T0136269

Combinatorics and Number Theory of Counting Sequences

Discrete Mathematics and Its Applications

Series Editors
Miklos Bona
Donald L. Kreher
Douglas West
Patrice Ossona de Mendez

Combinatorics of Compositions and Words
Silvia Heubach and Toufik Mansour

Handbook of Linear Algebra, Second Edition
Leslie Hogben

Combinatorics, Second Edition
Nicholas A. Loehr

Handbook of Discrete and Computational Geometry, Third Edition
C. Toth, Jacob E. Goodman and Joseph O'Rourke

Handbook of Discrete and Combinatorial Mathematics, Second Edition
Kenneth H. Rosen

Crossing Numbers of Graphs
Marcus Schaefer

Graph Searching Games and Probabilistic Methods
Anthony Bonato and Paweł Prałat

Handbook of Geometric Constraint Systems Principles
Meera Sitharam, Audrey St. John, and Jessica Sidman,

Volumetric Discrete Geometry
Karoly Besdek and Zsolt Langi

The Art of Proving Binomial Identities
Michael Z. Spivey

Combinatorics and Number Theory of Counting Sequences
István Mező

For more information about this series, please visit:
https://www.crcpress.com/Discrete-Mathematics-and-Its-Applications/book-series/CHDISMTHAPP?page=1&order=dtitle&size=12&view=list&status=published,forthcoming

Combinatorics and Number Theory of Counting Sequences

István Mező

CRC Press
Taylor & Francis Group
Boca Raton London New York

CRC Press is an imprint of the
Taylor & Francis Group, an **informa** business

A CHAPMAN & HALL BOOK

CRC Press
Taylor & Francis Group
6000 Broken Sound Parkway NW, Suite 300
Boca Raton, FL 33487-2742

**Visit the Taylor & Francis Web site at
http://www.taylorandfrancis.com**

**and the CRC Press Web site at
http://www.crcpress.com**

Para mi querida esposa, Pamela, és a kislányomnak Eszternek

Contents

Foreword

This book is, on one hand, an introduction to the theory of finite set partitions and to the enumeration of cycle decompositions of permutations; on the other hand, it is an extensive collection of references to newer works, and also, it contains number theoretical results on counting sequences of set partitions and permutations.

During writing, I tried to be as "combinatorial" as possible, prioritizing "enumerative" proofs wherever it is feasible. Still, there is a large number of analytic tools presented in the text, like generating functions (as there is no modern combinatorics without these), asymptotics (which provide valuable large-scale information on the studied sequences), and a bit of Riordan arrays, Hankel determinants, and orthogonal polynomials, to mention a few.

I suppose that some (or many) of the readers of this book are new to the topic. Therefore, I intended to be as elementary as possible so that a part of the text is available even on a high school level. Enthusiasts such as teachers and students can pick several identities from the book (especially Chapters 1, 6, 8–9), and can try to provide proof for them with a high probability of success. The self-training reader will find the exercises useful at the end of the chapters.

The book is useful for researchers also, as it collects vast information for many counting sequences (especially related to set partitions and permutations). Some parts of the text (mainly in Chapters 8–9, 13) appear for the first time in this book or in a unified way.

At the end of many chapters, there is an Outlook which guides the reader towards new topics and discusses the available results in the literature which do not fit in this book because of page limitations or because of the complexity of the tools and proofs. Also, the Outlooks contain results with respect to generalizations or some more particular subjects. These Outlooks, together with an extensive Bibliography and Tables at the end, make the book usable as a standard reference for combinatorialists working in the field.

An extremely useful and widely used tool in the everyday work of those who deal with integer sequences is the *On-line Encyclopedia of Integer Sequences* (OEIS). We shall occasionally cite entries from this encyclopedia via their IDs.

During the past centuries, a very large number of scientists have contributed to this area. For non-contemporary authors, we tried to give a short footnote with some very basic information so that the reader might guide himself or herself when the particular discovery and contribution was made.

No book is the sole contribution of its author. I would like to thank Prof. Miklós Bóna for his guidance during the whole process of the preparation of the text, and for his encouragement. I would also like to thank my colleague and friend, José Luis Ramírez Ramírez, for the great number of suggestions, corrections, and references. *¡Gracias, mi amigo!*

Many thanks also go to Ayhan Dil for the suggestions and for the careful reading of the text. I am thankful to Prof. Hacéne Belbachir for the discussion about the Kurepa conjecture.

András Bazsó and Ákos Pintér supplied a number of sources for the third part of the book. Gábor Nyul gave valuable suggestions and corrections.

I would like to also thank Jose Soto, Robert Ross, Paul Boyd, and Shashi Kumar at CRC Press for their professionalism shown during the preparation of the book.

Thank you, Pamela and Eszter, for supporting me during the writing of the book, and understanding how the work of a mathematician is. *¡Las amo mucho!*

Debrecen, 2009–2011. Nanjing, 2015–2019.

István Mező

About the Author

István Mező is a Hungarian mathematician who was born in Debrecen. He earned his PhD at the Doctoral School of Mathematics and Computer Science in 2010 at the University of Debrecen. He was working in this institute until 2014. After two years of Prometeo Professorship at the Escuela Politécnica Nacional (Quito, Ecuador) between 2012 and 2014, he moved to Nanjing, China where he is now a full-time research professor at the Nanjing University of Information Science and Technology, Nanjing, P. R. China.

Dr. Mező is currently working on Combinatorics and Special Function Theory.

Part I

Counting sequences related to set partitions and permutations

Chapter 1

Set partitions and permutation cycles

In this chapter we introduce the basic vocabulary that we will be using throughout the book. The notions to be introduced below are simple abstractions of everyday life questions like, "how many ways can six people be grouped into two groups?". The content of this chapter is entirely elementary, although for those who are not familiar with combinations, permutations and binomial coefficients, running over the Appendix might be useful.

1.1 Partitions and Bell numbers

Let us suppose that we have six people[1]. The most simple question might be the following: "in how many ways can we group these six people into some non-empty groups?". If we decide to label our people with the labels $1, 2, 3, 4, 5, 6$, then a possible grouping is the following:

$$1, 5|2, 3|4, 6;$$

meaning that the first and fifth person go into the first group, the second and third go into the second one, and the fourth and sixth are being put into the third group. The order of the groups does not count for now. Another possibility is

$$1|2, 6|3, 4, 5.$$

We feel that the number of different possibilities is not small, especially if there are more people to group. Concretely, there are 203 different groupings! Later on, we shall see how to get this exact number easily.

The problem of grouping or partitioning[2] objects is many hundred years old. Even more interestingly, the first appearance of this question is not in

---

[1] As the concrete entity that we are counting rarely matters, we can equally talk about "objects," or "elements" in general. What only matters is that our objects are distinguishable. This is expressed by the fact that we can put *labels* on our objects and say that this is the first, that is the second, and so on.

[2] We use the words "grouping" and "partitioning" interchangeably.

a mathematical treatise but in a Japanese novel, The Tale of Genji from the eleventh century. In mathematical work, this problem appears firstly in a 1796 paper of C. Kramp[3].

Let us study the question of partitioning in general and introduce the following definitions.

Definition 1.1.1. *Let an n-element set of objects A be given. If we have a non-empty collection of subsets of A such that*

- *the subsets in this collection are pairwise disjoint, and*

- *every object from A appears in one – and, by the previous point, in only one – subset,*

then we say that this collection of subsets is a partition *of A. The subsets in the collection are shortly called* blocks.

After introducing the notion of partitions, we arrive at a central definition of our topic, and we meet our first *counting sequence*.

Definition 1.1.2. *The number of all the possible partitions of the n element set A is the nth* Bell number, *denoted by B_n.*

E. T. Bell[4] wrote a comprehensive article on these numbers; therefore, later mathematicians named the Bell numbers after him.

For the sake of brevity, we shall often use the abbreviation "n-set" in place of "an n element set." As we have already remarked, the objects in the set A can be any kind of objects, like books, people, etc. As we assume them to be distinguishable, we do not lose the generality if we use a labeling of these objects, and consider the n-set A to be the label set $\{1, 2, \ldots, n\}$.

Let us determine the first few Bell numbers "by hand." Of course, one element can be grouped in only one way, hence $B_1 = 1$. For two objects $B_2 = 2$, and it still can easily be seen that $B_3 = 5$; all the partitions are listed here:

$$1, 2, 3 \quad 1|2, 3 \quad 2|1, 3, \quad 3|1, 2 \quad \text{and} \quad 1|2|3.$$

In our opening example, we took six people. Labeling these people with the labels $1, 2, \ldots, 6$, our set A is

$$A = \{1, 2, 3, 4, 5, 6\}.$$

This set – as we remarked – has 203 different partitions. Hence, $B_6 = 203$, as we shall see a bit later by concrete calculation – but without the necessity of

---

[3]Christian Kramp (1760-1826), French mathematician. He also published medical and physical papers.

[4]Eric Temple Bell (1883-1960), Scottish mathematician, writer and early science fiction author.

writing down all the partitions. We explicitly mentioned two partitions, which are

$$\{1,5\}, \ \{2,3\}, \ \{4,6\} \quad \text{and} \quad \{1\}, \ \{2,6\}, \ \{3,4,5\},$$

where now we are using set notations.

A recursion for the Bell numbers

As n grows, it is becoming harder and harder to calculate the Bell numbers by simply listing the partitions, so we must find a simpler way. To do this, we consider the following method. Let us fix an n-set A, and we pick out an arbitrary element a from that of n. If we consider all the partitions of A, there will be n possible cases: the element a is in a block of k elements, where k can be $1, 2, \ldots, n$. The first possibility, when $k = 1$, results in the block $\{a\}$. When a is in a two-element block, say, in $\{a, b\}$, then we have to choose the element b beside a. This can be done in $n - 1 = \binom{n-1}{1}$ ways. If a is contained in a block of three elements[5], then we have to choose two elements beside a on $\binom{n-1}{2}$ ways. And so on, until arriving $k = n$: if a is contained in an n-block (i.e., the partition contains one block and this is the whole A), then we have to choose $n - 1$ elements beside a. This is possible just in $\binom{n-1}{n-1} = 1$ way. After fixing the size k of the block of a, we can focus on the remaining elements not in the block of a. We see that the remaining $n - k$ elements can be partitioned arbitrarily and the total number of these partitions is B_{n-k} by the definition of the Bell numbers. Hence, for any single k there are $\binom{n-1}{k-1} B_{n-k}$ cases. We can sum up the possible values of k, because these cases are pairwise disjoint. Hence, we have that

$$B_n = 1 \cdot B_{n-1} + \binom{n-1}{1} \cdot B_{n-2} + \binom{n-1}{2} \cdot B_{n-3} + \cdots + \binom{n-1}{n-1} \cdot 1.$$

In the place of the last one we could write B_0 if we set $B_0 = 1$ to make the formula more consistent. Also, since $1 = \binom{n-1}{0}$ and we can fix $B_0 = 1$, this formula can be rewritten as

$$B_n = \binom{n-1}{0} \cdot B_{n-1} + \binom{n-1}{1} \cdot B_{n-2} + \binom{n-1}{2} \cdot B_{n-3} + \cdots + \binom{n-1}{n-1} \cdot B_0.$$

Shortly,

$$B_n = \sum_{k=0}^{n-1} \binom{n-1}{k} B_{n-k-1}.$$

For the symmetry $\binom{n}{k} = \binom{n}{n-k}$ of the binomial coefficients[6], we have that

---

[5] A k element block will also be called k-block for short.

[6] If we choose k elements from n, it is the same as choosing $n - k$ elements which we *do not want* to choose.

$\binom{n-1}{k} = \binom{n-1}{n-1-k}$, and hence

$$B_n = \sum_{k=0}^{n-1} \binom{n-1}{n-1-k} B_{n-1-k}.$$

Our formula can then be written in its final form, if we do the summation "from right to left":

$$B_n = \sum_{k=0}^{n-1} \binom{n-1}{k} B_k. \qquad (1.1)$$

This is a *recursive formula* or *recursion*, because it provides a method to calculate a member of a sequence by using the formerly calculated members.

Let us use our newly found formula! As we fixed, $B_0 = 1$. Moreover, as we have already seen, $B_1 = 1$, $B_2 = 2$, $B_3 = 5$. For B_4 we use our new recursion:

$$B_4 = \sum_{k=0}^{4-1} \binom{4-1}{k} B_k = \binom{3}{0} B_0 + \binom{3}{1} B_1 + \binom{3}{2} B_2 + \binom{3}{3} B_3 =$$

$$1 \cdot 1 + 3 \cdot 1 + 3 \cdot 2 + 1 \cdot 5 = 15.$$

B_5 and then B_6 can be calculated in a similar way, and now the reader can easily verify that there are 203 possible grouping of six elements, indeed.

At the end of the book in Tables, one can find a table of the Bell numbers up to B_{30}.

1.2 Partitions with a given number of blocks and the Stirling numbers of the second kind

After knowing how to determine the number of all possible groupings of a given number of objects, we continue our investigations with a new question: "how many partitions are there for an n-set which contain a *given number of blocks?*". Let us consider, for example, the following groupings:

$$1|2|3|4$$

$$1|2|3,4, \quad 1|3|2,4, \quad 1|4|2,3, \quad 2|3|1,4, \quad 2|4|1,3, \quad 3|4|1,2$$

$$1|2,3,4, \quad 2|1,3,4, \quad 3|1,2,4, \quad 4|1,2,3, \quad 1,2|3,4, \quad 1,3|2,4, \quad 1,4|2,3$$

$$1,2,3,4$$

All the $B_4 = 15$ possibilities are listed. It is obvious that there is only one partition which contains one block and only one exists with the maximal number of blocks. The cases in between are far more interesting. Let us introduce the following definition.

Definition 1.2.1. *Let A be an n element set. The partitions of A which contain exactly k blocks are called k-partitions of A. The number of k-partitions (on n elements) is denoted by $\left\{ {n \atop k} \right\}$, and called* Stirling number of the second kind *with parameters n and k*[7].

This name was given by Niels Nielsen[8] who gave this honor to James Stirling[9]. The $\left\{ {n \atop k} \right\}$ numbers are also called *Stirling partition numbers*. The reason why we classify these numbers as "second kind" will be clear soon. Nevertheless, one can easily guess before we reach there: there are first kind Stirling numbers also.

The notation $\left\{ {n \atop k} \right\}$ is similar to the binomial coefficient notation $\binom{n}{k}$[10], and it also tells us that we are talking about sets; sets are just limited by braces { and } in mathematical texts.

Our above example shows that

$$\left\{ {4 \atop 1} \right\} = 1, \quad \left\{ {4 \atop 2} \right\} = 7, \quad \left\{ {4 \atop 3} \right\} = 6, \quad \left\{ {4 \atop 4} \right\} = 1.$$

A table of the first several Stirling numbers can be found at the end of the book.

By the definition of $\left\{ {n \atop k} \right\}$ we won a new formula for the Bell numbers:

$$B_n = \left\{ {n \atop 1} \right\} + \left\{ {n \atop 2} \right\} + \cdots + \left\{ {n \atop n} \right\}. \tag{1.2}$$

Thus, we see additional evidence as to why $B_4 = 15$: $15 = 1 + 7 + 6 + 1$. The relation between the Bell and Stirling numbers of the second kind is obvious, because any partition is either a 1-partition, or a 2-partition, etc. Relation (1.2), however, does not help us at all to calculate B_n if we do not know how to calculate the individual Stirling number terms in the sum.

Special values of the Stirling numbers of the second kind

Certainly,

$$\left\{ {n \atop 1} \right\} = \left\{ {n \atop n} \right\} = 1.$$

---

[7]While "$\binom{n}{k}$" is pronounced as "n choose k", "$\left\{ {n \atop k} \right\}$" is read as "$n$ subset k".

[8]Niels Nielsen (1865-1931), Danish mathematician.

[9]James Stirling (1692-1770), Scottish mathematician. He calculated the Stirling numbers in his 1730 work *Methodus Differentialis* [539]. He studied the second kind of Stirling numbers first, and then he went to the first kind case. Exactly as we do in this chapter.

[10]In place of the widely used notation $\binom{n}{k}$, sometimes C_n^k is used. The former was introduced by Andreas Freiherr von Ettingshausen (1796-1878), Austrian mathematician and physicist in 1826. Ettingshausen developed some influential lecture notes at the University of Vienna, and he wrote a book [223] on combinatorial analysis in 1826 in which the mentioned notation for the binomial coefficients first appeared. The notation $\left\{ {n \atop k} \right\}$ for the Stirling numbers of the second kind is more than a century younger: Jovan Karamata (1902-1967), Serbian mathematician invented it in 1935. Therefore, we also call the $\left\{ {n \atop k} \right\}$ notation (and the later introduced notation $\left[{n \atop k} \right]$) as *Karamata notation*. See [227, 262, 336] on these questions but note that in [336] the credit was still given to I. Marx and A. Salmeri.

Let us try to find other special cases. We begin with $\left\{{n \atop 2}\right\}$. If we group n elements into 2 blocks, one block will contain the first element. For any other element from the $n - 1$, we have two possibilities: this element is contained in the block of the first element 1, or it is contained in the other block. These offer 2^{n-1} possibilities for the $n - 1$ elements. But it is not possible that all of the $n-1$ elements go to the block of 1, because this would be a 1-partition and not a 2-partition. Hence, we have to subtract this unique possibility. Thus,

$$\left\{{n \atop 2}\right\} = 2^{n-1} - 1. \tag{1.3}$$

Determining $\left\{{n \atop n-1}\right\}$ is also simple: n elements can go to $n - 1$ (always non-empty) blocks if all the blocks contain one element, but one block contains two elements. If we fix those two elements which go into this block, we are done. Two elements from n can be chosen in $\binom{n}{2}$ ways; therefore[11]

$$\left\{{n \atop n - 1}\right\} = \binom{n}{2}.$$

We can compare these results with the partition in the example. An n-set can have 2-partitions in the amount of $\left\{{4 \atop 2}\right\} = 2^{4-1} - 1 = 7$ and 3-partitions in the amount of $\left\{{4 \atop 3}\right\} = \binom{4}{2} = 6$.

Hence, we can easily calculate – one more time – that

$$B_4 = \left\{{4 \atop 1}\right\} + \left\{{4 \atop 2}\right\} + \left\{{4 \atop 3}\right\} + \left\{{4 \atop 4}\right\} = 1 + (2^{4-1} - 1) + \binom{4}{2} + 1 = 15.$$

Recursion of the Stirling numbers of the second kind

Let us go further to find more ways to calculate the second kind of Stirling numbers. If we have an n-set and a k-partition, there are two cases.

1. The first element is alone in its own 1-block[12]. Then the other $n - 1$ elements form $k - 1$ blocks in $\left\{{n-1 \atop k-1}\right\}$ ways.

2. The first element shares a block with other elements. The other $n - 1$ elements form a k-partition in $\left\{{n-1 \atop k}\right\}$ possible ways, and we put the first element in one of these. To perform this insertion, we have k options. This yields that in this case we have $k\left\{{n-1 \atop k}\right\}$ possibilities.

These two cases together give a simple calculation method for the Stirling numbers of the second kind:

$$\left\{{n \atop k}\right\} = \left\{{n - 1 \atop k - 1}\right\} + k\left\{{n - 1 \atop k}\right\}. \tag{1.4}$$

---

[11]Note that the order of the blocks does not count.
[12]1-blocks will also be called *singletons*.

As we already mentioned around (1.1), such formulas are called recursive.

To see how this recursion works in practice, let us try to answer the following question: "in how many ways can we group five elements into three groups?". In other words, we want to determine the value of $\left\{{5 \atop 3}\right\}$. By our new recursion we have that

$$\left\{{5 \atop 3}\right\} = \left\{{4 \atop 2}\right\} + 3\left\{{4 \atop 3}\right\}.$$

Luckily, we have already seen the Stirling numbers on the right: $\left\{{4 \atop 2}\right\} = 7$, and $\left\{{4 \atop 3}\right\} = 6$. Hence,

$$\left\{{5 \atop 3}\right\} = 7 + 3 \cdot 6 = 25.$$

We have not yet analyzed what numbers can go to the upper and lower position in $\left\{{n \atop k}\right\}$. The upper parameter n is always a number of objects, so it is a positive number; however, sometimes it is useful to permit $n = 0$. The lower parameter k is the number of blocks, so it cannot be greater than n. We also permit k to be zero, so $0 \le k \le n$, and $n \ge 0$. However, it is often useful to permit $k > n$ and in this case we simply fix

$$\left\{{n \atop k}\right\} = 0$$

There are some reasons we will meet later which suggest that we would better set $\left\{{0 \atop 0}\right\}$ to be one and not to be zero. We therefore fix

$$\left\{{0 \atop 0}\right\} = 1.$$

Yet another formula

Let us suppose that the block of the first element contains m objects and that there are k blocks in total. If we consider the partition of the rest of the elements, then they go to $k - 1$ partitions on $\left\{{n-m \atop k-1}\right\}$ possible ways. The $m - 1$ elements that are in the block of the first element can be chosen $\binom{n-1}{m-1}$ ways. Since m is arbitrary between 1 and n, we must sum up its possible values. The following important result follows:

$$\left\{{n \atop k}\right\} = \sum_{m=1}^{n} \binom{n-1}{m-1} \left\{{n-m \atop k-1}\right\}. \tag{1.5}$$

We now depart from the Stirling numbers of the second kind and we go to meet with the first kind Stirling numbers. These, at first sight, seem to be rather different from the second kind Stirlings, but there is a deep connection between the two. This connection will be seen in Section 1.5.

1.3   Permutations and factorials

After knowing how to partition objects into blocks, now we study how to *order* objects. If we have 30 books to put on a shelf (regardless of the alphabetical order of the titles, their sizes or colors, etc.) we can do this task as follows: we choose one book from the 30 and we put it the first on the shelf. Then we choose one from the remaining 29 books, and so on, until we have the last book to put. Therefore, we have $30 \cdot 29 \cdots 2 \cdot 1$ different cases in total. (This is a huge number with 33 digits.) This approach works for any number of books, so we introduce the following elementary notion.

Definition 1.3.1. *A fixed order of n elements is a* permutation *of these elements. The number of all the possible permutations is* $n \cdot (n-1) \cdot (n-2) \cdots 2 \cdot 1$, *which is denoted*[13] *by n! and we read it as n* factorial[14]. *n is also called the* length *of the permutation.*

Cycles

The permutations are often described by the following method. In the first line of a two-line table we put the labels of the elements, and in the second line we put their orders after *permuting* them. For example, if we consider the elements $1, 2, 3, 4$ and we put them in the order $2, 4, 1, 3$, then we have this table:

$$\begin{pmatrix} 1 & 2 & 3 & 4 \\ 2 & 4 & 1 & 3 \end{pmatrix}. \tag{1.6}$$

Now let us see a longer permutation. For instance:

$$\begin{pmatrix} 1 & 2 & 3 & 4 & 5 & 6 \\ 3 & 1 & 2 & 6 & 4 & 5 \end{pmatrix}.$$

Note that the first three and second three elements are "mixing" among themselves. In other words, this permutation can be rewritten as a "product" of two smaller permutations:

$$\begin{pmatrix} 1 & 2 & 3 \\ 3 & 1 & 2 \end{pmatrix} \quad \text{and} \quad \begin{pmatrix} 4 & 5 & 6 \\ 6 & 4 & 5 \end{pmatrix}.$$

A permutation which cannot be rewritten as a product of smaller permutations is called *cycle*, or sometimes *orbit*. The name "cycle" comes from the fact that the elements in a cycle permute cyclically: in the first permutation

---

[13]This notation was introduced by the already mentioned Christian Kramp in 1808.

[14]We shall see that $n!$ is the cousin (or, more fancily speaking, the dual) of the Bell numbers.

above 1 goes to the third place, 3 goes to the second place, and 2 goes to the first position, where 1 was before. For this reason it is very useful to introduce the following *cycle notation*:

$$\begin{pmatrix} 1 & 2 & 3 \\ 3 & 1 & 2 \end{pmatrix} = \begin{pmatrix} 1 & 3 & 2 \end{pmatrix}.$$

If we have more than one cycle, we just write them down one after another:

$$\begin{pmatrix} 1 & 2 & 3 & 4 & 5 & 6 \\ 3 & 1 & 2 & 6 & 4 & 5 \end{pmatrix} = \begin{pmatrix} 1 & 3 & 2 \end{pmatrix} \begin{pmatrix} 4 & 6 & 5 \end{pmatrix}. \tag{1.7}$$

What happens if an element stays in its position? Like 3 in the permutation

$$\begin{pmatrix} 1 & 2 & 3 & 4 \\ 2 & 4 & 3 & 1 \end{pmatrix}.$$

Now 3 forms a cycle of length one, the $1, 2, 4$ elements form another cycle:

$$\begin{pmatrix} 1 & 2 & 3 & 4 \\ 2 & 4 & 3 & 1 \end{pmatrix} = \begin{pmatrix} 1 & 2 & 4 \end{pmatrix} \begin{pmatrix} 3 \end{pmatrix}.$$

In this case we say that 3 is a *fixed point* of the permutation, since it stays in its original position.

We should note that in a cycle the order of the elements *do* matter, but the following cycles are still identical:

$$\begin{pmatrix} 1 & 2 & 3 \end{pmatrix} = \begin{pmatrix} 3 & 1 & 2 \end{pmatrix} = \begin{pmatrix} 2 & 3 & 1 \end{pmatrix}.$$

These cycles are related to each other by *right shifts*. In the initial cycle, we shift the elements to the right, and the outgoing element on the right comes back on the left.

One might note also that if we rewrite a permutation using this cycle notation (as in (1.7)), *the order of the cycles is indifferent.*

We introduce two special permutations: the *identical permutation*, or *identity* is a permutation in which every element is a fixed point. Moreover, a *transposition* is a permutation in which only two elements are interchanged. A transposition is therefore constituted by a cycle of length two and possibly some fixed points.

1.4 Permutation with a given number of cycles and the Stirling numbers of the first kind

There are $n!$ permutations and B_n partitions in total on n elements. In addition, there are $\left\{ {n \atop k} \right\}$ k-partitions on an n-set. These observations suggest that we can ask an analogous question: how many permutations are there with k cycles? The answer, as the reader might expect, is given by the Stirling numbers of the first kind.

Definition 1.4.1. *The* Stirling number of the first kind *with parameters n and k gives the number of permutations of length n with k cycles. This number is denoted by $\left[\begin{smallmatrix} n \\ k \end{smallmatrix}\right]$* [15].

The first kind Stirling numbers are sometimes called *Stirling cycle numbers* [16].

To know the $\left[\begin{smallmatrix} n \\ k \end{smallmatrix}\right]$ numbers a bit more profoundly, let us take an example. We would like to put four elements into 2 cycles. We have $\left[\begin{smallmatrix} 4 \\ 2 \end{smallmatrix}\right] = 11$ possibilities, because the following cases are all the possibilities we have:

$$\begin{pmatrix} 1 & 2 & 3 \end{pmatrix}\begin{pmatrix} 4 \end{pmatrix}, \quad \begin{pmatrix} 1 & 2 & 4 \end{pmatrix}\begin{pmatrix} 3 \end{pmatrix}, \quad \begin{pmatrix} 1 & 3 & 4 \end{pmatrix}\begin{pmatrix} 2 \end{pmatrix}, \quad \begin{pmatrix} 2 & 3 & 4 \end{pmatrix}\begin{pmatrix} 1 \end{pmatrix},$$

$$\begin{pmatrix} 2 & 1 & 3 \end{pmatrix}\begin{pmatrix} 4 \end{pmatrix}, \quad \begin{pmatrix} 2 & 1 & 4 \end{pmatrix}\begin{pmatrix} 3 \end{pmatrix}, \quad \begin{pmatrix} 3 & 1 & 4 \end{pmatrix}\begin{pmatrix} 2 \end{pmatrix}, \quad \begin{pmatrix} 3 & 2 & 4 \end{pmatrix}\begin{pmatrix} 1 \end{pmatrix},$$

$$\begin{pmatrix} 1 & 2 \end{pmatrix}\begin{pmatrix} 3 & 4 \end{pmatrix}, \quad \begin{pmatrix} 1 & 3 \end{pmatrix}\begin{pmatrix} 2 & 4 \end{pmatrix}, \quad \begin{pmatrix} 1 & 4 \end{pmatrix}\begin{pmatrix} 3 & 2 \end{pmatrix}.$$

Summing the Stirling numbers of the second kind we got the number of all the possible partitions and this was the B_n Bell number (see (1.2)). If we sum the first kind Stirling numbers, we must get the number of all the permutations which are – by the remark before Definition 1.3.1. – the factorials:

$$n! = \begin{bmatrix} n \\ 1 \end{bmatrix} + \begin{bmatrix} n \\ 2 \end{bmatrix} + \cdots + \begin{bmatrix} n \\ n \end{bmatrix}. \tag{1.8}$$

In this sense we meant above that $n!$ and B_n are duals of each other, compare (1.8) with (1.2).

Special values of the Stirling numbers of the first kind

For some special parameters, the first kind Stirling numbers can be calculated easily. For instance,

$$\begin{bmatrix} n \\ 1 \end{bmatrix} = (n-1)!,$$

because in the only one cycle in which all the elements must appear (in $n!$ different possible orders) the right shifts do not alter the permutation (see the end of the last section). Hence, $n!$ must be divided by the number of possible right shifts. The number of the latter is n, so $\left[\begin{smallmatrix} n \\ 1 \end{smallmatrix}\right] = n!/n = (n-1)!$, as we stated.

---

[15] This notation was also introduced by Karamata.

[16] The British Museum guards a manuscript of Thomas Harriot (1560-1621, English mathematician, ethnographer, translator; he was the first in depicting the moon by using a telescope) from the seventeenth century in which these numbers appear.

On the other hand, a permutation with n elements can have n cycles only if every element happens to be in its own cycle (so that these are fixed points):

$$\left[\begin{matrix} n \\ n \end{matrix}\right] = 1.$$

And, similarly as in the second kind case,

$$\left[\begin{matrix} n \\ n-1 \end{matrix}\right] = \binom{n}{2}.$$

Now let us see the parameters $n, n - 2$. n elements can go into $n - 2$ cycles in two ways:

1. We have $n - 3$ fixed points and three goes into one cycle. Then we have to choose these three elements: $\binom{n}{3}$ choice. (We can choose the $n - 3$ fixed points, but this is the same: $\binom{n}{n-3} = \binom{n}{3}$.) In the 3-cycle there are two possible arrangements of the elements, so we have to multiply $\binom{n}{3}$ by 2.

2. We can have two 2-cycles (transpositions). We have to choose two elements into each of these transpositions: to do this we have $\binom{n}{2}\binom{n-2}{2}$ choices. But the order of the two transpositions is indifferent, thus we divide by 2.

Altogether, we have that

$$\left[\begin{matrix} n \\ n-2 \end{matrix}\right] = 2\binom{n}{3} + \frac{1}{2}\binom{n}{2}\binom{n-2}{2}.$$

It can also be seen that

$$\left[\begin{matrix} n \\ k \end{matrix}\right] \geq \left\{\begin{matrix} n \\ k \end{matrix}\right\}.$$

This is so, because a k-partition on n elements always yields a k-cycle decomposition. For example, the 2-partition

$$\{1, 2, 3\},\ \{4, 5, 6\}$$

gives the decomposition

$$(1\quad 2\quad 3)(4\quad 5\quad 6),$$

but

$$(2\quad 1\quad 3)(5\quad 4\quad 6)$$

is a different permutation belonging to the partition

$$\{2, 1, 3\},\ \{5, 4, 6\},$$

which is still the same as before.

Recursion of the Stirling numbers of the first kind

After having some special values, let us try to find a recursive formula for the first kind Stirling numbers similar to (1.4) in the second kind case. It is worth it to read the proof of (1.4) again, because the basic idea can be applied here, too. Forming a k-cycle decomposition on n elements, we have two possibilities.

1. The first element is in a 1-cycle (i.e., it is a fixed point), and the rest of the elements form $k - 1$ cycles: $\begin{bmatrix} n-1 \\ k-1 \end{bmatrix}$ cases.

2. The first element is in a cycle longer than one. Then we already have k cycles on $n - 1$ elements: $\begin{bmatrix} n-1 \\ k \end{bmatrix}$. We have to insert the first element into one of the cycles: this can be done in $n - 1$ ways: at the first position of the cycles, or between the elements; if we insert it at the end of the cycles, we do not have new permutation by the shift invariance.

Hence, our recursion reads as:

$$\begin{bmatrix} n \\ k \end{bmatrix} = (n-1) \begin{bmatrix} n-1 \\ k \end{bmatrix} + \begin{bmatrix} n-1 \\ k-1 \end{bmatrix}. \tag{1.9}$$

The initial values for this recursion are $\begin{bmatrix} 0 \\ 0 \end{bmatrix} = 1$ and $\begin{bmatrix} n \\ 0 \end{bmatrix} = 0$ when $n \geq 1$.

A special value involving harmonic numbers

Having this recursion, we can find the special value $\begin{bmatrix} n \\ 2 \end{bmatrix}$ easily. By the recursion,

$$\begin{bmatrix} n \\ 2 \end{bmatrix} = (n-1) \begin{bmatrix} n-1 \\ 2 \end{bmatrix} + \begin{bmatrix} n-1 \\ 1 \end{bmatrix} = (n-1) \begin{bmatrix} n-1 \\ 2 \end{bmatrix} + (n-2)!$$

More concretely, if $n = 3$ or 4:

$$\begin{aligned}
\begin{bmatrix} 3 \\ 2 \end{bmatrix} &= 2 \begin{bmatrix} 2 \\ 2 \end{bmatrix} + \begin{bmatrix} 2 \\ 1 \end{bmatrix} = 2 \cdot 1 + 1! = 2! \left(1 + \frac{1}{2} \right) \\
\begin{bmatrix} 4 \\ 2 \end{bmatrix} &= 3 \begin{bmatrix} 3 \\ 2 \end{bmatrix} + \begin{bmatrix} 3 \\ 1 \end{bmatrix} = 3 \cdot 2! \left(1 + \frac{1}{2} \right) + 2! = 3! \left(1 + \frac{1}{2} + \frac{1}{3} \right),
\end{aligned}$$

and, in general,

$$\begin{bmatrix} n \\ 2 \end{bmatrix} = (n-1)! \left(1 + \frac{1}{2} + \frac{1}{3} + \cdots + \frac{1}{n-1} \right).$$

The sum on the right-hand side is denoted by H_{n-1} and it is called *harmonic number*[17]. Using this abbreviation, the above line can be written, after a reparametrization, as

$$\frac{1}{n!}\begin{bmatrix} n+1 \\ 2 \end{bmatrix} = H_n = \left(1 + \frac{1}{2} + \frac{1}{3} + \cdots + \frac{1}{n}\right). \tag{1.10}$$

We shall meet with the harmonic numbers later in Chapter 8.

1.5 Connections between the first and second kind Stirling numbers

There is more than just a similar naming convention between the first and second kind Stirling numbers. In this section, we reveal this connection.

Let us change the parameters and their signs in (1.9). What we get is the following:

$$\begin{bmatrix} -k \\ -n \end{bmatrix} = (n-1)\begin{bmatrix} -k \\ -n+1 \end{bmatrix} + \begin{bmatrix} -k+1 \\ -n+1 \end{bmatrix}.$$

Rearranging:

$$\begin{bmatrix} -k+1 \\ -n+1 \end{bmatrix} = \begin{bmatrix} -k \\ -n \end{bmatrix} + (-n+1)\begin{bmatrix} -k \\ -n+1 \end{bmatrix}.$$

We can realize an interesting fact. This recursion is the same as for the Stirling numbers of the second kind. The signs are a bit confusing, so let us take a concrete example. Let us substitute $n = 7$ and $k = 3$:

$$\begin{bmatrix} -2 \\ -6 \end{bmatrix} = \begin{bmatrix} -3 \\ -7 \end{bmatrix} + (-6)\begin{bmatrix} -3 \\ -6 \end{bmatrix}.$$

One can get the first term on the right-hand side, $\begin{bmatrix} -3 \\ -7 \end{bmatrix}$, from the left-hand side by decreasing the parameters in $\begin{bmatrix} -2 \\ -6 \end{bmatrix}$ by one. The Stirling number in the second term, $\begin{bmatrix} -3 \\ -6 \end{bmatrix}$, is multiplied by the lower parameter -6 and the upper parameter is decreased by one. Exactly as in (1.4). We therefore have the following important relation:

$$\begin{bmatrix} n \\ k \end{bmatrix} = \left\{ \begin{matrix} -k \\ -n \end{matrix} \right\}. \tag{1.11}$$

This relation says that one can get the Stirling numbers of the first kind by negating and interchanging the parameters in the second kind Stirling numbers during performing the recursion. This phenomenon is called *duality*.

---

[17]The name of the harmonic numbers comes from the following observation. A harmonic of a given tone is a tone with a frequency that is a positive integer multiple of the frequency of the original tone. The original tone is also called the first harmonic. As the wavelength is the reciprocal of the frequency, the nth harmonic of a given tone has the wavelength $1/n$-times the original.

1.6　Some further results with respect to the Stirling numbers

In this section, we give some new formulas and interpretations for the second kind Stirling numbers.

1.6.1　Rhyme schemes

If we rephrase the definition of $\left\{{n\atop k}\right\}$ by using rhyme schemes, we can have an interesting enumeration of schemes in strophes. Restricting us to two couplets in a strophe and a strophe is restricted to be of four lines[18], then the following possibilities can appear:

$$aabb, abab, abba, aaab, aaba, abaa, abbb$$

The first is called (double) couplet, the second is an alternate rhyme, the third is an enclosed rhyme, while the fifth is a rubaiyat[19].

Such rhymes can be related to set partitions by the following correspondences: elements $\leftrightarrow$ lines, and blocks $\leftrightarrow$ letters of rhymes:

$$1,2|3,4,\quad 1,3|2,4,\quad 1,4|2,3,\quad 1,2,3|4,\quad 1,2,4|3,\quad 1,3,4|2,\quad 1|2,3,4.$$

Similarly, one can generalize this for more lines and more rhyme schemes. At the end of the process and summing over all the possibilities, we have that the nth Bell number gives that how many rhyme schemes can be there in a poem of n lines.

1.6.2　Functions on finite sets

The second kind Stirling numbers are also related to functions on finite sets. First we ask a simple question: "how many functions are there on an n-set which attain at most k different values?". Suppose that these k values are $1, 2, \ldots, k$, while the n-set is simply $A = \{1, 2, \ldots, n\}$. The answer can be given easily: we map $1 \in A$ to one of the k elements, then map $2 \in A$ to one of the k elements again, and so on. Altogether we have $k \cdot k \cdots k = k^n$ possible functions.

The following question leads directly to the Stirling numbers: how many functions are there on A which map exactly to k elements? Such a function is called *surjective*. In this case $n \geq k$ is necessary. We can consider a surjective function as a partition: every element from A maps to one of the k numbers. This observation results in a k-partition on A, and, as we know, there are $\left\{{n\atop k}\right\}$

---

[18]Such strophes are called quatrains.
[19]More on these can be read in the Wikipedia article *Rhyme scheme*.

such partitions. However, the different orders of the blocks result in different functions; hence we have to multiply $\left\{{n \atop k}\right\}$ by $k!$ to get the total number of surjective functions.

We thus get that the number of *surjective* functions

$$f : \{1, 2, \ldots, n\} \to \{1, 2, \ldots, k\} \tag{1.12}$$

is

$$k!\left\{{n \atop k}\right\}.$$

With these preliminaries we can prove a nice identity. We have seen above that the total number of functions of the form (1.12) is k^n. It can also be seen that every such function is surjective on some subset $B \subset \{1, 2, \ldots, k\}$. The number of the elements of B is denoted by $|B|$. If we sum over the number of surjective functions on all the possible sets B (and this number is $|B|!\left\{{n \atop |B|}\right\}$), we get the number of all (not necessarily surjective) functions on the n elements: k^n. In symbols,

$$k^n = \sum_{B \subset \{1,2,\ldots,k\}} |B|!\left\{{n \atop |B|}\right\}.$$

B can be chosen in $\binom{k}{|B|}$ ways, hence instead of summing over B we can sum over $1 \le m = |B| \le k$:

$$k^n = \sum_{m=1}^{k} \binom{k}{m} m!\left\{{n \atop m}\right\} = \sum_{m=1}^{k} \frac{k!}{(k-m)!}\left\{{n \atop m}\right\} = \sum_{m=1}^{n} \left\{{n \atop m}\right\} k(k-1)(k-m+1).$$

(We can extend the sum to n if $m > k$, because the product $k(k-1)(k-m+1)$ will be zero.) That is,

$$\sum_{m=1}^{n} \left\{{n \atop m}\right\} k(k-1)(k-m+1) = k^n.$$

This formula will be of great importance later.

The pair of this identity exists for the first kind Stirling numbers:

$$k(k+1)\cdots(k+n-1) = \sum_{m=1}^{n} \left[{n \atop m}\right] k^m. \tag{1.13}$$

The proof is as follows. The left-hand side equals to

$$n!\binom{k+n-1}{n} = k(k+1)\cdots(k+n-1).$$

This expression counts the following selection process. We choose n balls, and we distribute them among k boxes such that the order of the balls in the boxes do count. Indeed: the first ball can go into one of the k boxes; k possibilities. The next ball can go into the $k - 1$ empty boxes or into the box where the previously selected ball is put, but here we have two options: the new ball goes to the left or to the right of the other one: $k - 1 + 2 = k + 1$ possibilities. The reader sees that this continues in a similar manner.

Next, let us see how this ball selection process is related to the right-hand side of (1.13). We put the balls in a given order (i.e., we make a permutation of the balls). If this permutation has m cycles ($1 \leq m \leq k$), then the number of such permutations is $\begin{bmatrix} n \\ m \end{bmatrix}$. Every cycle corresponds to a box, and the balls in the given cycle go into this box. There are k^m such mappings, as we saw earlier. Finally, we sum up the possible values of m to arrive at the identity (1.13).

1.7 d-regular partitions

After seeing how the Stirling numbers are related to the enumeration problem of functions, we study another question[20].

Once upon a time, a mad king of a small kingdom of five cities was facing a problem. The citizens of the capital did not want to pay his newly introduced air tax. The king knew that there were some rebels convincing the citizens not to pay the tax. He also knew that a group of six rebels were causing the problem who were living in adjacent houses. The king decided to separate the instigators: he announced that all of these rebels must be moved to five other cities. To make it even harder for them to form a group again, the king added that none of two adjacent neighbors can move to the same city. How many possibilities did the gendarmerie have to execute the king's order?

Since there are six rebellious instigators, but only five cities, two of the rebels will surely go to the same city. The first rebel can live together with the third, fourth, fifth, or sixth rebel, and the others are separated. These give four possibilities. Or, the second rebel can live together with one of the fourth, fifth, or sixth rebels, the third rebel together with the fifth or sixth (here we do not write the first-third pairing, because this configuration was counted before). Finally, the fourth rebel can live together with the sixth (and also with the already counted first and second) rebel. These give $4 + 3 + 2 + 1 = 10$ cases in total. The question has been answered.

How can we generalize this problem? The adjacent neighbors could not be put together. Therefore, a possible direction of generalization is that we restrict the citizens to live together only with even more distant neighbors. We introduce the following definition.

Definition 1.7.1. *If the individual blocks of a partition of the set $\{1, 2, \ldots, n\}$*

---

[20]This section is based on an article of Anisse Kasraoui [322].

contain elements with pairwise distances $\geq d$, *then this partition is called d-regular partition*[21].

As we saw, there are ten 2-regular partitions on six elements. The following partition is a 3-regular partition of a 7-set:

$$1, 4, 7|2|5|3, 6.$$

At first sight, we might think that the d-regular partitions offer result in a very different class of numbers from that of the $\left\{{n \atop k}\right\}$ numbers, but the truth is that the d-regular partitions are counted by the Stirling numbers of the second kind after a simple reparametrization. Indeed, we are going to show that the total number of d-regular k-partitions of an n-set is

$$\left\{{n - d + 1 \atop k - d + 1}\right\}. \tag{1.14}$$

This fact can be proven in an elegant and clever way. The proof is a *bijective proof*. This means that we try to establish a bijection (one-to-one correspondence) between two combinatorial arrangements; hence proving that in the two arrangements there are the same number of objects.

Let us imagine a triangle shaped table with $n - 1$ columns. In the first (leftmost) column there are $n - 1$ cells, in the second column there are $n - 2$ cells, and so on. The rightmost column contains only one cell. By reasons which will be clear later on, we lengthen the bottom and leftmost lines of the table, as it is seen below in the particular case when $n = 11$:

Now we recall the role of *rook*s in chess. The rooks can move horizontally and vertically[22]. On our triangle-shaped table, one can place $n - k$ rooks in exactly $\left\{{n \atop k}\right\}$ ways such that there is no pair of rooks in the same row or column (so that the rooks do not attack each other). A possible arrangement is shown here:

---

[21] The names *d-Fibonacci partition* and *non-consecutive partition* are also used.
[22] Rook arrangements often appear in enumerative combinatorics.

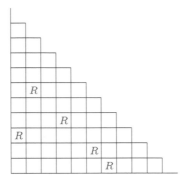

Next, number the lines from top to bottom and left to right, including the lengthened tiny lines:

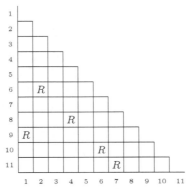

Now place the rooks applying the following rule: considering a k-partition on an n-set, we put a rook in the ith column and jth row if and only if i and j satisfy the following conditions:

1. i and j are in the same block,

2. $i < j$,

3. there is no third element between i and j in their blocks.

Such an i, j pair of elements is called *arc* of a partition. Let us take a 6-partition on 11 elements:

$$1, 9 | 2, 6, 10 | 3 | 4, 8 | 5 | 7, 11 \tag{1.15}$$

This partition has 5 arcs, namely,

$$1, 9; \quad 2, 6; \quad 6, 10; \quad 4, 8; \quad 7, 11$$

Hence, we have to place 5 rooks, exactly as in the configuration of the table above.

This example shows clearly that $n - k$ rooks belong to every k-partition.

This correspondence is unique because there is no arc between blocks (there are $k-1$ places between k blocks). If we list the elements in the blocks in increasing order, then there are at most $n-1$ arcs. That is, there are $n-1-(k-1) = n-k$ arcs (or, equivalently, rooks) to be placed. It can also be seen that from a given rook placement one can decipher the partition unambiguously.

The d-regularity will have a very nice interpretation if we apply it to rook placements. To see this interpretation in an example, we take the partition (1.15), which is $2, 3, 4$-regular. Set, say, $d = 4$. By d-regularity the smallest element in the block of a bigger than a must be $a + 4$. In the language of rook placements: the fourth column can contain rooks only in the positions $8, 9, 10, 11$. Repeating this for each element in the partition (or each column in the table), we can see that the first three cells counting from above are never used:

	1	2	3	4	5	6	7	8	9	10	11
1											
2	×										
3	×	×									
4	×	×	×								
5		×	×	×							
6		R	×	×	×						
7				×	×	×					
8				R	×	×	×				
9	R					×	×	×			
10						R	×	×	×		
11							R	×	×	×	

Leaving out the non-used cells, we get an equilateral triangle of size $n - (d-1) = n-d+1$ (together with the lengthened lines), where we can place the rooks. Since there is a bijection between the rook placements and partitions ($n-k$ rooks belong to k partitions), there are $k-(d-1)$ blocks on an $n-(d-1)$-set with the same number of rooks. Hence, we have proven that (1.14) is really the number of d-regular k-partitions on n elements.

Setting $n = 6$, $d = 2$, and $k = 5$ we see one more time that the gendarmerie of the mad king has $\left\{{6-2+1 \atop 5-2+1}\right\} = \left\{{5 \atop 4}\right\} = 10$ possibilities to separate the 6 rebels.

Finally, summing on k, we have that the total number of d-regular partitions on an n-set is the Bell number B_{n-d+1}.

1.8   Zigzag permutations

Applying some restriction on the combinatorial structures at hand often gives interesting constructions. We saw one example already, the d-regular partitions. In that instance, the restriction (that the elements must have some distance greater than one in between, resulted in no new number sequence because the d-regular partitions can be counted by the usual Stirling numbers of the second kind (see (1.14)).

Here we introduce an important subclass of permutations when the elements must be in a "zigzag" form. This definition will result in a new and interesting counting sequence. For example,

$$\begin{pmatrix} 1 & 2 & 3 & 4 & 5 \\ 3 & 1 & 4 & 2 & 5 \end{pmatrix}$$

is a "zigzag" permutation (more precisely, a down-up zigzag), because the elements in the bottom line can be ordered in a zigzag form:

$$3 > 1 < 4 > 2 < 5.$$

Here goes the general definition.

Definition 1.8.1. *A permutation*

$$\begin{pmatrix} 1 & 2 & 3 & \cdots & n \\ i_1 & i_2 & i_3 & \cdots & i_n \end{pmatrix}$$

is called up-down zigzag *if*

$$i_1 > i_2 < i_3 > \cdots.$$

Similarly, this permutation is down-up zigzag *if, instead, their elements satisfy the inequalities*

$$i_1 < i_2 > i_3 < \cdots.$$

The down-up zigzag permutations are also called *alternating*, while the up-down zigzags are often named as *reverse alternating*.

The reverse

$$\begin{pmatrix} 1 & 2 & 3 & \cdots & n \\ i_n & i_{n-1} & i_{n-2} & \cdots & i_1 \end{pmatrix}$$

of a down-up zigzag is an up-down zigzag, and vice versa. Therefore, it is enough to introduce only one sequence to count the number of (up-down or down-up) zigzag permutations. The number of up-down zigzag permutations on n elements is denoted by E_n and called the *Euler number sequence*. The fundamental properties of the zigzag permutations were studied by Désiré André[23] in 1879 [26], see also [27, 28].

On four elements there are $E_4 = 5$ down-up zigzags:

$$\begin{pmatrix} 1 & 2 & 3 & 4 \\ 2 & 1 & 4 & 3 \end{pmatrix}, \quad \begin{pmatrix} 1 & 2 & 3 & 4 \\ 3 & 2 & 4 & 1 \end{pmatrix}, \quad \begin{pmatrix} 1 & 2 & 3 & 4 \\ 3 & 1 & 4 & 2 \end{pmatrix}, \quad \begin{pmatrix} 1 & 2 & 3 & 4 \\ 4 & 2 & 3 & 1 \end{pmatrix},$$

and

$$\begin{pmatrix} 1 & 2 & 3 & 4 \\ 4 & 1 & 3 & 2 \end{pmatrix}.$$

---

[23] Désiré André (1840-1917), French mathematician.

Let us see a simple recursion satisfied by these numbers.

A basic recurrence for the Euler numbers

For any $n \geq 1$

$$2E_{n+1} = \sum_{k=0}^{n} \binom{n}{k} E_k E_{n-k}. \tag{1.16}$$

To see the validity of this recursion, choose an arbitrary subset of k elements from the set $\{1, 2, \ldots, n\}$ (in $\binom{n}{k}$ ways). Then choose a π_1 up-down zigzag on these k elements, and a π_2 down-up zigzag on the remaining $n - k$ elements. From the former, there are E_k; from the latter, there are E_{n-k}, so up to this point we are given $\binom{n}{k} E_k E_{n-k}$ possibilities. If we construct a permutation π by *concatenating* the π_1^r reverse of π_1, the fixed point one element permutation $n+1$, and π_2, we see that π is either a down-up or an up-down zigzag on $n+1$ elements. To see this in an example, let $n = 7$ and we choose three elements for π_1. Let them be, say, $2, 5, 7$; while the rest goes into π_2. Then we form an up-down on $2, 5, 7$: $5, 7, 2$, and a down-up on the other elements: $3, 1, 6, 4$. Concatenating these with 8 in between, π will be $2, 7, 5, 8, 3, 1, 6, 4$.

In fact, π uniquely encodes π_1 and π_2; these can be *decoded* from π, with the aid of $n+1$. In our example $n+1 = 8$, so we split up π: $2, 7, 5$ and $3, 1, 6, 4$. The first is the reverse of π_1, the second is π_2.

This uniqueness gives that we have constructed all two types of zigzag permutations (letting k run from zero to n); altogether there are $2E_{n+1}$ of these. This gives that the recursion is indeed correct.

1.8.1 Zigzag permutations and trees

The Euler numbers appear in many combinatorial contexts, not only in the problem of zigzags. We refer to [573] for a survey. Here we work out one specific example which has multiple connections to our combinatorial numbers.

Let us take a zigzag permutation:

$$\pi = \begin{pmatrix} 1 & 2 & 3 & 4 & 5 & 6 & 7 & 8 & 9 \\ 6 & 3 & 4 & 1 & 5 & 2 & 8 & 7 & 9 \end{pmatrix}$$

In this subsection we write out only the second line and we left the parenthesis. This shortens our formulas, and we do not lose any information. So π will look like[24] $\pi = 634152879$. We split this permutation into two parts, according to its minimal element: 634 and 52879. Splitting these again with respect to their minimal elements, we get 6, 4, 5, and 879. Finally, the only splittable subpermutation is 879 which splits into 8 and 9. This splitting process together with the minimal elements and the final remainders can be arranged in a shape which we call *tree*:

---

[24]This short notation is called *word representation*.

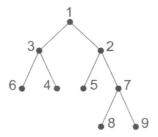

More precisely, such a tree is called *labeled increasing plane 2-tree*. Labeled, because the edges (nodes) of the tree have their labels; increasing[25], because the labels increase on any path going downwards; plane, because the order of the elements stemming to the left or right (the *children*) from a given edge (like 3 and 4 from 2) matters; 2-tree, because every edge must contain two children if it is not a terminal node. A terminal node is called *leaf.*

We have to note, however, that this correspondence (which is, we shall see immediately, bijective!) between zigzags and 2-trees works only when the number of elements in the permutation (and the number of edges in the tree) is odd. It is easy to see that there is no labeled non-decreasing plane 2-tree of even edges. How can we recover the bijective relation between zigzags and trees? We consider a zigzag on an odd number of elements: $i_1 i_2 \cdots i_{n-1}$, then try to insert n. In order to keep the zigzag property, n cannot be inserted wherever we want. We are going to show that we have only one option. n must be inserted somewhere where we have $i_k < i_{k+1}$, so that $i_k < n > i_{k+1}$ (obviously, n cannot be inserted in a place where $i_k > i_{k+1}$, because it would immediately break the zigzag property). This seems to be okay as far as we do not consider the element next to i_{k+1}: $i_k < i_{k+1} > i_{k+2}$ now turns to be $i_k < n > i_{k+1} > i_{k+2}$, which is not acceptable. The only way out is that i_{k+1} is the *last* element in the permutation. In the corresponding tree, this yields that n appears as the leaf stemming out from the *rightmost* edge (to the left). To illustrate this, insert $n = 10$ into the above permutation $\pi = 634152879$. The resulting tree is

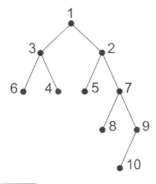

---

[25] We admit that adding both adjectives "labeled" and "increasing" is somewhat super-fluous, because the increasing property is defined only for labeled trees.

confirming our above observation.

Collecting the above facts, our statement is the following. If n is odd, E_n counts the number of labeled increasing plane 2-trees on n edges. If n is even, E_n counts the labeled increasing plane trees which have two children except the rightmost edge which has only one child (stemming to the left).

A similar bijective proof can be found in [112].

Exercises

1. Calculate the fifth and sixth Bell number by using the recursion (1.1).

2. Write down all the 4-partition of $\{1, 2, 3, 4, 5\}$. (There are 10.)

3. Show that $\left\{{n \atop 3}\right\} = \frac{9}{2}3^{n-3} - 4 \cdot 2^{n-3} + 1/2$ $(n \geq 3)$.

4. Prove that
$$\left\{{n+1 \atop k+1}\right\} = \sum_{m=0}^{n} \binom{n}{m}\left\{{m \atop k}\right\}.$$
   (Note that this is only another form of (1.5).)

5. By using the recursion (1.4), calculate the first six lines in the second kind Stirling triangle.

6. Decompose the permutation
$$\begin{pmatrix} 1 & 2 & 3 & 4 & 5 & 6 & 7 & 8 \\ 8 & 1 & 4 & 2 & 7 & 6 & 5 & 3 \end{pmatrix}$$
   into cycles.

7. Give all the 4-partitions of a 5-set. (There are 10.)

8. How many configurations are there if we would like 10 persons to sit around four *round* tables such that none of the tables can be empty?

9. Show that the number of ways a product of n distinct primes can be factored is B_n.

10. Prove the pair of (1.5):
$$\left[{n \atop k}\right] = \sum_{m=0}^{n} \left[{n \atop m}\right]\binom{m}{k-1}.$$

11. Compute the first six lines of the first kind Stirling triangle.

12. Write a computer algorithm which calculates the Stirling numbers of the first kind applying the recursion (1.9).

13. In how many ways can we give n coins to k people, if each person receives at least one coin?

14. How many functions are there which map n elements into k elements, such that none of the k images appears more than once (such function is called *invertible* or *injective*)?

15. Look for the arcs in the partition

$$1, 2, 6, 7, 8|3, 4, 9|5.$$

16. What partition belongs to the below rook placement?

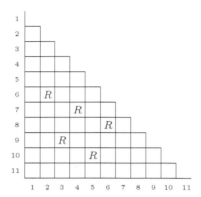

What is the maximal level d of regularity?

17. In how many ways can we place rooks on an 8×8 chessboard (avoiding attacks)?

18. Solve the previous problem for a general $n \times m$ chessboard and k rooks. (The answer is $k!\binom{n}{k}\binom{m}{k}$, where, of course, k cannot be greater than the minimum of n and m.)

19. Determine the tree corresponding to the zigzag 725491638.

20. Decipher the zigzag permutation encoded in the below tree.

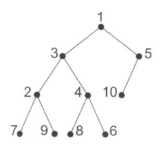

21. The increasing non-plane $1 - 2$ trees are trees such that the non-leaf edges can have one or two children, and the order of the children does not matter (if there is only one child, it can go to the left or right; this makes no difference). Prove that the number of such trees is E_n. (So E_n counts at least two different tree classes.) (Hint: the combinatorial

proof is not so easy, it was done by R. Donaghey [207]. A simpler proof can be done by generating a function technique. This technique will be detailed in Chapter 2. The proof via generating a function method can be found in [536] (Theorem 3.2) or in [83] (Exercises (14-16) on p. 471, and solutions on p. 478-479.)

Outlook

1. The harmonic numbers are related to overhanging deck packings. See [566] for a good source of the description and discussion of the problem.

2. Relation (1.10) can be proven combinatorially, without relying on recursion (as we did). Not only (1.10), but a wealth of other identities for the Stirling numbers of the first kind and harmonic numbers are presented in [63]. We can learn from this paper, for example, that the expected number of cycles in a randomly and uniformly chosen permutation on n elements is H_n. This is equivalent to the pretty formula

$$\frac{1}{n!} \sum_{k=1}^{n} k \begin{bmatrix} n \\ k \end{bmatrix} = H_n.$$

   The reader is called to verify this statement by himself/herself. The expected number of blocks in a partition is among the exercises of Chapter 3.

3. Some recent papers are devoted to study the number of permutations and partitions where the smallest or largest fixed point or singleton is given. See [191] for permutations and [547, 549] for partitions.

4. The Stirling and Bell numbers can be extended to graphs. Let $S(G, k)$ be the number of partitions of the vertices of the graph G into k nonempty, independent sets. Then, quite obviously, for the graph E_n with n vertices and no edges we have that $S(E_n, k) = \begin{Bmatrix} n \\ k \end{Bmatrix}$. According to the research of D. Galvin and Do T. Thanh [241], the quantity $S(G, k)$ was introduced by I. Tomescu [562]. Reasonably, the $S(G, k)$ numbers are called *graphical Stirling numbers* by B. Duncan and R. Peele [209]. The sequence $(S(G, k))_{k \geq 0}$ is related to colorings, and it is called the chromatic vector of G [256]. See [46, 331] and the literature therein for more sources, and also [407] for related investigations.

5. The Stirling and Bell numbers can be applied in the enumeration of certain types of forests (forests are collections of trees) [581].

6. Symbolic summation methods are algorithms which automatize the finding of recursions for sequences which are defined by finite sums with terms of specific type. Such methods are available when the summands are hypergeometric, holonomic, nested sum-product expressions, and some others. Recently, symbolic summation methods were developed by M. Kauers and C. Schneider for Stirling numbers. See [326, 327] for the details, and for a collection of citations for the other above-mentioned algorithms.

7. There is a surprising integral representation for the Bell numbers which was discovered in 1885 by Cesàro [131] and holds for all $n \geq 1$:

$$B_n = \frac{2n!}{\pi e} \int_0^\pi e^{e^{\cos\theta}\cos(\sin\theta)} \sin\left(e^{\cos\theta}\sin(\sin\theta)\right) \sin n\theta \, d\theta.$$

This integral appeared in an editorial remark to a Monthly problem [102]. An easily available proof is due to Callan [111].

8. The Stirling numbers of the first kind can be represented by integrals [7, 473], by moments of some properly defined random variable [6], and by the hypergeometric function [3] (see also the Outlook of Chapter 8).

9. A simple *Mathematica* code was written by Robert M. Dickau [194] to construct lists of all the partitions of an n element set.

10. Those who would like to know more about computer-aided combinatorics should consult the book of S. Pemmaraju and S. Skiena [458] in which the use of the combinatorics features of the *Mathematica* software are described in great detail.

11. See a reprint of Stirling's tables and a short introduction to Stirling numbers in [93].

12. The nth power of the differential operator $xD = x\frac{d}{dx}$ is a linear combination of $x^k D^k$ with Stirling number coefficients:

$$(xD)^n = \sum_{k=0}^n \left\{ n \atop k \right\} x^k D^k.$$

This is a result from 1823 by H. Scherk [509]. In general, the possibility to write down polynomials of xD as a sum of the monomials $x^k D^k$ is known as the *normal ordering problem*. This problem was generalized by Scherk himself, and later further generalized by a number of authors. Schwatt [513] shows the utility of $(xD)^n$ in the summation of powers and like series. Carlitz, inspired by the work of Schwatt [128, 129] studied, among others, $(x^\lambda D^\lambda)^n$, Comtet [159] and Lang [355] considered $(x^r D)^n$. All these operators are expressible as a linear combination of $x^k D^k$, only the coefficients alter. The above-mentioned studies, however, did not consider the combinatorial properties of the arising coefficients. The first who gave a presentation on a possible interpretation of these numbers was Navon [451] in 1973. He did actually even more, considered an arbitrary finite word w over the alphabet $\{x, D\}$, and showed that w can be expressed as a linear combination of $x^k D^k$ with Stirling-like number coefficients, denoted by $S_w(k)$, and gave a description of these numbers for arbitrary w (the parameter n does not appear explicitly, because the length n of the linear combination depends on the length of the word w). Codara et al. [154] connected the $S_w(k)$ numbers with graph coloring in the particular case when $w = (x^s D^s)^n$. The newest graph theoretic achievements with respect to the normal ordering problem are detailed in [214]. Interpretation and calculations of $S_w(k)$ via tree structures was achieved by a number of authors, see [37, 80, 356, 402]. Relations to Feynman diagrams was discovered by Solomon et. al. [526]. The operator $(x^\lambda D^\lambda)^n$ was studied in [325] for non-integer λ.

See also [394] for a modern treatise where historical remarks on the Stirling and related numbers can also be found. A so-called q-analogue of the Stirling numbers was defined by Garsia and Remmel [242], from which a whole theory of q-Stirling numbers has been growing out. See again [394], and [512]. Also, the book [392] of T. Mansour contains a great deal of historical information about set partitions, and it is a good introduction to the so-called pattern avoidance.

13. The Euler numbers are very widely studied in the literature, see the survey [536] for references and [282] for some simple proofs of basic properties of the Euler numbers and some refinements.

14. For an interesting connection between permutations and RNA, see [145].

15. Let (a_n) and (b_n) be two sequences such that $a_1 = b_1$, and

$$b_n + a_1 b_{n-1} + \cdots + a_{n-1} b_1 = n a_n \quad (n > 1).$$

We then say that the pair (a_n, b_n) is a *Newton-Euler pair*. Further, we say that (b_n) is a *Newton-Euler sequence* if a_n is integer (positive, negative, or zero) for all $n \geq 1$. T. Lengyel proved that $(k!\{{n \atop k}\})$ is a Newton-Euler sequence [368].

16. Henry Gould (an American mathematician, born in 1928) has a famous and huge register of combinatorial formulas. These were compiled into a book [474] by J. Quaintance. The Foreword and Preface paints a lovely story about the workstyle and deep knowledge of Gould. Also, [474] contains a large number of formulas not only for the Stirling numbers but for the binomial coefficients, Bernoulli numbers, and many other combinatorially interesting number sequences. The first chapters introduce methods that are worth to be known by everyone who is interested in the *symbolic derivation* of combinatorial formulas.

Chapter 2

Generating functions

The method of generating functions is extremely useful in enumeration problems. With the aid of generating functions we can discover a host of new identities, divisibility properties and asymptotical behavior[1] of our counting sequences. This chapter needs some knowledge from the reader's side in analyis, especially differential and integral calculus, and some basics from differential equations. In a non-essential way, some matrix algebra will also be used. An excellent book which contains a detailed discussion of generating functions is [262][2].

2.1   On the generating functions in general

For a while we leave our study of partitions and permutations. Later – having the strong tool of generating functions in our hand – we turn back to combinatorics.

Examples

Let a_n be a sequence of (real) numbers (with $n = 0, 1, 2, \ldots$). Then we can form a function

$$f(x) = a_0 x^0 + a_1 x^1 + a_2 x^2 + a_3 x^3 + \cdots$$

which is called the *generating function* of the sequence a_n.

In particular, all the polynomials are generating functions. For example, the polynomial

$$f(x) = 2 + 4x - 3x^2 + 5x^3$$

---

[1] Asymptotic investigations reveal how rapidly grows (or decreases) a sequence of numbers.

[2] [262] is a good source also for the basics of Stirling numbers. The books [597] and [235] are other excellent sources for those who want to study the interrelationship between combinatorics and analysis. A good introduction to generating functions is Chapter 8 in [82], but the whole book [82] is worth reading for a wide scope introduction to combinatorics.

is the generating function of the sequence $a_0 = 2, a_1 = 4, a_2 = -3, a_3 = 5$, and $a_n = 0$ if $n > 3$.

In other words, the *polynomials* are generating functions of *finite sequences*. Now let $a_0 = a_1 = a_2 = \cdots = a_n = 1$. Then

$$f(x) = 1 + x + x^2 + x^3 + \cdots + x^n.$$

Let us multiply both sides with x:

$$xf(x) = x + x^2 + x^3 + \cdots + x^{n+1}.$$

Subtracting this from the above, we see that only two members survive on the right-hand side:

$$f(x) - xf(x) = 1 - x^{n+1}.$$

This can be rewritten as

$$f(x) = \frac{1 - x^{n+1}}{1 - x}. \tag{2.1}$$

We have proven a handy calculation formula for sums of increasing powers of a fixed number. Let us take an example. Let $x = 3$ and $n = 10$. Then

$$f(3) = 1 + 3 + 3^2 + 3^3 + 3^4 + 3^5 + 3^6 + 3^7 + 3^8 + 3^9 + 3^{10} = 88\,573.$$

Applying (2.1) we have that

$$f(3) = \frac{1 - 3^{11}}{1 - 3} = 88\,573.$$

Let us move on and take infinite sequences. For instance, let *every* $a_n = 1$. Since $f(x) = \frac{1 - x^{n+1}}{1 - x}$ for any n, (2.1) is valid for every n. But if we try to apply it to $x = 3$, the numerator becomes larger and larger as n grows. At the end, we get that the *infinite* sum of the powers of 3 is *infinity* – an obvious fact.

But if x is small, like $x = \frac{1}{2}$, then the powers of x in (2.1) become smaller and smaller. In this case, if n tends to infinity, this argument results that x^{n+1} is negligibly small in the expression of $f(x)$. Hence,

$$f(x) = 1 + x + x^2 + x^3 + x^4 + \cdots = \frac{1}{1 - x}. \tag{2.2}$$

We can shed more light on this argument by taking another example. Let $x = \frac{1}{10}$. Then

$$
\begin{aligned}
f\left(\frac{1}{10}\right) &= 1 + \frac{1}{10} + \frac{1}{10^2} + \frac{1}{10^3} + \frac{1}{10^4} + \cdots \\
&= 1 + 0,1 + 0,01 + 0,001 + 0,0001 + \cdots \\
&= 1,111111\ldots \\
&\overset{(2.2)}{=} \frac{1}{1 - \frac{1}{10}} = \frac{10}{9}.
\end{aligned}
$$

This is really true, because $1.\overline{11} = 1\frac{1}{9} = \frac{10}{9}$, a high school fact.

Complicating a bit, we set $a_n = 1$, if n is even and 0 if n is odd. Thus,

$$f(x) = 1 + x^2 + x^4 + x^6 + \cdots = (x^2)^0 + (x^2)^1 + (x^2)^3 + \cdots = \frac{1}{1 - x^2},$$

if we write x^2 in place of x in (2.2).

Exponential generating functions

We introduce an important class of generating functions. If a_n is a sequence of numbers, then the function

$$f(x) = \frac{a_0}{0!}x^0 + \frac{a_1}{1!}x^1 + \frac{a_2}{2!}x^2 + \frac{a_3}{3!}x^3 + \cdots$$

is called the *exponential generating function*[3] of a_n (or, which is the same, f is the generating function of $a_n/n!$). Hence, for example, the exponential generating function of $a_n = n!$ is $\frac{1}{1-x}$. In the simplest case, when $a_n = 1$, the exponential generating function is

$$f(x) = \frac{1}{0!}x^0 + \frac{1}{1!}x^1 + \frac{1}{2!}x^2 + \frac{1}{3!}x^3 + \cdots \qquad (2.3)$$

which is denoted by e^x, and called *exponential function* (this is where the name "exponential generating function" comes). Here $e \approx 2,718$ is the base of the *natural logarithm*[4].

2.2 Operations on generating functions

2.2.1 Addition and multiplication

One can introduce operations on generating functions to construct new generating functions from older ones.

If we have two sequences like a_n and b_n, and their generating functions are $f(x)$ and $g(x)$, respectively, then the generating function of $a_n + b_n$ is simply $f(x) + g(x)$, because

$$\begin{aligned}
(a_0 + b_0)x^0 + (a_1 + b_1)x^1 + \cdots \ &= \ a_0x^0 + b_0x^0 + a_1x^1 + b_1x^1 + \cdots = \\
&\quad\ a_0x^0 + a_1x^1 + \cdots + b_0x^0 + b_1x^1 + \cdots = \\
&\quad\ f(x) + g(x).
\end{aligned}$$

---

[3] By convention $0! = 1$.

[4] The letter "e" is chosen to remember Leonhard Euler (1707-1783), Swiss mathematician and physicist, one of the greatest scientists in the history of science.

It can similarly be seen that if the generating function of a_n is $f(x)$, then the generating function of $c \cdot a_n$ is simply $c \cdot f(x)$ (here c is an arbitrary constant).

If we need the generating function of $c^n a_n$ just take $f(cx)$ in place of $f(x)$:

$$f(cx) = \sum_{n=0}^{\infty} a_n (cx)^n = \sum_{n=0}^{\infty} (c^n a_n) x^n.$$

Life is not so easy if we would like to construct the generating function of $a_n b_n$ from $f(x)$ and $g(x)$. Let us first consider the finite case. For example, if the only non-zero terms are a_0, a_1, a_2 and b_0, b_1, b_2, then

$$\begin{aligned} f(x)g(x) =& (a_0 + a_1 x + a_2 x^2)(b_0 + b_1 x + b_2 x^2) \\ =& (a_0 b_0 + a_0 b_1 x + a_0 b_2 x^2) + (a_1 b_0 x + a_1 b_1 x^2 + a_1 b_2 x^3) \\ & + (a_2 b_0 x^2 + a_2 b_1 x^3 + a_2 b_2 x^4). \end{aligned} \quad (2.4)$$

(From now on, in place of x^0 we simply write 1.)

From the application point of view, this product is not so useful as another product in which we "forget" the terms with degree greater than the degree of the factors. Applying this rule to the above product, we leave the terms of degree ≥ 2:

$$(a_0 b_0 + a_0 b_1 x + a_0 b_2 x^2) + (a_1 b_0 x + a_1 b_1 x^2) + (a_2 b_0 x^2) =$$

$$a_0 b_0 + (a_0 b_1 + a_1 b_0)x + (a_0 b_2 + a_1 b_1 + a_2 b_0)x^2.$$

In a more compact form:

$$\sum_{n=0}^{2} x^n \left(\sum_{k=0}^{n} a_k b_{n-k} \right).$$

This is useful also because it can be generalized easily for arbitrary (not only finite) a_n, b_n sequences. The product of the generating functions of these sequences is then

$$f(x)g(x) = \sum_{n=0}^{\infty} x^n \left(\sum_{k=0}^{n} a_k b_{n-k} \right). \quad (2.5)$$

(From now on, we frequently use the sum symbol, hence the formulas will be more readable and compact.) This product is called *Cauchy*[5] *product*. In what follows, we will always use the Cauchy product and never will use products like (2.4). Hence, if we talk about the product of two generating functions, the coefficients in the product are always taken to be as Cauchy's product dictates.

---

[5] Augustin Louis Cauchy (1789-1857), French mathematician.

We demonstrate in a simple example how Cauchy product works in practice. Let $a_n = b_n = 1$. Then $f(x) = g(x) = \frac{1}{1-x}$, and

$$f(x)g(x) = \frac{1}{(1-x)^2}.$$

On the other hand, by (2.5)

$$\sum_{n=0}^{\infty} x^n \left(\sum_{k=0}^{n} 1 \cdot 1 \right) = \sum_{n=0}^{\infty} (n+1)x^n. \qquad (2.6)$$

Hence, the Cauchy product of the generating functions of $a_n = b_n = 1$ is $c_n = n + 1$.

Now we study the exponential generating function of the same sequences a_n and b_n. Their exponential generating functions are e^x (see (2.3)).

$$e^x \cdot e^x = \sum_{n=0}^{\infty} x^n \left(\sum_{k=0}^{n} \frac{1}{k!} \frac{1}{(n-k)!} \right).$$

The members of the inner sum are $\frac{1}{n!} \binom{n}{k}$, we carry out $\frac{1}{n!}$. Hence, the entire sum equals to

$$\sum_{n=0}^{\infty} \frac{1}{n!} x^n \left(\sum_{k=0}^{n} \binom{n}{k} \right).$$

It is known that the inner sum equals to 2^n (see the binomial theorem in the Appendix for a simple proof); thus the final form is

$$\sum_{n=0}^{\infty} \frac{1}{n!} 2^n x^n = \sum_{n=0}^{\infty} \frac{1}{n!} (2x)^n = e^{2x}.$$

(This shows that the function e^x behaves as we expect: $e^x \cdot e^x = e^{x+x} = e^{2x}$.)

It is now easy to find a formula for the Cauchy product of two exponential generating functions in general. Let

$$f(x) = \sum_{n=0}^{\infty} \frac{a_n}{n!} x^n \quad \text{and} \quad g(x) = \sum_{n=0}^{\infty} \frac{b_n}{n!} x^n.$$

Then their Cauchy product is

$$f(x)g(x) = \sum_{n=0}^{\infty} \left(\sum_{k=0}^{n} \frac{a_k}{k!} x^k \cdot \frac{b_{n-k}}{(n-k)!} x^{n-k} \right).$$

Here x^k and x^{-k} cancel and the resulting x^n can be carried out from the inner sum. Moreover, $\frac{1}{k!} \frac{1}{(n-k)!} = \frac{1}{n!} \binom{n}{k}$, and $\frac{1}{n!}$ can also be carried out. The result is

$$f(x)g(x) = \sum_{n=0}^{\infty} \frac{x^n}{n!} \left(\sum_{k=0}^{n} \binom{n}{k} a_k b_{n-k} \right). \qquad (2.7)$$

The inner sum $\sum_{k=0}^{n} \binom{n}{k} a_k b_{n-k}$ will play a very important role, see Section 2.3.

2.2.2   Some additional transformations

We describe three important observations that help in the determination of generating functions.

1. If $f(x)$ is multiplied by x, we get the generating function of a_{n-1}:

$$xf(x) = \sum_{n=0}^{\infty} a_n x^{n+1} = \sum_{n=1}^{\infty} a_{n-1} x^n,$$

2. If $f(x)$ is divided by x, we get the generating function of a_{n+1}:

$$\frac{f(x)}{x} = \sum_{n=1}^{\infty} a_n x^{n-1} = \sum_{n=0}^{\infty} a_{n+1} x^n.$$

(To eliminate negative powers of x, we supposed that $a_0 = 0$.)

3. Let a_n be a sequence and $f(x)$ be its generating function. Multiplying $f(x)$ by $\frac{1}{1-x}$ (which belongs to $b_n = 1$) we have that

$$\frac{f(x)}{1-x} = \sum_{n=0}^{\infty} x^n \left(\sum_{k=0}^{n} a_k \cdot 1 \right).$$

Therefore, $\frac{f(x)}{1-x}$ is the generating function of the *partial sums* of a_n:

$$\frac{f(x)}{1-x} = a_0 + (a_0 + a_1)x + (a_0 + a_1 + a_2)x^2 + (a_0 + a_1 + a_2 + a_3)x^3 + \cdots.$$

These observations are valid for ordinal generating functions. In the exponential case, we get that $xf(x)$ belongs to $\frac{a_{n-1}}{(n-1)!}$, and $f(x)/x$ belongs to $\frac{a_{n+1}}{(n+1)!}$.

In the third case, the exponential generating function of $b_n = 1$ is e^x, and we have to apply (2.7):

$$f(x)e^x = \sum_{n=0}^{\infty} \frac{x^n}{n!} \left(\sum_{k=0}^{n} \binom{n}{k} a_k \right). \tag{2.8}$$

The coefficient in the sum on the right-hand side is the *binomial transform* of a_n. This is an important transformation, we will turn back to this notion in Section 2.3.

2.2.3   Differentiation and integration

In this subsection we need to know a bit about differential and integral calculus.

Manipulations of the ordinary generating function

Let a_n be an arbitrary sequence and $f(x)$ be its generating function. By using differentiation and integration we can introduce some additional tricks in order to construct new generating functions from older ones.

1. The derivative of $f(x)$ is[6]

$$f'(x) = \left(\sum_{n=0}^{\infty} a_n x^n\right)' = \sum_{n=0}^{\infty} a_n (x^n)' = \sum_{n=1}^{\infty} a_n n x^{n-1}.$$

Multiplying by x we have that the generating function of na_n is $xf'(x)$.

2. Integrating and dividing by x, we have the generating function of $\frac{a_n}{n+1}$:

$$\frac{1}{x}\int f(x)dx = \frac{1}{x}\int \left(\sum_{n=0}^{\infty} a_n x^n\right) dx = \frac{1}{x}\sum_{n=0}^{\infty} a_n \int x^n dx =$$

$$\frac{1}{x}\sum_{n=0}^{\infty} a_n \frac{x^{n+1}}{n+1} = \sum_{n=0}^{\infty} \frac{a_n}{n+1} x^n.$$

3. If we divide by x *before* integrating, we get the generating function of $\frac{a_n}{n}$:

$$\int \frac{f(x)}{x}dx = \int \left(\sum_{n=0}^{\infty} a_n x^{n-1}\right) = \sum_{n=0}^{\infty} a_n \int x^{n-1}dx = \sum_{n=0}^{\infty} \frac{a_n}{n} x^n.$$

Manipulations of the exponential generating function

1. The derivative of $f(x)$:

$$f'(x) = \sum_{n=0}^{\infty} \frac{a_n}{n!}(x^n)' = \sum_{n=1}^{\infty} \frac{a_n}{(n-1)!}x^{n-1} = \sum_{n=0}^{\infty} \frac{a_{n+1}}{n!}x^n.$$

Hence, $f'(x)$ is the exponential generating function of the sequence $b_n = a_{n+1}$.

2. By integration, we get the exponential generating function of $b_n = a_{n-1}$ $(n > 0)$, and $b_0 = 0$:

$$\int f(x)dx = \sum_{n=0}^{\infty} \frac{a_n}{n!}\int x^n dx = \sum_{n=0}^{\infty} \frac{a_n}{(n+1)!}x^{n+1} = \sum_{n=1}^{\infty} \frac{a_{n-1}}{n!}x^n.$$

---

[6]It is not true in general that the derivative of an infinite sum of functions can be taken termwise. The series must converge uniformly. However, inside the radius of convergence this always satisfies.

3. If we divide by x before integration we have the exponential generating function of $\frac{a_n}{n}$:

$$\int \frac{f(x)}{x} dx = \sum_{n=0}^{\infty} \frac{a_n}{n!} \int x^{n-1} dx = \sum_{n=0}^{\infty} \frac{a_n}{n} \frac{x^n}{n!}.$$

Of course, one could add more and more rules dividing or multiplying by x^2, differentiating two times, and so on. We will practice these tricks many times in the course of this book, so here we mention just one additional example: let us find the generating function of $b_n = n$. We can take, for instance, the sequence $a_n = 1$ and then na_n will exactly be b_n. The first transformation of this section says that the generating function of na_n is $xf'(x)$. Now $f(x) = \frac{1}{1-x}$, so

$$\sum_{n=1}^{\infty} nx^n = x \left(\frac{1}{1-x} \right)' = \frac{x}{(1-x)^2}.$$

Thus, the generating function of $b_n = n$ is $\frac{x}{(1-x)^2}$.

2.2.4   Where do the name generating functions come from?

After introducing the generating functions and their most basic operations, we show where their name comes from.

Let us take the function

$$f(x) = \sum_{n=0}^{\infty} a_n x^n,$$

and the nth term of the sum: $a_n x^n$. If we take the derivative of this function n times, the terms with powers smaller than n disappear. In addition, the first not disappearing term, $a_n x^n$, turns to be $a_n n x^{n-1}$ after the first derivation, then $a_n n(n-1)x^{n-2}$ after the second derivation, and finally $a_n n(n-1)\cdots 2 \cdot 1x^0 = a_n n!$ after taking the nth derivative. If we now substitute $x = 0$ into the nth derivative, all the terms with power higher than 0 disappear, and the value of this derivative at zero will be $n!a_n$. Dividing by $n!$, we have the fundamental fact that

$$\frac{1}{n!} f^{(n)}(x) \Big|_{x=0} = a_n. \tag{2.9}$$

Here the vertical line and $x = 0$ in the lower index means that we substitute $x = 0$ into the derivative. If $f(x)$ is an exponential generating function, then it is equally easy to see that we do not divide by $n!$:

$$f^{(n)}(x) \Big|_{x=0} = a_n. \tag{2.10}$$

To take an example, we look for the sequence a_n for which the sine function is the generating function. a_0 must simply be the zeroth derivative at $x = 0$, divided by $0! = 1$:

$$a_0 = \frac{1}{0!} \sin^{(0)}(x)\Big|_{x=0} = \frac{1}{0!} \sin 0 = 0.$$

Similarly,

$$a_1 = \frac{1}{1!} \sin^{(1)}(x)\Big|_{x=0} = \frac{1}{1!} \cos 0 = 1.$$

$$a_2 = \frac{1}{2!} \sin^{(2)}(x)\Big|_{x=0} = \frac{1}{2!} - \sin 0 = 0.$$

$$a_3 = \frac{1}{3!} \sin^{(3)}(x)\Big|_{x=0} = \frac{1}{3!} - \cos 0 = -\frac{1}{3!}.$$

$$a_4 = \frac{1}{4!} \sin^{(4)}(x)\Big|_{x=0} = \frac{1}{4!} \sin 0 = 0.$$

$$a_5 = \frac{1}{5!} \sin^{(5)}(x)\Big|_{x=0} = \frac{1}{5!} \cos 0 = \frac{1}{5!}.$$

We see that, in general, $a_{2n+1} = \frac{(-1)^n}{(2n+1)!}$, and the even indexed a_ns are all zero[7]. We have the following statement for the sine function:

$$\sin x = \sum_{n=0}^{\infty} \frac{(-1)^n}{(2n+1)!} x^{2n+1}.$$

From this nice expansion[8] – and from the geometrical fact that $\sin \frac{\pi}{2} = 1$ – comes the nice infinite sum identity:

$$\sum_{n=0}^{\infty} \frac{(-1)^n}{2^{2n+1}(2n+1)!} \pi^{2n+1} = 1.$$

Many classical functions from analysis (like some hyperbolic functions, and the trigonometric functions in general) are generating functions of sequences having combinatorial meaning. This branch of mathematics is on the border of combinatorics and analysis and it is called *combinatorial trigonometry*[9]

---

[7] If the reader is familiar with the notion of odd and even functions, she/he maybe knows that sine is an *odd* function. This is so, because the even powers of x disappear in its expansion. In generating function vocabulary, the even indexed members of the generating sequence are zero.

[8] This is also called the *Taylor-expansion* of the sine function.

[9] A nice comprehensive treatment of this narrow area of mathematics is due to Arthur T. Benjamin. His presentation can be read here: math.nist.gov/mcsd/Seminars/2008/2008-01-11-Benjamin.html. See also [60, 236].

2.3   The binomial transformation

As we noted during deducing (2.8), the binomial transform of a sequence a_n is defined by

$$b_n = \sum_{k=0}^{n} \binom{n}{k} a_k$$

We also know that if the exponential generating function of a_n is $f(x)$, then the exponential generating function of b_n is $e^x f(x)$.

We had actually met with a binomial transform at the very beginning. This was the recursion (1.1). This recursion says that for the Bell numbers the binomial transform is $b_n = B_{n+1}$. In other words, the binomial transformation acts on the Bell number sequence as an *index shifting*.

Let us take another example. The binomial theorem (see the Appendix) says that

$$(x + y)^n = \sum_{k=0}^{n} \binom{n}{k} x^k y^{n-k}.$$

If here we substitute $y = 1$ we have that

$$(x + 1)^n = \sum_{k=0}^{n} \binom{n}{k} x^k.$$

In this case $a_n = x^n$ and $b_n = (x + 1)^n$. The binomial transform is naturally applicable on polynomial sequences.

We can *invert* the transform to get back a_n from b_n in the above example. We just need to substitute $x + 1$ in place of x and -1 in place of y:

$$(x + 1 - 1)^n = x^n = \sum_{k=0}^{n} \binom{n}{k} (x+1)^k (-1)^{n-k}.$$

A similar approach works in general (the one line proof is seen below). If

$$b_n = \sum_{k=0}^{n} \binom{n}{k} a_k, \tag{2.11}$$

then

$$a_n = \sum_{k=0}^{n} \binom{n}{k} (-1)^{n-k} b_k. \tag{2.12}$$

This is the *binomial inversion theorem*. Applying this, say, to the Bell numbers we readily get a new formula for them:

$$B_n = \sum_{k=0}^{n} \binom{n}{k} (-1)^{n-k} B_{k+1}.$$

The proof of the binomial inversion theorem is just one line if we apply exponential generating functions. If we note that the exponential generating function of the transformed sequence is always $e^x f(x)$, so, to get back $f(x)$ we simply multiply by e^{-x}.

The matrix form of the binomial transformation

The results in the above section can be rephrased in matrix form; this approach can result in more comprehensive formulas and might be easier to generalize.

Looking at the right-hand side of (2.11), we see immediately that it is, in fact, a matrix multiplication. Considering the so-called *Pascal*[10] *matrix*

$$
\mathcal{B}_n = \begin{pmatrix}
\binom{0}{0} & 0 & 0 & \cdots & 0 \\
\binom{1}{0} & \binom{1}{1} & 0 & \cdots & 0 \\
\binom{2}{0} & \binom{2}{1} & \binom{2}{2} & \cdots & 0 \\
\vdots & \vdots & \ddots & \ddots & \vdots \\
\binom{n}{0} & \binom{n}{1} & \cdots & \cdots & \binom{n}{n}
\end{pmatrix}
\tag{2.13}
$$

and the vector

$$
v = \begin{pmatrix} a_0 \\ a_1 \\ \vdots \\ a_n \end{pmatrix}
$$

then their product $\mathcal{B}_n v$ is the right-hand side of (2.11), indeed. Hence, one can write (2.11) as

$$
\begin{pmatrix} b_0 \\ b_1 \\ \vdots \\ b_n \end{pmatrix} = \mathcal{B}_n \begin{pmatrix} a_0 \\ a_1 \\ \vdots \\ a_n \end{pmatrix}.
$$

The binomial inversion theorem is nothing else, but the realization that the multiplication by the inverse of $\mathcal{B}_n$ gives v:

$$
\begin{pmatrix} a_0 \\ a_1 \\ \vdots \\ a_n \end{pmatrix} = \mathcal{B}_n^{-1} \begin{pmatrix} b_0 \\ b_1 \\ \vdots \\ b_n \end{pmatrix}.
$$

---

[10]Blaise Pascal (1623-1662), French philosopher, physicist, and mathematician.

The inversion theorem yields that the inverse of the matrix $\mathcal{B}_n$ is

$$
\mathcal{B}_n^{-1} = \begin{pmatrix}
\binom{0}{0} & 0 & 0 & \cdots & 0 \\
-\binom{1}{0} & \binom{1}{1} & 0 & \cdots & 0 \\
\binom{2}{0} & -\binom{2}{1} & \binom{2}{2} & \cdots & 0 \\
\vdots & \vdots & & \ddots & \vdots \\
\binom{n}{0} & -\binom{n}{1} & \cdots & \cdots & \binom{n}{n}
\end{pmatrix}.
$$

(Here we supposed that n is even, hence the last line begins with a positive number. The sign of $\binom{n}{k}$ is always $(-1)^{n-k}$ in the inverse matrix).

The binomial transform and ordinary generating functions – Euler's theorem

It was easy to get the exponential generating function of the binomial transformed sequence. Can we deduce the ordinary generating function for b_n in general? This question was answered by Euler in 1755. Euler stated[11] that if $f(x)$ is the generating function of a_n, then the generating function of the binomial transform is

$$
\frac{1}{1-x} f\left(\frac{x}{1-x}\right) \tag{2.14}
$$

Let us verify this. We rewrite the above function in more details:

$$
\frac{1}{1-x} f\left(\frac{x}{1-x}\right) = \frac{1}{1-x} \sum_{n=0}^{\infty} a_n \frac{x^n}{(1-x)^n} = \sum_{n=0}^{\infty} a_n \frac{x^n}{(1-x)^{n+1}}.
$$

The denominator of $\frac{x^n}{(1-x)^{n+1}}$ is the generating function of a well-known sequence of the rising factorials. We will meet these later and introduce them in more details. For the moment, we just remark that

$$
\frac{1}{(1-x)^n} = \sum_{k=0}^{\infty} n^{\overline{k}} \frac{x^k}{k!},
$$

where $n^{\overline{k}} = n(n+1)(n+2)\cdots(n+k-1)$. See Exercise 23 for the details. We can now proceed as follows:

$$
\frac{1}{(1-x)^{n+1}} = \sum_{k=0}^{\infty} (n+1)^{\overline{k}} \frac{x^k}{k!} = \sum_{k=0}^{\infty} (n+1)(n+2)\cdots(n+k) \frac{x^k}{k!} =
$$

$$
\sum_{k=0}^{\infty} \frac{(n+k)!}{n!} \frac{x^k}{k!} = \sum_{k=0}^{\infty} \binom{n+k}{n} x^k.
$$

---

[11]The collected works of Euler appeared in a sequence of books entitled *Opera Omnia* (this term is not referring only to Euler's work; it is a Latin term meaning "Complete works"). This result is contained in the 10th volume of the first series. The whole collection contains 79 volumes! See the Euler archive http://eulerarchive.maa.org/ for the compilation.

Writing this back to the former expression, we get

$$\sum_{n=0}^{\infty} a_n \frac{x^n}{(1-x)^{n+1}} = \sum_{n=0}^{\infty} a_n \sum_{k=0}^{\infty} \binom{n+k}{n} x^{n+k}.$$

The double sum can be rewritten if we note that every index and power is of the form n or $n+k$. We keep the outer summation index n, then the inner index is $m = n - k$:

$$\sum_{n=0}^{\infty} a_n \sum_{k=0}^{\infty} \binom{n+k}{n} x^{n+k} = \sum_{n=0}^{\infty} x^n \sum_{m=n-k} \binom{n}{m} a_m.$$

The right-hand side is the binomial transform of a_n, so we have verified the statement of Euler.

2.4 Applications of the above techniques

2.4.1 The exponential generating function of the Bell numbers

We are now endowed with sufficient knowledge to turn back to our combinatorial numbers and study them from a new point of view. Many new identities can be deduced with the generating function technique.

In the first chapter, firstly we met with the Bell numbers. Recall (1.1) and (2.8). Based on these, and letting $f(x)$ be the generating function of B_n,

$$f(x)e^x = \sum_{n=0}^{\infty} \frac{x^n}{n!} \left(\sum_{k=0}^{n} \binom{n}{k} B_k \right) = \sum_{n=0}^{\infty} B_{n+1} \frac{x^n}{n!}.$$

We also know that the derivative of $f(x)$ gives the exponential generating function of B_{n+1}, that is,

$$f(x)e^x = f'(x). \tag{2.15}$$

Which function can be $f(x)$? We can easily see that $f(x) = e^{e^x}$ is a suitable function satisfying the above differential equation. Equation (2.15) can be multiplied by any real number c, $cf(x)$ is also a solution. What distinguishes the actual exponential generating function of the Bell numbers in the whole class $cf(x)$? We know from the theory of generating functions that $f(0)$ is the zeroth coefficient in the series expansion of $f(x)$ (see (2.9)). The zeroth coefficient is $B_0 = 1$ in our case. Since $ce^{e^0} = ce$, we infer that $c = \frac{1}{e}$. We determined the exponential generating function of the Bell numbers:

$$\sum_{n=0}^{\infty} B_n \frac{x^n}{n!} = \frac{1}{e} e^{e^x}. \tag{2.16}$$

This result has a nice and immediate consequence that shows why the generating functions are so useful. This consequence is the Dobiński formula.

2.4.2 Dobiński's formula

A beautiful expression from the year 1877 for the Bell numbers is presented now. This is called the *Dobiński[12] formula*. The following transformations are made:

$$\frac{1}{e}e^{e^x} = \frac{1}{e}\sum_{k=0}^{\infty}\frac{(e^x)^k}{k!} = \frac{1}{e}\sum_{k=0}^{\infty}\frac{e^{kx}}{k!} = \frac{1}{e}\sum_{k=0}^{\infty}\frac{1}{k!}\left(\sum_{n=0}^{\infty}\frac{(kx)^n}{n!}\right).$$

Changing the order of the summation,

$$\frac{1}{e}e^{e^x} = \sum_{n=0}^{\infty}\frac{x^n}{n!}\frac{1}{e}\left(\sum_{k=0}^{\infty}\frac{k^n}{k!}\right).$$

On the left-hand side, there is the exponential generating function of the Bell numbers; therefore, on the right-hand side the coefficient of $\frac{x^n}{n!}$ must be B_n[13]. We deduced the formula of Dobiński:

$$B_n = \frac{1}{e}\sum_{k=0}^{\infty}\frac{k^n}{k!}. \tag{2.17}$$

Therefore, the Bell numbers can be represented by infinite sums. If, in particular, we take $n = 4$, we have

$$\frac{1}{e}\left(\frac{0^4}{0!} + \frac{1^4}{1!} + \frac{2^4}{2!} + \frac{3^4}{3!} + \frac{4^4}{4!} + \frac{5^4}{5!} + \frac{6^4}{6!} + \frac{7^4}{7!}\right) \approx 14,9548$$

One more term gives $14,992$. Remember: the exact value is $B_4 = 15$.

Interestingly, the Dobiński formula can be deduced via probability theory [467, 498].

2.4.3 The exponential generating function of the second kind Stirling numbers

We continue with the determination of the exponential generating function of the second kind Stirling numbers. Let us denote this function by

$$f_k(x) = \sum_{n=0}^{\infty}\left\{{n \atop k}\right\}\frac{x^n}{n!}.$$

---

[12]G. Dobiński – We could not find information on him. The paper in which his formula appears is [205]. The Internet Archive [206] contains a digitalized copy of Dobiński's original work. No affiliation is visible there.

[13]If two generating functions are equal, they must belong to the very same sequence. This can be seen by successive derivation.

For every k we have a separate exponential generating function. Summing on k

$$\sum_{k=0}^{\infty} f_k(x) = \sum_{k=0}^{\infty} \sum_{n=0}^{\infty} \left\{ {n \atop k} \right\} \frac{x^n}{n!} = \sum_{n=0}^{\infty} \frac{x^n}{n!} \left(\sum_{k=0}^{\infty} \left\{ {n \atop k} \right\} \right). \qquad (2.18)$$

The inner sum is the nth Bell number. Note that the sum on k is a *finite* sum, because if $k > n$ the members of the sum are all zero. So

$$\sum_{k=0}^{\infty} f_k(x) = \sum_{n=0}^{\infty} B_n \frac{x^n}{n!}.$$

On the right-hand side, we have the exponential generating function of the Bell numbers (after carrying the factor $1/e$ to the power):

$$\sum_{k=0}^{\infty} f_k(x) = e^{e^x - 1}.$$

By the (2.3) definition of the exponential function

$$e^{e^x - 1} = \sum_{k=0}^{\infty} \frac{(e^x - 1)^k}{k!},$$

which helps us make a guess about the final form of $f_k(x)$. It is

$$f_k(x) = \frac{(e^x - 1)^k}{k!} = \sum_{n=0}^{\infty} \left\{ {n \atop k} \right\} \frac{x^n}{n!}. \qquad (2.19)$$

This is actually not a proof, because there might be many function sequences summing to $e^{e^x - 1}$. To justify this conjecture, we recall the recursion (1.4):

$$\left\{ {n \atop k} \right\} = \left\{ {n-1 \atop k-1} \right\} + k \left\{ {n-1 \atop k} \right\}.$$

Multiplying by x^n and dividing by $n!$ and then summing over n:

$$\sum_{n=0}^{\infty} \left\{ {n \atop k} \right\} \frac{x^n}{n!} = \sum_{n=0}^{\infty} \left\{ {n-1 \atop k-1} \right\} \frac{x^n}{n!} + k \sum_{n=0}^{\infty} \left\{ {n-1 \atop k} \right\} \frac{x^n}{n!}.$$

Considering the operations for the exponential generating functions (especially the fact that decreasing the index is done by integration) this is equivalent to the equation

$$f_k(x) = \int f_{k-1}(x) + k \int f_k(x).$$

Or, taking the derivative on both sides,

$$f_k'(x) = f_{k-1}(x) + k f_k(x).$$

Substituting (2.19), we get that our conjecture was correct, indeed:

$$
\begin{aligned}
\left(\frac{(e^x-1)^k}{k!}\right)' &= \frac{(e^x-1)^{k-1}}{(k-1)!} + k\frac{(e^x-1)^k}{k!} \\
k\frac{(e^x-1)^{k-1}}{k!}e^x &= \frac{(e^x-1)^{k-1}}{(k-1)!} + k\frac{(e^x-1)^k}{k!} \\
\frac{(e^x-1)^{k-1}}{(k-1)!}e^x &= \frac{(e^x-1)^{k-1}}{(k-1)!} + \frac{(e^x-1)^k}{(k-1)!} \\
e^x &= 1 + e^x - 1 \\
e^x &= e^x
\end{aligned}
$$

As for the Bell numbers, it is possible to give another representation for the Stirling numbers. Applying the binomial theorem from the Appendix:

$$
\frac{(e^x-1)^k}{k!} = \frac{1}{k!}\sum_{l=0}^{k}\binom{k}{l}(e^x)^l(-1)^{k-l}.
$$

We know that $(e^x)^l = e^{lx}$ is the exponential generating function of l^n:

$$
\frac{1}{k!}\sum_{l=0}^{k}\binom{k}{l}(e^x)^l(-1)^{k-l} = \frac{1}{k!}\sum_{l=0}^{k}\binom{k}{l}(-1)^{k-l}\sum_{n=0}^{\infty}\frac{l^n}{n!}x^n.
$$

Carrying everything under the sum over n:

$$
\frac{1}{k!}\sum_{l=0}^{k}\binom{k}{l}(-1)^{k-l}\sum_{n=0}^{\infty}\frac{l^n}{n!}x^n = \sum_{n=0}^{\infty}\frac{x^n}{n!}\frac{1}{k!}\sum_{l=0}^{k}\binom{k}{l}(-1)^{k-l}l^n.
$$

Since we started with the exponential generating function of $\left\{{n \atop k}\right\}$, on the right-hand side we must have these numbers as the coefficients of $\frac{x^n}{n!}$. Our result is therefore that

$$
\left\{{n \atop k}\right\} = \frac{1}{k!}\sum_{l=0}^{k}\binom{k}{l}l^n(-1)^{k-l}. \tag{2.20}
$$

This formula not only provides an easy calculation of the second kind Stirling numbers, but it shows one more thing. The *inverse* binomial transform of l^n is $k!\left\{{n \atop k}\right\}$ (see (2.12)). Hence, (2.11) yields that

$$
k^n = \sum_{m=0}^{n}\binom{k}{m}m!\left\{{n \atop m}\right\} = \sum_{m=0}^{n}\left\{{n \atop m}\right\}k(k-1)\cdots(k-m+1).
$$

This is nothing other than our former identity on page 17 which we deduced on a combinatorial way there!

An alternative proof of Dobiński's formula

Having (2.20) at our disposal, we can give an alternative proof of Dobiński's formula. As B_n is the sum of the Stirling numbers, we have that

$$B_n = \sum_{k=0}^{n} \frac{1}{k!} \sum_{l=0}^{k} \binom{k}{l} l^n (-1)^{k-l} = \sum_{l=0}^{n} \frac{l^n}{l!} \sum_{k=l}^{n} \frac{1}{(k-l)!} (-1)^{k-l}.$$

Since $k - l$ runs from 0 to $n - l$, we can write the following:

$$B_n = \sum_{l=0}^{n} \frac{l^n}{l!} \sum_{k=0}^{n-l} \frac{(-1)^k}{k!}.$$

Note that l can run from any number greater or equal to n if we modify the inner sum such that it runs from 0 to this number minus l. Thus, it remains true that

$$B_n = \sum_{l=0}^{n+m} \frac{l^n}{l!} \sum_{k=0}^{n+m-l} \frac{(-1)^k}{k!} \quad (m \geq 0). \tag{2.21}$$

This formula is interesting in itself, and it can even be used to deduce the Dobiński formula; just take $m \to \infty$:

$$B_n = \sum_{l=0}^{\infty} \frac{l^n}{l!} \sum_{k=0}^{\infty} \frac{(-1)^k}{k!} = \frac{1}{e} \sum_{l=0}^{\infty} \frac{l^n}{l!}.$$

Identity (2.21) appears in the following sources: [467, Eq. (21)], [597, Section 1.6].

2.4.4 The ordinary generating function of the second kind Stirling numbers

The recursion for the Stirling numbers will be helpful in the determination of the generating function of $\left\{{n \atop k}\right\}$. Multiplying by x^n and summing over n we have

$$\sum_{n=0}^{\infty} \left\{{n \atop k}\right\} x^n = \sum_{n=0}^{\infty} \left\{{n-1 \atop k-1}\right\} x^n + k \sum_{n=0}^{\infty} \left\{{n-1 \atop k}\right\} x^n.$$

Hence, the generating function must satisfy the equation

$$f_k(x) = x f_{k-1}(x) + kx f_k(x);$$

see the operations on p. 38. Thus,

$$f_k(x) = \frac{1}{1-kx} x f_{k-1}(x).$$

This is a recursion; it expresses f_k in terms of f_{k-1}. Knowing the first function $f_0(x)$ we know all the others. $f_0(x)$ is simply

$$f_0(x) = \sum_{n=0}^{\infty} \left\{{n \atop 0}\right\} x^n = 1$$

(this is so, because $\left\{ {n \atop 0} \right\} = 0$ if $n > 0$). Hence,

$$f_1(x) = \frac{x}{1-x},$$

$$f_2(x) = \frac{x^2}{(1-x)(1-2x)},$$

and in general

$$f_k(x) = \frac{x^k}{(1-x)(1-2x)\cdots(1-kx)}.$$

The ordinary generating function of the Stirling numbers of the second kind is then:

$$\sum_{n=0}^{\infty} \left\{ {n \atop k} \right\} x^n = \frac{x^k}{(1-x)(1-2x)\cdots(1-kx)}. \tag{2.22}$$

2.4.5 The generating function of the Bell numbers and a formula for $\left\{ {n \atop k} \right\}$

The above leads to the generating function of the Bell numbers – at least to a preliminary form of it. Remembering the step we made in (2.18):

$$\sum_{n=0}^{\infty} B_n x^n = \sum_{n=0}^{\infty} \left(\sum_{k=0}^{\infty} \left\{ {n \atop k} \right\} \right) x^n = \sum_{k=0}^{\infty} \left(\sum_{n=0}^{\infty} \left\{ {n \atop k} \right\} x^n \right) = \sum_{k=0}^{\infty} f_k(x).$$

Thus, the function we were looking for is

$$\sum_{n=0}^{\infty} B_n x^n = \sum_{k=0}^{\infty} \frac{x^k}{(1-x)(1-2x)\cdots(1-kx)}.$$

The function on the right-hand side seems to be too complicated but can be simplified by some tools we have not yet known. In Section 8.4 we turn back to this question and offer a satisfying answer.

A consequence of (2.22)

Knowing (2.22) we can deduce yet another formula for the $\left\{ {n \atop k} \right\}$ numbers. To do a trick, we pretend that we know the binomial transform of $\left\{ {n \atop k} \right\}$, and we look for the original sequence. If we would do the reverse we got back (1.5) (see the exercises at the end of the chapter and exercise 4 in Chapter 1).

Let k be fixed and we look for a_n with generating function $f(x)$, supposing that the binomial transform of a_n is $\left\{ {n \atop k} \right\}$. Upon this setting, we know that the recently deduced generating function equals to the Euler-transformed form of $f(x)$ (see (2.14)):

$$\frac{x^k}{(1-x)(1-2x)\cdots(1-kx)} = \frac{1}{1-x} f\left(\frac{x}{1-x} \right).$$

Multiplying by $\frac{1}{1-x}$ and substituting $\frac{x}{1+x}$ in place of x^{14}, we get that

$$
f(x) = \left(1 - \frac{x}{1+x}\right) \frac{\left(\frac{x}{1+x}\right)^k}{\left(1 - \frac{x}{1+x}\right)\left(1 - 2\frac{x}{1+x}\right)\cdots\left(1 - k\frac{x}{1+x}\right)}
$$

$$
= \frac{1}{1+x} \frac{\frac{x^k}{(1+x)^k}}{\frac{1+x-x}{1+x}\frac{1+x-2x}{1+x}\cdots\frac{1+x-kx}{1+x}}
$$

$$
= \frac{1}{1+x} \frac{x^k}{(1-x)(1-2x)\cdots(1-(k-1)x)}
$$

$$
= \frac{x}{1+x} \frac{x^{k-1}}{(1-x)(1-2x)\cdots(1-(k-1)x)}
$$

$$
= \frac{x}{1+x} \sum_{n=0}^{\infty} \left\{ {n \atop k-1} \right\} x^n.
$$

The first factor is almost the geometric series (2.2):

$$
\frac{x}{1+x} = \sum_{n=0}^{\infty} (-1)^n x^{n+1},
$$

hence

$$
f(x) = \left(\sum_{n=0}^{\infty} (-1)^n x^{n+1}\right) \left(\sum_{n=0}^{\infty} \left\{ {n \atop k-1} \right\} x^n\right)
$$

$$
= \sum_{n=0}^{\infty} \left(\sum_{m=0}^{n} (-1)^m x^{m+1} \left\{ {n-m \atop k-1} \right\} x^{n-m}\right)
$$

$$
= \sum_{n=0}^{\infty} x^{n+1} \left(\sum_{m=0}^{n} (-1)^m \left\{ {n-m \atop k-1} \right\}\right),
$$

by the definition of the Cauchy product. It follows that $f(x)$ is the generating function of

$$
\sum_{m=0}^{n-1} (-1)^m \left\{ {n-1-m \atop k-1} \right\}
$$

(if $n = 0$ this equals 0 – empty sums are always zero). Its binomial transform is $\left\{ {n \atop k} \right\}$:

$$
\left\{ {n \atop k} \right\} = \sum_{l=0}^{n} \binom{n}{l} \sum_{m=0}^{l-1} (-1)^m \left\{ {l-1-m \atop k-1} \right\}.
$$

This is the formula we wanted to prove.

---

[14] $\frac{x}{1+x}$ is the functional inverse of $\frac{1}{1-x}$.

2.4.6 The exponential generating function of the first kind Stirling numbers

To deduce the exponential generating function of the first kind Stirling numbers, we can follow the same approach as in the second kind case. Recall Equation (2.18). Let us suppose that for the exponential generating function belonging to the fixed lower parameter k is $f_k(x)$:

$$\sum_{n=0}^{\infty} \begin{bmatrix} n \\ k \end{bmatrix} \frac{x^n}{n!} = f_k(x).$$

Summing on k – and remembering (1.8) – we have that

$$\sum_{k=0}^{\infty}\sum_{n=0}^{\infty} \begin{bmatrix} n \\ k \end{bmatrix} \frac{x^n}{n!} = \sum_{n=0}^{\infty} \frac{x^n}{n!} \sum_{k=0}^{\infty} \begin{bmatrix} n \\ k \end{bmatrix} = \sum_{n=0}^{\infty} x^n = \sum_{k=0}^{\infty} f_k(x).$$

The penultimate sum is $\frac{1}{1-x}$, so $f_k(x)$ must be an expression such that its sums to $\frac{1}{1-x}$. Of course $f_k(x) = x^k$ could be a candidate, but this is obviously not the generating function of the Stirling number sequence. Another candidate is

$$f_k(x) = \frac{1}{k!} \ln \left(\frac{1}{1-x} \right)^k. \tag{2.23}$$

This function sequence already gives the correct answer. We leave the verification as an exercise for the reader. (Hint: show that f_k satisfies a recursion corresponding to the basic recursion (1.9) of the $\begin{bmatrix} n \\ k \end{bmatrix}$ numbers.)

2.4.7 Some particular lower parameters of the first kind Stirling numbers

We have already seen that the $\begin{bmatrix} n \\ 2 \end{bmatrix}$ Stirling numbers can be expressed by the harmonic numbers. In this section, this observation will be generalized and it will be shown that not only the $k = 2$ lower parameter leads to harmonic numbers but *any* k.

Our first approach is based on the exponential generating function, the second approach (what we postpone until p. 63) will be based on the horizontal generating function to be introduced in the next section.

It is easy to see by integration that the function $\ln \left(\frac{1}{1-x} \right) = -\ln(1-x)$ is the generating function of $a_0 = 0$, $a_n = 1/n$ $(n > 0)$.

Hence, for its square,

$$\sum_{n=0}^{\infty} \begin{bmatrix} n \\ 2 \end{bmatrix} \frac{x^n}{n!} = \frac{1}{2!} \ln \left(\frac{1}{1-x} \right)^2 = \frac{1}{2} \left(\sum_{n=1}^{\infty} \frac{x^n}{n} \right)^2 =$$

$$\frac{1}{2} \sum_{n=1}^{\infty} x^n \sum_{k=1}^{n-1} \frac{1}{k} \frac{1}{n-k}.$$

Comparing the coefficients we get that

$$\begin{bmatrix} n \\ 2 \end{bmatrix} = \frac{n!}{2} \sum_{k=1}^{n-1} \frac{1}{k} \frac{1}{n-k}$$

holds whenever $n \geq 2$.

The reader can easily prove that

$$\sum_{k=1}^{n-1} \frac{1}{k} \frac{1}{n-k} = \frac{2}{n} H_{n-1}, \tag{2.24}$$

hence the more compact identity

$$\begin{bmatrix} n \\ 2 \end{bmatrix} = (n-1)! H_{n-1}$$

follows. We have already proved this on page 14.

For $k = 3$ we have that

$$\sum_{n=0}^{\infty} \begin{bmatrix} n \\ 3 \end{bmatrix} \frac{x^n}{n!} = \frac{1}{3!} \ln \left(\frac{1}{1-x} \right)^3.$$

The function on the right-hand side can be decomposed as

$$\ln \left(\frac{1}{1-x} \right)^3 = \ln \left(\frac{1}{1-x} \right) \ln \left(\frac{1}{1-x} \right)^2.$$

We know that the two factors are the generating functions of $a_n = 1/n$ $(n \geq 1)$ and $b_n = \frac{2}{n} H_{n-1}$ (see (2.24)), respectively. Then by Cauchy's product, it is not hard to see that

$$\begin{bmatrix} n \\ 3 \end{bmatrix} = \frac{n!}{3} \sum_{k=1}^{n-1} \frac{H_{k-1}}{k(n-k)}. \tag{2.25}$$

This is valid for $n \geq 3$, for smaller n these Stirling numbers are zero.

We leave to the reader to verify rigorously the special value

$$\begin{bmatrix} n \\ 4 \end{bmatrix} = \frac{n!}{6} \sum_{k=1}^{n-1} \frac{H_{k-1}}{k} \frac{H_{n-k-1}}{n-k}.$$

See [582] for similar relations.

2.5 Additional identities coming from the generating functions

In the previous sections, we have not extracted all the information on $\begin{bmatrix} n \\ k \end{bmatrix}$ and $\begin{Bmatrix} n \\ k \end{Bmatrix}$ from the generating functions. In this section, we gain more identities and we encounter additional similarities between $\begin{bmatrix} n \\ k \end{bmatrix}$ and $\begin{Bmatrix} n \\ k \end{Bmatrix}$.

Our first results are the following identities.

$$\begin{bmatrix} n \\ k \end{bmatrix} = \frac{n!}{k!} \sum_{j_1+j_2+\cdots+j_k=n} \frac{1}{j_1 j_2 \cdots j_k}, \tag{2.26}$$

$$\begin{Bmatrix} n \\ k \end{Bmatrix} = \frac{n!}{k!} \sum_{j_1+j_2+\cdots+j_k=n} \frac{1}{j_1! j_2! \cdots j_k!}. \tag{2.27}$$

(The j_is are positive; hence the denominator is never zero.)
In particular,

$$\begin{Bmatrix} n \\ 2 \end{Bmatrix} = \frac{n!}{2} \sum_{j_1+j_2=n} \frac{1}{j_1! j_2!} = \frac{n!}{2} \sum_{k=1}^{n-1} \frac{1}{k!(n-k)!} =$$

$$\frac{1}{2} \sum_{k=1}^{n-1} \frac{n!}{k!(n-k)!} = \frac{1}{2} \sum_{k=1}^{n-1} \binom{n}{k} = \frac{1}{2}(2^n - 2) = 2^{n-1} - 1.$$

Two other representations will be proven:

$$\begin{bmatrix} n \\ k \end{bmatrix} = \sum_{\substack{j_1+2j_2+\cdots+nj_n=n \\ j_1+j_2+\cdots+j_n=k}} \frac{n!}{j_1! j_2! \cdots j_n!} \left(\frac{1}{1}\right)^{j_1} \left(\frac{1}{2}\right)^{j_2} \cdots \left(\frac{1}{n}\right)^{j_n}, \tag{2.28}$$

$$\begin{Bmatrix} n \\ k \end{Bmatrix} = \sum_{\substack{j_1+2j_2+\cdots+nj_n=n \\ j_1+j_2+\cdots+j_n=k}} \frac{n!}{j_1! j_2! \cdots j_n!} \left(\frac{1}{1!}\right)^{j_1} \left(\frac{1}{2!}\right)^{j_2} \cdots \left(\frac{1}{n!}\right)^{j_n} \tag{2.29}$$

The proof of these can be worked out as follows. By (2.23) we have that

$$\sum_{n=0}^{\infty} \begin{bmatrix} n \\ k \end{bmatrix} \frac{x^n}{n!} = \frac{1}{k!} \ln\left(\frac{1}{1-x}\right)^k.$$

Since $-\ln(1-x)$ is the generating function of $\frac{1}{n}$,

$$\sum_{n=0}^{\infty} \begin{bmatrix} n \\ k \end{bmatrix} \frac{x^n}{n!} = \frac{1}{k!} \left(\sum_{j=1}^{\infty} \frac{x^j}{j}\right)^k.$$

First let $k = 2$. Then

$$\left(\sum_{j=1}^{\infty} \frac{x^j}{j}\right) \left(\sum_{j=1}^{\infty} \frac{x^j}{j}\right) = \sum_{j=1}^{\infty} \sum_{l=1}^{j-1} \frac{x^l}{l} \frac{x^{j-l}}{j-l} =$$

$$\sum_{j=1}^{\infty} x^j \sum_{l=1}^{j-1} \frac{1}{l(j-l)} = \sum_{j=1}^{\infty} x^j \sum_{j_1+j_2=j} \frac{1}{j_1 j_2}.$$

This is already enough to see how we can get the general case (2.26) for arbitrary k.

In the second kind case, we have the powers of $e^x - 1$, and this is the exponential generating function of $\frac{1}{n!}$ (if $n \geq 1$); therefore, we see why the factorials appear in (2.27) – the other steps are the same.

The other couple of identities ((2.28)-(2.29)) can be proven similarly but now in place of the Cauchy product we use the polynomial theorem from the Appendix. Again from (2.23) we know that

$$\sum_{n=0}^{\infty} \begin{bmatrix} n \\ k \end{bmatrix} \frac{x^n}{n!} = \frac{1}{k!} \left(\frac{x}{1} + \frac{x^2}{2} + \cdots + \frac{x^n}{n} + \cdots \right)^k.$$

We look for the coefficient of x^n. It is

$$\frac{1}{k!} \sum_{j_1+j_2+\cdots+j_n=k} \frac{k!}{j_1!j_2!\cdots j_n!} \left(\frac{x^1}{1} \right)^{j_1} \left(\frac{x^2}{2} \right)^{j_2} \cdots \left(\frac{x^n}{n} \right)^{j_n}.$$

In this expression, the power of x is $1j_1 + 2j_2 + \cdots + nj_n$, and this sum must be equal to n to match the coefficients on the two sides. This gives the other condition for the sum in (2.28).

The representation (2.29) can be proven similarly.

As a consequence, we can easily have a new sum representation for the Bell numbers. We just have to sum over k in (2.29). The terms with $k > n$ are zero, so

$$B_n = \sum_{\substack{j_1+2j_2+\cdots+nj_n=n \\ j_1+j_2+\cdots+j_n \geq 0}} \frac{n!}{j_1!j_2!\cdots j_n!} \left(\frac{1}{1!} \right)^{j_1} \left(\frac{1}{2!} \right)^{j_2} \cdots \left(\frac{1}{n!} \right)^{j_n}. \tag{2.30}$$

We offer a combinatorial proof for this result. By definition, B_n counts the partitions of an n-set. Any partition can be constructed by the following algorithm. We put the n elements in an order (this is a permutation) and then we count the different blocks with i elements ($1 \leq i \leq n$). Let j_i be the number of the i-blocks in the partition. It must hold true that

$$j_1 + 2j_2 + \cdots + nj_n = n.$$

Let a $j_1, \ldots, j_n$ sequence be fixed, and we group the first $1 \cdot j_1$ elements in a separate class (in which all elements form singletons). Then put the next $2 \cdot j_2$ elements in another class, where two-two elements are in a block, and so on. In general, there are $i \cdot j_i$ elements in the ith class, and there are j_i blocks of size i:

$$\underbrace{i|i|\cdots|i}_{j_i}.$$

These classes can be ordered in $j_i!$ ways. The individual elements in the classes have $i!$ different orders, and there are j_i blocks so the total number of orders is $(i!)^{j_i}$.

For example, if $j_2 = 3$ then two-two-two (6 in total) elements are put in three blocks. There are $3! = j_2!$ order of the blocks, and in the three individual blocks the order of the elements is $2!$ and $2!2!2! = (2!)^{j_2}$ in total.

In general, the number of the possibilities is

$$j_1!(1!)^{j_1} \cdot j_2!(2!)^{j_2} \cdots j_n!(n!)^{j_n},$$

so, if we divide $n!$ (the possible orders of all the elements) with this quantity we get the number of the $(j_1 + \cdots + j_n)$-partitions. This is nothing else but $\left\{ {n \atop j_1 + \cdots + j_n} \right\}$. This is the combinatorial proof for (2.29). If $j_1 + \cdots + j_n$ is arbitrary, we get the relation (2.30) for the Bell numbers.

Partition patterns and permutation patterns

The above considerations use the notion of *patterns*[15] If we look for a grouping of n elements such that we want to prescribe how many 1-blocks, 2-blocks etc. are there, we then prescribe a partition pattern. If we decide that there must be j_1 pieces of 1-blocks, j_2 pieces of 2-blocks, etc. then the number of partitions with this pattern is

$$\frac{n!}{j_1!(1!)^{j_1} \cdot j_2!(2!)^{j_2} \cdots j_n!(n!)^{j_n}}. \tag{2.31}$$

Of course the condition $1j_1 + 2j_2 + \cdots + nj_n = n$ must be satisfied.

In other words, a $(j_1, j_2, \ldots, j_n)$ number sequence identifies a partition pattern.

Patterns can be introduced for permutations, too. The only one difference is that in permutation cycles the order of the elements is not indifferent; however, right shifts do not alter the cycle. A cycle of length i can be shifted i times. Hence, we must divide by i and not by $i!$ as we did for partitions. Hence, the number of permutations with a given pattern $(j_1, j_2, \ldots, j_n)$ is

$$\frac{n!}{j_1!1^{j_1} \cdot j_2!2^{j_2} \cdots j_n!n^{j_n}}. \tag{2.32}$$

This leads to the combinatorial proof of (2.28).

2.6   Orthogonality

If we take a closer look at the generating functions of the two kinds of Stirling numbers, we can see an interesting fact: the two functions are almost

---

[15] Pattern is often called shape or type.

inverses of each other. If, in place of

$$\ln\left(\frac{1}{1-x}\right)$$

we take

$$-\ln\left(\frac{1}{1+x}\right) \tag{2.33}$$

then this is already the inverse of $e^x - 1$:

$$e^{-\ln\left(\frac{1}{1+x}\right)} - 1 = e^{\ln(1+x)} - 1 = 1 \mid x - 1 = x. \tag{2.34}$$

The function (2.33) is the generating function of a sequence which is easy to find (see the exercises). Now we need to use the function

$$\frac{1}{k!}\left[-\ln\left(\frac{1}{1+x}\right)\right]^k. \tag{2.35}$$

If we consider the identity

$$\sum_{n=0}^{\infty}\begin{bmatrix}n\\k\end{bmatrix}\frac{x^n}{n!} = \frac{1}{k!}\ln\left(\frac{1}{1-x}\right)^k$$

with $(-1)^k$ and in place of x we write $-x$ we get (2.35):

$$\sum_{n=0}^{\infty}(-1)^k\begin{bmatrix}n\\k\end{bmatrix}\frac{(-x)^n}{n!} = \sum_{n=0}^{\infty}(-1)^{n-k}\begin{bmatrix}n\\k\end{bmatrix}\frac{x^n}{n!} = \frac{1}{k!}\left[-\ln\left(\frac{1}{1+x}\right)\right]^k. \tag{2.36}$$

The numbers $(-1)^{n-k}\begin{bmatrix}n\\k\end{bmatrix}$ are called *signed Stirling numbers of the first kind*. It will be very convenient to introduce a notation for the signed Stirling numbers of the first kind:

$$\left\lceil\begin{matrix}n\\k\end{matrix}\right\rceil = (-1)^{n-k}\begin{bmatrix}n\\k\end{bmatrix}.$$

Why do we need this change in sign? Considering (2.34) again, and knowing that $\frac{1}{k!}(e^x - 1)^k$ is the exponential generating function of the second kind Stirlings we have

$$\frac{x^k}{k!} = \frac{1}{k!}\left(e^{-\ln\left(\frac{1}{1+x}\right)} - 1\right)^k = \sum_{n=0}^{\infty}\left\{\begin{matrix}n\\k\end{matrix}\right\}\frac{1}{n!}\left[-\ln\left(\frac{1}{1+x}\right)\right]^n. \tag{2.37}$$

Applying (2.36):

$$\frac{x^k}{k!} = \sum_{n=0}^{\infty}\left\{\begin{matrix}n\\k\end{matrix}\right\}\frac{1}{n!}\left[-\ln\left(\frac{1}{1+x}\right)\right]^n = \sum_{n=0}^{\infty}\left\{\begin{matrix}n\\k\end{matrix}\right\}\sum_{m=0}^{\infty}\left\lceil\begin{matrix}m\\n\end{matrix}\right\rceil\frac{x^m}{m!} =$$

$$\sum_{m=0}^{\infty} \frac{x^m}{m!} \left(\sum_{n=0}^{\infty} \left\{ {n \atop k} \right\} \overline{\left[{m \atop n} \right]} \right).$$

In the inner sum n can run between k and m:

$$\sum_{m=0}^{\infty} \frac{x^m}{m!} \left(\sum_{n=k}^{m} \left\{ {n \atop k} \right\} \overline{\left[{m \atop n} \right]} \right) = \frac{x^k}{k!}.$$

The right-hand side is the exponential generating function of the sequence $a_n = 1$ if $k = n$ and $a_n = 0$ otherwise. Hence, on the left-hand side, the coefficient of $\frac{x^m}{m!}$ is 1 if $m = k$ and 0 otherwise. This is indeed a very simple sum of one single non-zero term! We have the fundamental identity

$$\sum_{n=k}^{m} \overline{\left[{m \atop n} \right]} \left\{ {n \atop k} \right\} = \begin{cases} 1, & \text{if } m = k; \\ 0, & \text{otherwise} \end{cases} \tag{2.38}$$

This property is called *orthogonality* and the formula is called *inversion formula*.

The orthogonality in matrix form

The above inversion formula looks like a matrix multiplication written out term wise. Indeed, if $A = (a_{i,j})$ and $B = (b_{i,j})$ are two $m \times m$ matrices, then the $c_{i,j}$ item of AB is given by

$$c_{i,j} = \sum_{n=0}^{m} a_{i,n} b_{n,j}.$$

The left-hand side of the inversion formula (2.38) is an item of the matrix product of two matrices formed by the Stirling numbers:

$$\mathcal{S}_1 = \begin{pmatrix} \overline{\left[{0 \atop 0} \right]} & 0 & 0 & 0 & 0 & \cdots \\ \overline{\left[{1 \atop 0} \right]} & \overline{\left[{1 \atop 1} \right]} & 0 & 0 & 0 & \cdots \\ \overline{\left[{2 \atop 0} \right]} & \overline{\left[{2 \atop 1} \right]} & \overline{\left[{2 \atop 2} \right]} & 0 & 0 & \cdots \\ \vdots & \vdots & \ddots & \vdots & \vdots & \\ \overline{\left[{n \atop 0} \right]} & \overline{\left[{n \atop 1} \right]} & \cdots & \overline{\left[{n \atop n} \right]} & 0 & \cdots \\ \vdots & & & & & \end{pmatrix},$$

$$\mathcal{S}_2 = \begin{pmatrix} \left\{ {0 \atop 0} \right\} & 0 & 0 & 0 & 0 & \cdots \\ \left\{ {1 \atop 0} \right\} & \left\{ {1 \atop 1} \right\} & 0 & 0 & 0 & \cdots \\ \left\{ {2 \atop 0} \right\} & \left\{ {2 \atop 1} \right\} & \left\{ {2 \atop 2} \right\} & 0 & 0 & \cdots \\ \vdots & \vdots & \ddots & \vdots & \vdots & \\ \left\{ {n \atop 0} \right\} & \left\{ {n \atop 1} \right\} & \cdots & \left\{ {n \atop n} \right\} & 0 & \cdots \\ \vdots & & & & & \end{pmatrix}.$$

By the orthogonality, these matrices are *inverses* of each other. Since $S_1 S_2 = S_2 S_1 = E$, another inversion formula comes:

$$\sum_{k=m}^{n} \left\{ {n \atop k} \right\} \overline{\left[{k \atop m} \right]} = \left\{ \begin{array}{ll} 1, & \text{if } n = m; \\ 0, & \text{otherwise} \end{array} \right. .$$

The Stirling transform

Orthogonality permits us to introduce an interesting notion, which is similar to the binomial transform. This is the *Stirling transformation*. Let $a_0, a_1, \ldots$ be a given sequence. If we construct $b_0, b_1, \ldots$ via

$$\begin{pmatrix} b_0 \\ b_1 \\ \vdots \end{pmatrix} = S_2 \begin{pmatrix} a_0 \\ a_1 \\ \vdots \end{pmatrix},$$

then, by orthogonality, one can recover $a_0, a_1, \ldots$ from $b_0, b_1, \ldots$.

$$\begin{pmatrix} a_0 \\ a_1 \\ \vdots \end{pmatrix} = S_2^{-1} \begin{pmatrix} b_0 \\ b_1 \\ \vdots \end{pmatrix} = S_1 \begin{pmatrix} b_0 \\ b_1 \\ \vdots \end{pmatrix}.$$

Considering sums in place of matrix equations, we have the following connection between two arbitrary sequences a_n and b_n.

$$b_n = \sum_{k=0}^{n} \left\{ {n \atop k} \right\} a_k, \tag{2.39}$$

$$a_n = \sum_{k=0}^{n} \overline{\left[{n \atop k} \right]} b_k. \tag{2.40}$$

The sequence $b_0, b_1, \ldots$ is called the *Stirling transform* of $a_0, a_1, \ldots$.

An easy particular case is $a_i = 1$. Then (2.39) gives the sequence of Bell numbers: $b_n = B_n$. In other words, the Bell number sequence is the Stirling transform of the sequence $a_n = 1$.

Orthogonality will be used in the following section.

2.7 Horizontal generating functions and polynomial identities

So far we have studied generating functions formed by sums on the upper parameter of the Stirling numbers. If we consider the Stirling triangles, fix the parameter k (the index of columns) and run n, we move downwards in the triangle. The generating functions for a fixed k can then be called *vertical*. In turn, if we fix n (the row) and we form the generating functions of the arising numbers, we get the so-called *horizontal* generating functions. In a given row there are only finitely many elements, thus these generating functions will be *polynomials*.

Beginning with the second kind Stirling numbers, we fix n and run k:

$$\sum_{k=0}^{n} \left\{ {n \atop k} \right\} x^k.$$

These polynomials have their own name and we dedicate a whole chapter to them. At the moment we only mention that these polynomials are called *Bell polynomials* and they are denoted by $B_n(x)$. From the (1.2) definition of Bell numbers, we have that $B_n(1) = B_n$ – hence the name.

The polynomials with first kind Stirling number coefficients can also be determined by a small effort.

$$\sum_{k=0}^{n} \left[{n \atop k} \right] x^k = x(x+1)(x+2)\cdots(x+n-1). \tag{2.41}$$

The proof is by induction[16]. It is obvious that for $n = 1$ the statement is true:

$$\sum_{k=0}^{1} \left[{1 \atop k} \right] x^k = 0 \cdot 1 + 1 \cdot x = x.$$

We suppose that (2.41) holds up to n and we try to prove it for $n + 1$ by induction:

$$x(x+1)(x+2)\cdots(x+n-1)(x+n) = \left(\sum_{k=0}^{n} \left[{n \atop k} \right] x^k \right)(x+n) =$$

$$\sum_{k=0}^{n} \left[{n \atop k} \right] x^{k+1} + \sum_{k=0}^{n} n \left[{n \atop k} \right] x^k =$$

$$\sum_{k=1}^{n+1} \left[{n \atop k-1} \right] x^k + \sum_{k=0}^{n} n \left[{n \atop k} \right] x^k =$$

$$\sum_{k=1}^{n+1} \left(\left[{n \atop k-1} \right] + n \left[{n \atop k} \right] \right) x^k =$$

$$\sum_{k=1}^{n+1} \left[{n+1 \atop k} \right] x^k = \sum_{k=0}^{n+1} \left[{n+1 \atop k} \right] x^k.$$

---

[16] In a recent paper [136], this formula was proven by using the notion of right-to-left minima. We will encounter left-to-right minima in Section 8.8. The two are easily transformed into each other. Four other proofs can be found in [534], and another in the online note [110].

Hence, (2.41) is indeed correct. As an example, we take the polynomial $x(x+1)(x+2)(x+3)$:

$$x(x+1)(x+2)(x+3) = x^4 + 6x^3 + 11x^2 + 6x,$$

and

$$\begin{bmatrix} 4 \\ 4 \end{bmatrix} = 1, \quad \begin{bmatrix} 4 \\ 3 \end{bmatrix} = 6, \quad \begin{bmatrix} 4 \\ 2 \end{bmatrix} = 11, \quad \begin{bmatrix} 4 \\ 1 \end{bmatrix} = 6,$$

as it must be. By (2.41) one can deduce the horizontal generating function of the signed Stirling numbers of the first kind:

$$\sum_{k=0}^{n} \overline{\begin{bmatrix} n \\ k \end{bmatrix}} x^k = x(x-1)(x-2)\cdots(x-n+1). \tag{2.42}$$

By using orthogonality again, this latter result offers an identity for the second kind Stirling numbers. On one hand,

$$\sum_{n=0}^{m} \begin{Bmatrix} m \\ n \end{Bmatrix} \left(\sum_{k=0}^{n} \overline{\begin{bmatrix} n \\ k \end{bmatrix}} x^k \right) = \sum_{n=0}^{m} \begin{Bmatrix} m \\ n \end{Bmatrix} x(x-1)(x-2)\cdots(x-n+1).$$

On the other hand, the second inversion formula yields

$$\sum_{n=0}^{m} \begin{Bmatrix} m \\ n \end{Bmatrix} \left(\sum_{k=0}^{n} \overline{\begin{bmatrix} n \\ k \end{bmatrix}} x^k \right) = \sum_{k=0}^{n} x^k \left(\sum_{n=0}^{m} \begin{Bmatrix} m \\ n \end{Bmatrix} \overline{\begin{bmatrix} n \\ k \end{bmatrix}} \right) = x^m.$$

Therefore

$$\sum_{k=0}^{n} \begin{Bmatrix} n \\ k \end{Bmatrix} x(x-1)(x-2)\cdots(x-k+1) = x^n. \tag{2.43}$$

This is just the same what we got by a combinatorial argument on page 17. We got the same result as there. Remember that the same result was obtained in a third way before (see p. 48.).

Polynomials of the form $x(x-1)(x-2)\cdots(x-n+1)$ and $x(x+1)(x+2)\cdots(x+n-1)$ will appear so frequently that we introduce a new notation for them. Let

$$x(x-1)(x-2)\cdots(x-n+1) = x^{\underline{n}}.$$

This polynomial is called *falling factorial*, while

$$x(x+1)(x+2)\cdots(x+n-1) = x^{\overline{n}},$$

is called *rising factorial*. This latter is often denoted by $(x)_n$ and called *Pochhammer*[17] *symbol*. We set $x^{\underline{0}} = x^{\overline{0}} = 1$.

---

[17]Leo August Pochhammer (1841-1920), Prussian mathematician.

Using these new notations, we rewrite the results of the section one more time.

$$\sum_{k=0}^{n} \left[\begin{matrix} n \\ k \end{matrix} \right] x^k \;=\; x^{\overline{n}}, \tag{2.44}$$

$$\sum_{k=0}^{n} \overline{\left[\begin{matrix} n \\ k \end{matrix} \right]} x^k \;=\; x^{\underline{n}}, \tag{2.45}$$

$$\sum_{k=0}^{n} \left\{ \begin{matrix} n \\ k \end{matrix} \right\} x^{\underline{k}} \;=\; x^n. \tag{2.46}$$

The polynomials like x^n, $x^{\overline{n}}$ or $x^{\underline{n}}$ are connected by the Stirling numbers, as we see here. Using fancy vocabulary, we say that the Stirling numbers are the *connecting coefficients* among the appropriate monomials, falling and rising factorials[18].

Note that the dual of (2.46) is (2.45) and not (2.44).

The second and third *polynomial identities* can be rewritten by the matrices introduced above:

$$\mathcal{S}_1 \begin{pmatrix} x^0 \\ x^1 \\ \vdots \\ x^n \end{pmatrix} \;=\; \begin{pmatrix} x^{\underline{0}} \\ x^{\underline{1}} \\ \vdots \\ x^{\underline{n}} \end{pmatrix},$$

$$\mathcal{S}_2 \begin{pmatrix} x^{\underline{0}} \\ x^{\underline{1}} \\ \vdots \\ x^{\underline{n}} \end{pmatrix} \;=\; \begin{pmatrix} x^0 \\ x^1 \\ \vdots \\ x^n \end{pmatrix}.$$

(Of course, in $\mathcal{S}_1$ and $\mathcal{S}_2$ we must take the first n lines only.)

Note that if we apply $\mathcal{S}_2$ one more time on both sides of the second identity we get the Bell polynomials.

It is also worth to note that x^n can be expanded by the rising (or falling) factorials and vice versa. Based on this observation, we might think that the rising and falling factorials can be expanded by each other with some connecting coefficients, too. This is so, and we deal with this question in Section 2.8.

---

[18]The polynomials with real coefficients form an infinite dimensional vector space. This means that polynomials can be arbitrarily added or can be multiplied by constants, the result will be a polynomial. In every vector space there are bases, such that every vector (polynomial in this case) is a linear combination of base vectors. It is quite obvious that any polynomial sequence $p_n(x)$ such that $\deg p_n = n$ ($n \geq 0$) forms a base – so the rising and falling factorials as well as the monomials x^n ($n \geq 0$) are bases. The bases can be transformed into each other, that is, one base vector of a given base can be expressed in terms of some base vectors from another base. This expressibility and the concrete forms of the expressions are the contents of the formulas (2.44)-(2.46). We see more examples when we turn to the study of the Lah numbers.

But first, we extract some information from the horizontal generating function of $\begin{bmatrix} n \\ k \end{bmatrix}$.

2.7.1 Special values of the Stirling numbers of the first kind - second approach

After having known the horizontal generating function of the $\begin{bmatrix} n \\ k \end{bmatrix}$ numbers, we show another technique to deduce low parameter special values for these numbers. Recall (2.41) and perform the following transformations:

$$\sum_{k=0}^{n} \begin{bmatrix} n+1 \\ k+1 \end{bmatrix} = n! \left(1 + \frac{x}{1}\right)\left(1 + \frac{x}{2}\right)\cdots\left(1 + \frac{x}{n}\right) =$$

$$n! \exp\left(\log\left(1 + \frac{x}{1}\right) + \log\left(1 + \frac{x}{2}\right) + \cdots + \log\left(1 + \frac{x}{n}\right)\right) =$$

$$n! \exp\left(\sum_{j=1}^{n}\sum_{k=1}^{\infty} \frac{(-1)^{k+1}}{k}\left(\frac{x}{j}\right)^k\right) =$$

$$n! \exp\left(\sum_{k=1}^{\infty} \frac{(-1)^{k+1}}{k}x^k \sum_{j=1}^{n} \frac{1}{j^k}\right) = n! \exp\left(\sum_{k=1}^{\infty} \frac{(-1)^{k+1}}{k}x^k H_{n,k}\right).$$

Here $H_{n,k}$ denotes the *generalized harmonic number*:

$$H_{n,k} = \sum_{j=1}^{n} \frac{1}{j^k}. \tag{2.47}$$

These numbers appear very often in analysis, so it is good to know about them. Here we express $\begin{bmatrix} n \\ k \end{bmatrix}$ in terms of generalized harmonic numbers. Actually, it can already be seen by the above transformation of the horizontal generating function that it is enough to extract the corresponding power of x in

$$n! \exp\left(\sum_{k=1}^{\infty} \frac{(-1)^{k+1}}{k}x^k H_{n,k}\right)$$

to get the wanted relations. Recall that the series of the exponential function begins as

$$\exp(x) = 1 + x + \frac{x^2}{2!} + \frac{x^3}{3!} + \cdots,$$

and we do not need the whole sum under the above exponent to determine the smaller lower parameter first kind Stirling numbers: for $\begin{bmatrix} n \\ k \end{bmatrix}$ one needs only $k-1$ terms. Truncate the series after the third term and write the much simpler

$$n! \exp\left(H_{n,1}x - \frac{1}{2}H_{n,2}x^2 + \frac{1}{3}H_{n,3}x^3\right)$$

we get that

$$\begin{bmatrix} n+1 \\ 1 \end{bmatrix} = n!,$$

$$\begin{bmatrix} n+1 \\ 2 \end{bmatrix} = n!H_n,$$

$$\begin{bmatrix} n+1 \\ 3 \end{bmatrix} = \frac{n!}{2}(H_n^2 - H_{n,2}),$$

$$\begin{bmatrix} n+1 \\ 4 \end{bmatrix} = \frac{n!}{6}(H_n^3 - 3H_nH_{n,2} + 2H_{n,3}).$$

The reader is kindly asked to verify that these coincide with the values given by the first approach in Section 2.4.7.

2.8   The Lah numbers

As we noted before, one can expand the rising factorial with respect to the falling factorial and vice versa by some connecting coefficients. In this section we are going to determine the connecting coefficients, denoted by $\left\lfloor \begin{smallmatrix} n \\ k \end{smallmatrix} \right\rfloor$, such that

$$x^{\overline{n}} = \sum_{k=0}^{n} \left\lfloor \begin{matrix} n \\ k \end{matrix} \right\rfloor x^{\underline{k}}, \tag{2.48}$$

and

$$x^{\underline{n}} = \sum_{k=0}^{n} \overline{\left\lfloor \begin{matrix} n \\ k \end{matrix} \right\rfloor} x^{\overline{k}}. \tag{2.49}$$

These numbers are called *Lah numbers*[19]. The notation $\left\lfloor \begin{smallmatrix} n \\ k \end{smallmatrix} \right\rfloor$ is a *Karamata-Knuth notation*[20].

How to begin to find the exact form of these coefficients? We shall need a useful identity which is the counterpart of the binomial theorem for falling factorials. Namely,

$$(x+y)^{\underline{n}} = \sum_{k=0}^{n} \binom{n}{k} x^{\underline{k}} y^{\underline{n-k}}. \tag{2.50}$$

This can be proven as follows. Divide by $n!$ on both sides. Then the left-hand

---

[19]Ivo Lah (1896-1979), Slovenian mathematician. He introduced these numbers in 1954 in [350], see also [351].

[20]Knuth and his co-authors propagated the notations of Karamata in the famous book *Concrete Mathematics* [262]. Therefore, such notations are called Karamata-Knuth.

side becomes

$$\frac{(x+y)^{\underline{n}}}{n!} = \binom{x+y}{n},$$

and the right-hand side is

$$= \sum_{k=0}^{n} \frac{x^{\underline{k}}}{k!} \frac{y^{\underline{n-k}}}{(n-k)!} = \sum_{k=0}^{n} \binom{x}{k}\binom{y}{n-k}.$$

It can be a bit peculiar at first sight, but we can write *any* real x in the binomial coefficient, if it is defined as

$$\binom{x}{k} = \frac{x(x-1)(x-2)\cdots(x-k+1)}{k!} = \frac{x^{\underline{k}}}{k!}. \tag{2.51}$$

Hence, our transformed relation reads as

$$\binom{x+y}{n} = \sum_{k=0}^{n} \binom{x}{k}\binom{y}{n-k}. \tag{2.52}$$

This is a famous theorem of *Vandermonde*[21], called *Vandermonde convolution*[22]. The proof of (2.52) is as follows. Fix a set of x women and y men. The left-hand side $\binom{x+y}{n}$ gives the number of ways we can choose n person from the whole group of $x+y$ members. This selection can be performed by selecting k women from x and $n-k$ men from y, where k is between 0 and n. For a fixed k this offers

$$\binom{x}{k}\binom{y}{n-k}$$

possibilities. Summing over k, we get the Vandermonde convolution formula.

Having (2.50) in our hand, we continue with the determination of the Lah numbers. Note that

$$x^{\overline{n}} = (x+n-1)^{\underline{n}}.$$

On the right-hand side, we apply (2.50):

$$(x+n-1)^{\underline{n}} = \sum_{k=0}^{n} \binom{n}{k} x^{\underline{k}} (n-1)^{\underline{n-k}}.$$

Since

$$x^{\overline{n}} = \sum_{k=0}^{n} \binom{n}{k} x^{\underline{k}} (n-1)^{\underline{n-k}} = \sum_{k=0}^{n} \left\lfloor \begin{matrix} n \\ k \end{matrix} \right\rfloor x^{\underline{k}},$$

---

[21] Alexandre-Théophile Vandermonde (1735-1796), French mathematician, violin player and chemist.

[22] We remark that this identity was known much before Vandermonde. In 1303 the Chinese Chu Shih-Chieh (in modern transcription: Zhu ShiJie), who lived between 1249 and 1314 used it: therefore, (2.52) is often called *Chu-Vandermonde identity*. For the history, see [38, pp. 59-60].

we get a simple expression for $\left\lfloor {n \atop k} \right\rfloor$:

$$\left\lfloor {n \atop k} \right\rfloor = \binom{n}{k}(n-1)^{\underline{n-k}}.$$

This can be rewritten:

$$\binom{n}{k}(n-1)^{\underline{n-k}} = \binom{n}{k}(n-1)(n-2)\cdots(n-1-(n+k)+1) =$$

$$\binom{n}{k}(n-1)(n-2)\cdots k.$$

(Remember that $0 \le k \le n$.) One more step to arrive at a compact expression:

$$\left\lfloor {n \atop k} \right\rfloor = \binom{n}{k}(n-1)(n-2)\cdots k = \frac{n!}{k!}\binom{n-1}{k-1}. \qquad (2.53)$$

Finally, since $(-1)^n(-x)^{\overline{n}} = x^{\underline{n}}$, $\left\lceil {n \atop k} \right\rceil$ can be expressed easily:

$$\left\lceil {n \atop k} \right\rceil = (-1)^{n-k}\left\lfloor {n \atop k} \right\rfloor.$$

In the case of Lah numbers, the first and second kinds are equal up to sign. Hence, in spite of talking about first and second kind Lah numbers, we say that $\left\lceil {n \atop k} \right\rceil$ is a *signed Lah number*. Note that this kind of notation is compatible with the notation for the signed first kind Stirling numbers.

2.8.1 The combinatorial meaning of the Lah numbers

Ordered lists

The Stirling numbers of both kinds are *connecting coefficients* of certain polynomials, and, in the same time they have a well-defined combinatorial meaning. This – and the presence of the binomial coefficients in (2.53) – suggest that the $\left\lfloor {n \atop k} \right\rfloor$ Lah numbers also have some combinatorial interpretation. We dedicate this section to explain this meaning in more details. We begin with a definition.

Definition 2.8.1. *Let a k-partition of an n-set be given. If we take into account the order of the elements in the individual blocks, we say that this partition is an* ordered list.

For example, in the usual sense, the following 2-partitions of $\{1,2,3,4,5\}$ are identical:

$$1,2|3,4,5 \quad \text{and} \quad 2,1|4,5,3$$

This is so because the order of the elements does not matter in a set. But if we consider ordered lists, they are no longer identical. Hence, the question comes: how many ordered lists are there on n elements with k blocks? The answer can be guessed: $\left\lfloor \begin{smallmatrix} n \\ k \end{smallmatrix} \right\rfloor$.

To prove this, we construct all the ordered lists on n elements with k blocks one by one. This can be established as follows. Choose k elements from n which will be the first elements in their blocks. Then from the remaining $n - k$ elements we choose one after another and put them down in the possible places. The first element can be put down in k ways (between the k elements and after the last). The next element has $k + 1$ positions to go, and the last has $k + (n - k - 1) = n - 1$ positions. This means that altogether we have

$$\binom{n}{k} k(k+1) \cdots (n-1) = \frac{n!}{k!} \binom{n-1}{k-1}$$

different ordered lists with k blocks on n elements. This expression is the same as (2.53).

Two identities for the Lah numbers

The basic recursion follows from the above combinatorial definition:

$$\left\lfloor \begin{matrix} n+1 \\ k \end{matrix} \right\rfloor = (n+k) \left\lfloor \begin{matrix} n \\ k \end{matrix} \right\rfloor + \left\lfloor \begin{matrix} n \\ k-1 \end{matrix} \right\rfloor. \tag{2.54}$$

An ordered list of $n + 1$ elements with k blocks can be constructed from an ordered list of n elements. (1) The last element can form a singleton and the remaining n elements form an ordered list with $(k-1)$-blocks in $\left\lfloor \begin{smallmatrix} n \\ k-1 \end{smallmatrix} \right\rfloor$ ways. (2) If the last element is not in a singleton, then first we form an ordered list of k blocks on the n other elements in $\left\lfloor \begin{smallmatrix} n \\ k \end{smallmatrix} \right\rfloor$ ways and then we insert the last element somewhere between the n elements ($n-1$ positions), or before the first element, or after the last one ($1 + 1$ options). We need to be careful: between the last element of a block and the first element of the next block, there are two distinct places. So we altogether have $n - 1 + 1 + 1 + (k - 1) = n + k$ places to insert the last element. This gives the $(n + k) \left\lfloor \begin{smallmatrix} n \\ k \end{smallmatrix} \right\rfloor$ term.

Having this combinatorial interpretation for the Lah numbers, another nice expression can be deduced. Namely,

$$\left\lfloor \begin{matrix} n \\ k \end{matrix} \right\rfloor = \sum_{j=0}^{n} \left[\begin{matrix} n \\ j \end{matrix} \right] \left\{ \begin{matrix} j \\ k \end{matrix} \right\}. \tag{2.55}$$

The construction of an ordered list with k blocks on n elements can be done in a different way than above, so that we arrive at (2.55). If we put the n elements into a permutation with j cycles and then we group these cycles into k groups, we get an ordered list with k blocks. Summing over j we get the formula.

Note that (2.55) is like the orthogonality relation (2.38) without the alternating signs; and it can actually be proven by the matrix symbolism we introduced before. We continue to use $\mathcal{S}_2$, but in place of $\mathcal{S}_1$ we use the matrix with the unsigned Stirlings of the first kind; which we denote by $\mathcal{S}_{1,\mathrm{us}}$. From (2.46) we know that $\mathcal{S}_2$ maps the vector $\left(x^{\underline{0}} \quad x^{\underline{1}} \quad x^{\underline{2}} \quad \cdots\right)^T$ to $\left(x^0 \quad x^1 \quad x^2 \quad \cdots\right)^T$, and (2.44) says that $\mathcal{S}_{1,\mathrm{us}}$ maps the latter vector to $\left(x^{\overline{0}} \quad x^{\overline{1}} \quad x^{\overline{2}} \quad \cdots\right)^T$. Here T stands for the transpose. If we construct the matrix $\mathcal{L}$ with the Lah numbers as items, it maps the vector of falling factorials to the vector of rising factorials. Thus, we get that

$$\mathcal{L} = \mathcal{S}_{1,\mathrm{us}}\mathcal{S}_2,$$

which is the matricial equivalent of (2.55).

More information on the linear algebraic properties of the Lah matrices is accessible in [557].

2.9 The total number of ordered lists and the horizontal sum of the Lah numbers

Once we have the Lah number triangle (see also the table at the end of the book), we can form the horizontal sum of these numbers as we did for the Stirlings to obtain $n!$ and B_n.

Definition 2.9.1. *Let*

$$L_n = \sum_{k=0}^{n} \left\lfloor \begin{matrix} n \\ k \end{matrix} \right\rfloor \quad (n = 0, 1, 2, \dots). \tag{2.56}$$

These numbers enumerate those partitions in which the order of the elements in the blocks does *matter[23]; that is, L_n is the total number of ordered lists on n elements.*

See the table at the end of the book for the first thirty values of this sequence.

The analogue of the (1.1) recursion of the Bell numbers follows easily:

$$L_{n+1} = \sum_{k=0}^{n} \binom{n}{k}(k+1)!L_{n-k}.$$

To form a partition counted by L_{n+1} we can choose k elements to put in the

---

[23] As far as we know, these numbers do not have widely accepted names.

ordered block of $n+1$ ($\binom{n}{k}$ choices), then we order this block in $(k+1)!$ ways. The rest of the elements can go into another ordered list in L_{n-k} ways.

Later we will define the polynomial version of L_n, see Section 4.4.

A relation among B_n, L_n and $\left\{{n \atop k}\right\}$

As another application of orthogonality, we can find a relation between B_n, L_n and $\left\{{n \atop k}\right\}$ based on (2.55). By using the matrix machinery again, (2.55) can be rewritten in the compact form

$$\mathcal{L} = \mathcal{S}_{1,\mathrm{us}}\mathcal{S}_2,$$

where, as before, "us" means unsigned, and $\mathcal{L}$ is the (unsigned) Lah matrix. Thus,

$$\mathcal{S}_2 = \mathcal{S}_{1,\mathrm{us}}^{-1}\mathcal{L},$$

The inverse $\mathcal{S}_{1,\mathrm{us}}^{-1}$ is the *signed* second kind Stirling matrix as it follows from orthogonality. Writing this latter relation out,

$$\left\{{n \atop k}\right\} = \sum_{j=0}^{n}(-1)^{n-j}\left\{{n \atop j}\right\}\left\lfloor{j \atop k}\right\rfloor.$$

Summing over k we get the Bell numbers on the left and L_j in the sum on the right-hand side. Thus, we arrive at an interesting formula:

$$B_n = \sum_{j=0}^{n}(-1)^{n-j}\left\{{n \atop j}\right\}L_j.$$

This result was proved recently in [472].

A Spivey-like formula for L_n

In 2008 M. Spivey [531] proved the following identity with respect to the Bell numbers:

$$B_{n+m} = \sum_{k=0}^{n}\sum_{j=0}^{m}j^{n-k}\left\{{m \atop j}\right\}\binom{n}{k}B_k.$$

This formula will be proven in Section 11.7 where we will explicitly use it. Why we mention it now is because there is a formula of the same flavor with respect to the L_n numbers. Namely,

$$L_{n+m} = \sum_{k=0}^{n}\sum_{j=0}^{m}\left\lfloor{m \atop j}\right\rfloor (m+j)^{\overline{n-k}}\binom{n}{k}L_k, \qquad (2.57)$$

for all $n, m \geq 0$. The proof goes as follows. On the left-hand side, L_{n+m} counts the sets of lists of $n + m$ elements. We can construct any of these as follows: we pick k element from n in $\binom{n}{k}$ ways and put these into an ordered list: L_k possibilities. From the other m elements, we form j ordered lists in $\left\lfloor \genfrac{}{}{0pt}{}{m}{j} \right\rfloor$ ways.

However, there are $n - k$ elements already out of the ordered lists. We put these among the m elements. The first element, say a, can go to $m+j$ positions: among the m elements, there are $m - 1$ positions but we must be careful: there are *two* positions between two elements if they separate blocks. Among j blocks there are $j-1$ places. Hence, up to now, we have $m-1+j-1 = m+j-2$ places. In addition, there are two more places at the beginning of the list and at the end. Thus, in total, there are $m + j - 2 + 2 = m + j$ positions, as we stated.

Then the next element can be placed into $m+j+1$ positions, because the former element a has created a new space. And so on. Finally, if we would like to put down the last $(n - k)$th element, we have $(m + j)(m + j + 1) \cdots (m + j + n - k - 1)$ possibilities.

Summing over the possible values of k and j, we are done.

Identity (2.57) was given in a slightly different form by Xu and Cen [605, 606].

2.10   The Hankel transform

The so-called Hankel transform of sequences leads to a nice theory. As this transform is based on generating functions, and is related to the binomial transform, we include it in this chapter. The content of this section is not indispensable to understand the later chapters.

The *Hankel[24]-matrices* of the sequence a_n are

$$
H_n = \begin{pmatrix}
a_0 & a_1 & a_2 & \cdots & a_n \\
a_1 & a_2 & a_3 & \cdots & a_n \\
\vdots & & & & \\
a_n & a_{n+1} & a_{n+2} & \cdots & a_{2n}
\end{pmatrix}
$$

$(n + 1) \times (n + 1)$ matrices $(n = 0, 1, 2, \dots)$.

The indices of the rows and columns start from zero; hence we get the simple correspondence:

$$(H_n)_{i,j} = a_{i+j}.$$

For instance, the H_2 Hankel matrix of $a_n = n$ is

$$
H_2 = \begin{pmatrix}
0 & 1 & 2 \\
1 & 2 & 3 \\
2 & 3 & 4
\end{pmatrix}.
$$

---

[24]Hermann Hankel (1839-1873), German mathematician.

The determinants $\det(H_n)$ of the Hankel matrices are called *Hankel determinants* and denoted by h_n. Hence, for any sequence there is a corresponding sequence of the Hankel determinants. The sequence of the Hankel determinants is the *Hankel transform* of the given sequence. The Hankel transform is not unique: many sequences have the same sequence of Hankel determinants – we shall see examples.

The question we are interested in is, of course, the following: what are the Hankel transforms of the sequences we have studied so far? One parameter sequences must be considered, so the Bell numbers come first. To be able to determine the Hankel transform easily, we need to know the so-called Euler-Seidel matrices.

2.10.1 The Euler-Seidel matrices

An interesting thing happens when we multiply a Hankel matrix with a Pascal matrix (see (2.13)):

$$[\mathcal{B}_n H_n]_{i,j} = \left[\begin{pmatrix} \binom{0}{0} & 0 & 0 & 0 \\ \binom{1}{0} & \binom{1}{1} & 0 & 0 \\ \binom{2}{0} & \binom{2}{1} & \binom{2}{2} & 0 \\ \vdots & \vdots & \ddots & \vdots \\ \binom{n}{0} & \binom{n}{1} & \cdots & \binom{n}{n} \end{pmatrix} \begin{pmatrix} a_0 & a_1 & a_2 & \cdots & a_n \\ a_1 & a_2 & a_3 & \cdots & a_{n+1} \\ \vdots & & & \ddots & \\ a_n & a_{n+1} & a_{n+2} & \cdots & a_{2n} \end{pmatrix} \right]_{i,j} =$$

$$\sum_{l=0}^{i} \binom{i}{l} a_{l+j}.$$

For a given sequence the $\mathcal{B}_n H_n$ matrices are called *Euler-Seidel*[25] *matrices*. Thus, the (i,j)-th element of the $n \times n$ Euler-Seidel matrix is

$$\sum_{l=0}^{i} \binom{i}{l} a_{l+j}.$$

Note that in the first column ($j = 0$) the binomial transform of a_n appears. The Euler-Seidel matrices are very useful proving new identities. We have already seen one example when we talked about the Euler formula (2.14).

The invariance of the Hankel transform under the binomial transformation

There is an important fact we need later: the Hankel transform is invariant under the binomial transform. In other words, a sequence and its binomial transform have the same Hankel determinant sequence[26].

---

[25] Philipp Ludwig von Seidel (1821-1896), German mathematician who also worked on optical lenses and astronomy.

[26] This is a theorem of J. W. Layman [358] from 2001.

Taking $\mathcal{B}_n H_n \mathcal{B}_n^T$, where, as before, T stands for the transpose, we have that

$$
\begin{pmatrix}
\binom{0}{0} & 0 & 0 & 0 \\
\binom{1}{0} & \binom{1}{1} & 0 & 0 \\
\binom{2}{0} & \binom{2}{1} & \binom{2}{2} & 0 \\
\vdots & \vdots & \ddots & \vdots \\
\binom{n}{0} & \binom{n}{1} & \cdots & \binom{n}{n}
\end{pmatrix}
\begin{pmatrix}
a_0 & a_1 & \cdots & a_n \\
a_1 & a_2 & \cdots & a_{n+1} \\
\vdots & & \ddots & \\
a_n & a_{n+1} & \cdots & a_{2n}
\end{pmatrix}
\begin{pmatrix}
\binom{0}{0} & \binom{1}{0} & \binom{2}{0} & \cdots & \binom{n}{0} \\
0 & \binom{1}{1} & \binom{2}{1} & \cdots & \binom{n}{1} \\
0 & 0 & \binom{2}{2} & \cdots & \binom{n}{2} \\
\vdots & \vdots & \ddots & \cdots & \vdots \\
0 & 0 & \cdots & \cdots & \binom{n}{n}
\end{pmatrix}.
$$

According to the rules of matrix multiplication, and taking into account that $(H_n)_{i,j} = a_{i+j}$, we have

$$
(\mathcal{B}_n H_n \mathcal{B}_n^T)_{i,j} = \sum_{k=0}^{i} \binom{i}{k} \sum_{m=0}^{j} a_{k+m} \binom{j}{m} = \sum_{k=0}^{i} \binom{i}{k} \sum_{n=m+k} a_n \binom{j}{n-k} =
$$

$$
\sum_{n=0}^{i+j} a_n \sum_{k=0}^{i} \binom{i}{k} \binom{j}{n-k} = \sum_{n=0}^{i+j} a_n \binom{i+j}{n}.
$$

The steps were the following: first we took the product, in the second step we introduced a new index n, and in the last step we used the Vandermonde convolution. Finally we note that the last sum is nothing else but the binomial transform of a_n. The multiplication theorem for determinants now provides the invariance of the Hankel transform under the binomial transformation, because

$$
\det\left(\mathcal{B}_n H_n \mathcal{B}_n^T\right) = \det \mathcal{B}_n \det H_n \det \mathcal{B}_n^T = \det H_n,
$$

knowing that

$$
\det \mathcal{B}_n = \det \mathcal{B}_n^T = 1.
$$

Moreover, as the Euler-Seidel matrix is just the product $\mathcal{B}_n H_n$, we also get that

$$
h_n = \det(\mathcal{B}_n H_n \mathcal{B}_n^T) = \det(\mathcal{E}_n \mathcal{B}_n^T) = \det(\mathcal{E}_n) \cdot 1.
$$

This means that the $(n+1) \times (n+1)$ Euler-Seidel matrix $\mathcal{E}_n$ has the same determinant sequence as a_n itself: h_n.

We remark that a more general statement is also true. If a_n is a sequence, then not only the

$$
b_n = \sum_{k=0}^{n} \binom{n}{k} a_k
$$

sequence has the same Hankel transform, but also the more general sequence

$$
b_n(x) = \sum_{k=0}^{n} \binom{n}{k} a_k x^{n-k}. \tag{2.58}
$$

This statement can be proven by the reader with a high probability of success. Also, see Proposition 1 in [311] for a short proof.

It often happens that a sequence we are studying is the binomial transform or the more general transform (2.58) of a simpler sequence. If we can determine the Hankel transform of this simpler sequence, we know the Hankel transform of our sequence, too. An example of this observation is the sequence of r-Bell polynomials in Section 8.3.

In the following two subsections, we show two tools, a simpler and a more advanced one, to find the Hankel transform.

2.10.2 A generating function tool to calculate the Hankel transform

Let a_n be a sequence with exponential generating function $f(x)$. If there is a sequence $f_k(x)$ of functions and a d_k sequence of numbers such that

$$f_k^{(n)}(0) = 0 \text{ if } k > n,$$

and

$$\sum_{k=0}^{\infty} d_k f_k(x) f_k(y) = f(x+y), \tag{2.59}$$

then the Hankel determinants of a_n are determined as

$$h_n = \prod_{k=0}^{n} d_k [f_k^{(k)}(0)]^2. \tag{2.60}$$

Here $f_k^{(k)}(0)$ is the kth derivative of f_k at $x = 0$.

The proof of (2.60)

The proof of this theorem is not hard[27]. First we have that

$$f(x+y) = \sum_{n=0}^{\infty} a_n \frac{(x+y)^n}{n!} = \sum_{n=0}^{\infty} \frac{a_n}{n!} \left(\sum_{k=0}^{n} \binom{n}{k} x^k y^{n-k} \right) =$$

$$\sum_{n=0}^{\infty} a_n \left(\sum_{k=0}^{n} \frac{x^k}{k!} \frac{y^{n-k}}{(n-k)!} \right) = \sum_{n=0}^{\infty} \sum_{m=0}^{\infty} a_{n+m} \frac{x^n}{n!} \frac{y^m}{m!}.$$

On the other hand, by (2.9):

$$f_k(x) = \sum_{n=0}^{\infty} \frac{f_k^{(n)}(0)}{n!} x^n.$$

Hence,

$$\sum_{k=0}^{\infty} d_k f_k(x) f_k(y) = \sum_{k=0}^{\infty} \sum_{n,m=0}^{\infty} d_k f_k^{(n)}(0) f_k^{(m)}(0) \frac{x^n}{n!} \frac{y^m}{m!}.$$

---

[27] The statement, together with the proof can be found in [311].

By our assumptions, this must be equal to $f(x+y)$. Comparing the coefficients we have that

$$a_{n+m} = \sum_{k=0}^{\infty} d_k f_k^{(n)}(0) f_k^{(m)}(0).$$

Note that this is a finite sum. And, in addition, the sum can be considered as a product of three matrices. These matrices are, by our assumptions, lower triangular, diagonal, and upper triangular matrices, respectively. The first and third are a transpose of each other[28]. On the left-hand side, we have an element of the Hankel matrix, and the determinant formed by these elements is the same as the determinant formed by the elements on the right-hand side. The latter is easy to calculate because of the structure of the matrices – the determinant of a triangular or diagonal matrix is the product of the diagonal elements. Hence, (2.60) is proven.

We note that we often have generating functions with two variables. This is the case if we consider the exponential generating function of the polynomials $p_n(x)$. We suppose that the sequence of these polynomials has its exponential generating function as $f(x, y)$. It is now useful to introduce the two variable functions $f_k(x, y)$. Now the d_k numbers turn to be functions; moreover, we suppose that $f_k(x, y) = \partial_x^k f(x, y)$. With these assumptions (2.59) turns into

$$\sum_{k=0}^{\infty} d_k(x) \partial_x^k f(x, y) \partial_x^k f(x, z) = f(x, y+z). \tag{2.61}$$

The Hankel determinants (which now depend on x) take the form:

$$h_n(x) = \prod_{k=0}^{n} d_k(x) [\partial_x^k p_k(x)]^2.$$

Applications of (2.60)

Considering the Bell numbers and polynomials, we show how this machinery works in practice. For the moment it suffices to know that the $B_n(x)$ Bell polynomials have the exponential generating function $\exp(x(e^y - 1))$. (See the proof of (3.3).) In this case then

$$f(x, y) = \exp(x(e^y - 1)),$$

and (2.61) becomes:

$$\sum_{k=0}^{\infty} d_k(x) \partial_x^k f(x, y) \partial_x^k f(x, z) = \sum_{k=0}^{\infty} d_k(x) e^{x(e^y-1)} (e^y - 1)^k e^{x(e^z-1)} (e^z - 1)^k =$$

---

[28]This decomposition is called LDL^T *decomposition* in numerical mathematics.

$$e^{x(e^y-1)}e^{x(e^z-1)}\sum_{k=0}^{\infty}d_k(x)[e^{y+z}-e^y-e^z+1]^k$$

This must be equal to

$$f(x,y+z)=e^{x(e^{y+z}-1)}.$$

Dividing by the exponential factor in the penultimate line, we get

$$\sum_{k=0}^{\infty}d_k(x)[e^{y+z}-e^y-e^z+1]^k=e^{x(e^{y+z}-1)-x(e^y-1)-x(e^z-1)}=e^{x(e^{y+z}-e^y-e^z+1)}.$$

This equality satisfies if we choose

$$d_k(x)=\frac{x^n}{n!}.$$

The Hankel determinants thus become

$$h_n(x)=\prod_{k=0}^{n}\frac{x^k}{k!}k!^2=\prod_{k=0}^{n}x^kk!=x^{0+1+2+\cdots+n}\prod_{k=0}^{n}k!=x^{n(n+1)/2}\prod_{k=0}^{n}k!.$$

Setting $x=1$, we get a beautiful result for the Bell numbers:

$$\det\begin{pmatrix}B_0 & B_1 & \cdots & B_n \\ B_1 & B_2 & \cdots & B_{n+1} \\ \vdots & & \ddots & \\ B_n & B_{n+1} & \cdots & B_{2n}\end{pmatrix}=\prod_{k=0}^{n}k!. \tag{2.62}$$

The sequence on the right-hand side is called *superfactorial*[29]. This result appears in Aigner [12], in [243, p. 35], and the polynomial form

$$\det\begin{pmatrix}B_0(x) & B_1(x) & \cdots & B_n(x) \\ B_1(x) & B_2(x) & \cdots & B_{n+1}(x) \\ \vdots & & \ddots & \\ B_n(x) & B_{n+1}(x) & \cdots & B_{2n}(x)\end{pmatrix}=x^{\binom{n+1}{2}}\prod_{k=0}^{n}k!.$$

was proven in [477]. See also [211], where a combinatorial connection between the Hankel determinants and specific permutation-partition pairs is given.

2.10.3 Another tool to calculate the Hankel transform involving orthogonal polynomials

We show another useful but more advanced way to calculate Hankel determinants. This approach uses the theory of orthogonal functions – a theory

---

[29]The superfactorials are the special values of the *Barnes G function*. If we take the kth powers of k and not its factorials, we get the *hyperfactorial function*.

which we cannot develop in this book, so this subsection will not be self-contained. However, it is easy for anyone to apply this theory without being familiar with the otherwise rich topic of orthogonal functions (in fact, here only orthogonal polynomials will be used). For a good general introduction, see the book of Szegő [556]. The theorems that we are going to present here are taken from [345] (see also the complement [346]).

This tool is usable when we can write the generating function of our sequence μ_n in a *continued fraction* form:

$$\sum_{n=0}^{\infty} \mu_n x^n = \cfrac{a_0}{1 + a_0 x - \cfrac{b_1 x^2}{1 + a_1 x - \cfrac{b_2 x^2}{1 + a_2 x - \ddots}}}. \qquad (2.63)$$

If this is possible, then the Hankel determinant of size $n \times n$ equals

$$\mu_0^n \prod_{k=1}^{n-1} b_k^{n-k}. \qquad (2.64)$$

The continued fraction might seem to be scary and too special for our purposes: until now we have not seen any combinatorial sequence which generating function would have continued fraction expansion. This is not entirely true, however, as we shall see soon. The a_n and b_n sequences in the continued fraction can be determined with a small effort in many cases. We *will* find such continued fraction expansion for some of our sequences[30]. The bridge between the two sides of (2.63) is the theory of orthogonal polynomials.

A sequence $(p_n(x))_{n \geq 0}$ of polynomials is called *orthogonal* if the degree of $p_n(x)$ is n, and there exists a linear functional[31] L such that

$$L(p_n(x)p_m(x)) = \begin{cases} c_n, & \text{if } n = m; \\ 0, & \text{if } n \neq m \end{cases}$$

with some non-zero sequence c_n. Favard[32] proved [226] that a polynomial sequence $(p_n(x))_{n \geq 0}$ is orthogonal if and only if for all $n \geq 1$ these polynomials satisfy a second order recurrence

$$p_{n+1}(x) = (a_n + x)p_n(x) - b_n p_{n-1}(x) \qquad (2.65)$$

with some sequences $(a_n)_{n \geq 1}$ and $(b_n)_{n \geq 1}$, with the initial condition $p_0(x) = 1$ and $p_1(x) = x + a_0$. (See also [579, Theorem 50.1].)

---

[30] The general theory of continued fractions connected to generating functions can be found in Chapter XI of [579].

[31] Roughly speaking, a linear functional map functions to numbers such that $L(cf(x)) = cL(f(x))$, and $L(f(x)+g(x)) = L(f(x))+L(g(x))$. $L_{a,b}(f(x)) = \int_a^b f(t)dt$ is a good example of a linear functional, where f is integrable on $[a, b]$.

[32] Jean Favard (1902-1965), French mathematician, a former president of the French Mathematical Society.

We have two sets of quantities in (2.63): our sequence μ_n, and the pair of sequences a_n and b_n constituting the continued fraction. The statement is that *if*

$$\mu_n = L(x^n), \tag{2.66}$$

then a_n and b_n are the sequences in Favard's theorem, and the (2.63) expansion holds true. This extraordinary fact is proven in [579], see Theorem 51.1 there.

So, what do we need to do to find the Hankel determinants of a sequence? First, we need to find a suitable sequence of orthogonal polynomials (usually from a database that we will discuss below) for which our sequence satisfies (2.66). Second, look at the recurrence of these polynomials, particularly the sequence b_n is important, because a_n does not appear in (2.64). Third, calculate the Hankel determinants via (2.64).

How this works in practice is shown in an example. When we will know the so-called Fubini numbers, we present another example in Subsection 6.1.5.

Let $\mu_n = n!$, and we want to calculate the determinants

$$h_n = \begin{vmatrix} 0! & 1! & 2! & \cdots & n! \\ 1! & 2! & 3! & \cdots & (n+1)! \\ \vdots & & & & \\ n! & (n+1)! & (n+2)! & \cdots & (2n)! \end{vmatrix} \quad (n = 0, 1, 2, \dots).$$

First we are looking for a sequence of polynomials such that

$$\mu_n = n! = L(x^n).$$

The basic source of the database of common orthogonal polynomials is [338]. In Section 1.11 of [338] we find the basic properties of the *Laguerre polynomials* $L_n^{(\alpha)}(x)$. Equation (1.11.2) says that

$$\int_0^\infty e^{-x} x^\alpha L_n^{(\alpha)}(x) L_m^{(\alpha)}(x) dx = \frac{\Gamma(n+\alpha+1)}{n!} \delta_{mn} \quad (\alpha > -1),$$

where $\delta_{mn} = 1$ if $n = m$ and zero otherwise. This exactly means that $L_n^{(\alpha)}(x)$ is an orthogonal polynomial sequence with respect to the linear functional

$$L(f(x)) = \int_0^\infty e^{-t} t^\alpha f(t) dt.$$

Why is this good for us? In the particular case when $f(x) = x^n$ and $\alpha = 0$ we have that

$$L(x^n) = \int_0^\infty e^{-t} t^n dt.$$

This integral[33] is calculable by partial integration:

$$L(x^n) = \int_0^\infty e^{-t} t^n dt = n! = \mu_n \quad (n \geq 0),$$

---

[33] This integral will be more extensively used in Subsection 7.3.1, where more information is given.

exactly as we wish.

The first step is thus done. In the second step we look at the recurrence of these polynomials: (1.11.3) in [338] could be used, but the normalized recurrence (1.11.4), which is simpler, does the job. It refers to the polynomial

$$p_n(x) = n! L_n^{(\alpha)}(x)$$

but it makes no difference since the $p_n(x)$ polynomials are orthogonal with respect to the same functional L. When $\alpha = 0$ (1.11.4) reads

$$p_{n+1}(x) = (x - (2n+1))p_n(x) - n^2 p_{n-1}(x),$$

whence

$$a_n = -(2n+1), \quad \text{and} \quad b_n = n^2$$

by (2.65). The second step is performed.

In the third step we are already going to calculate the determinants. Recall (2.64). Being careful with the size of the matrix and the indices:

$$h_n = \begin{vmatrix} 0! & 1! & 2! & \cdots & n! \\ 1! & 2! & 3! & \cdots & (n+1)! \\ \vdots & & & & \\ n! & (n+1)! & (n+2)! & \cdots & (2n)! \end{vmatrix} = \prod_{k=1}^{n} (k^2)^{n+1-k} \quad (n = 1, 2, \dots).$$

Nice result, is this not it?

What about the continued fraction expansion (2.63)? It is obvious that the sum

$$\sum_{n=0}^{\infty} n! x^n$$

converges nowhere except at the origin, so it makes not much sense to compare it to the continued fraction. This, however, does not alter the correctness of our result.

Exercises

1. Determine the generating function of the following sequences:

$$
\begin{aligned}
a_n &= \begin{cases} 1, & \text{if } n \text{ odd;} \\ 0, & \text{if } n \text{ even} \end{cases}, \\
b_n &= 1 + 2 + 3 + \cdots + n, \\
c_n &= H_n = \frac{1}{1} + \frac{1}{2} + \frac{1}{3} + \cdots + \frac{1}{n}, \\
d_n &= n, \\
e_n &= c^n \ (c \text{ is an arbitrary real number}), \\
f_n &= n2^n, \\
g_n &= n^2, \\
h_n &= \frac{1}{n}, \\
i_n &= \frac{(-1)^{n+1}}{n}.
\end{aligned}
$$

2. Give the generating function of the recursive sequence $F_0 = 0$, $F_1 = 1$, $F_n = F_{n-1} + F_{n-2}$. (The sequence F_n is the famous *Fibonacci*[34] *sequence.*)

3. Calculate the exact values of the sums

$$
\sum_{n=1}^{\infty} \frac{(-1)^{n+1}}{n}, \qquad \sum_{n=0}^{\infty} \frac{n+1}{3^n}.
$$

4. To which sequence does the generating function $\cos(x)$ belong?

5. To which sequence does the exponential generating functions $-\ln(1-x)$ and $-\ln(1+x)$ belong?

6. Let us suppose that the generating function of a_n is $f(x)$. What is the generating function of $\frac{a_n}{n+2}$?

7. If $f(x)$ is the generating function of a_n, then for which sequence does the generating function $f(x^2)$ belong?

8. Verify the

$$
\sum_{n=0}^{\infty} \begin{bmatrix} n \\ k \end{bmatrix} \frac{x^n}{n!} = \frac{1}{k!} \ln\left(\frac{1}{1-x}\right)^k
$$

generating function identity by using the standard recursion for $\begin{bmatrix} n \\ k \end{bmatrix}$.

---

[34] Fibonacci ($\approx$1170-$\approx$1250), Leonardo de Pisa or Leonardo Pisano, Italian mathematician.

9. Show the validity of the following sum formula for the generalized harmonic numbers:

$$\sum_{k=1}^{n} H_{k,m} = (n+1)H_{n,m} + H_{n,m-1} \quad (m \geq 1).$$

(The definition of the generalized harmonic numbers is on p. 63.)

10. Show without referring to the Stirling numbers that

$$\frac{n!}{2}(H_n^2 - H_{n,2}) = \frac{n!}{3}\sum_{k=1}^{n-1}\frac{H_{k-1}}{k(n-k)}$$

11. Calculate $\left\{{4 \atop 2}\right\}$ by taking the derivative of the exponential generating function (2.19).

12. Show that the exponential generating function of the E_n Euler numbers is

$$\sum_{n=0}^{\infty} E_n \frac{x^n}{n!} = \sec x + \tan x.$$

This is a famous result of D. André [26], and it is actually very simple once we have the recurrence (1.16). This recurrence, together with the initial condition $E_0 = 1$, is equivalent to the differential equation $2f'(x) = f^2(x) + 1$ with $f(0) = 1$. The solution of this equation with the said initial value is $\sec x + \tan x$.) Note that $\sec x$ is an even function, and $\tan x$ is an odd function. So the even-indexed Euler numbers are determined only by $\sec x$, and the odd-indexed Euler numbers occur only in $\tan x$. By this very reason, sometimes the numbers E_{2n} are called *secant numbers* and E_{2n+1} are called *tangent numbers*.

13. Show by explicit calculation that the 3×3 Stirling matrices are inverses of each other, as it follows from the orthogonality.

14. Write (1.5) into matrix form.

15. Use the (2.14) Euler formula to obtain (1.5).

16. Express the determinant of $\mathcal{B}_n$ in terms of the superfactorial.

17. Verify (2.42) for the signed first kind Stirling numbers.

18. Calculate

$$\sum_{k=0}^{n}(-1)^k \left[{n \atop k}\right].$$

19. Let a_n be a sequence such that

$$\sum_{n=0}^{\infty} a_n \frac{x^n}{n!} = f(x),$$

and let b_n be the Stirling transform of a_n. Show that the exponential generating function of b_n is $f(e^x - 1)$. (For more information on transformation sequences, see the paper of M. Bernstein, N. J. A. Sloane, [72].)

20. Show that

$$\sum_{k=1}^{n} \left\{ {n \atop k} \right\} \frac{(-1)^k}{(k-1)!} = 0 \quad (n \geq 2).$$

21. Let the pattern $(j_1, j_2, \ldots, j_8) = (0, 2, 2, 0, 0, 0, 0, 0)$ be given. How many permutations are there on 8 elements with this given pattern?

22. Solve the previous exercise for partitions.

23. Prove that the exponential generating function of the $x^{\overline{n}}$ rising factorial is

$$\sum_{n=0}^{\infty} x^{\overline{n}} \frac{y^n}{n!} = \left(\frac{1}{1-y} \right)^x.$$

(Hint: apply the sum

$$\sum_{n=0}^{\infty} \left(\sum_{k=0}^{n} \left[{n \atop k} \right] x^k \right) \frac{y^n}{n!} .)$$

24. Prove that the exponential generating function of the $x^{\underline{n}}$ falling factorial is

$$\sum_{n=0}^{\infty} x^{\underline{n}} \frac{y^n}{n!} = (1+y)^x.$$

(Now use the exponential generating function of the signed first kind Stirling numbers: (2.36).)

25. Express $n!$ in terms of falling and rising factorials.

26. Prove the binomial theorem for rising factorials:

$$(x+y)^{\overline{n}} = \sum_{k=0}^{n} \binom{n}{k} x^{\overline{k}} y^{\overline{n-k}}$$

(Apply the $x^{\overline{n}} = (-1)^n (-x)^{\underline{n}}$ transformation both to x and y, then use (2.50).)

27. Calculate the Hankel transform of $a_n = 1$.

28. Prove that the vertical generating function of the Lah numbers is

$$\sum_{n=k}^{\infty} \left\lfloor {n \atop k} \right\rfloor \frac{x^n}{n!} = \frac{x^k}{k!} \frac{1}{(1-x)^k}.$$

29. Let L_n be defined as in (2.56). Prove that the exponential generating function of L_n is

$$\sum_{n=0}^{\infty} L_n \frac{x^n}{n!} = e^{\frac{x}{1-x}}.$$

30. Prove the following recurrence for L_n:

$$L_n = (2n-1)L_{n-1} - (n-1)(n-2)L_{n-2} \quad (n \geq 2)$$

with $L_0 = L_1 = 1$.

31. Prove that the Hankel transform of the sequence $L_{n+1}/(n+1)!$ is

$$\frac{(-1)^{a(n)}}{\prod_{i=0}^{n-1} \frac{(n+i)!}{i!}} \quad (n \geq 1),$$

where

$$a(n) = \begin{cases} 1, & \text{if } n \equiv 1,2 \pmod 4; \\ 0, & \text{if } n \equiv 0,3 \pmod 4. \end{cases}$$

(Hint: first note that by the definition of L_n and the exact binomial expression for the Lah numbers, it is true that

$$\frac{L_{n+1}}{(n+1)!} = \sum_{k=0}^{n} \binom{n}{k} \frac{1}{(k+1)!}.$$

Then use Layman's theorem which says that the binomial transform does not modify the Hankel determinants. So it is enough to find the Hankel determinant of $\frac{1}{(k+1)!}$. This result is taken from [430].)

Outlook

1. A formula similar to (2.26) was presented by V. Adamchik [3]:

$$\begin{bmatrix} n \\ k \end{bmatrix} = \sum_{j_1 < j_2 < \cdots < j_{k-1}} \frac{(n-1)!}{j_1 j_2 \cdots j_{k-1}}.$$

   This was later generalized and extended in [324].

2. P. Flajolet [234] established continued fraction representations of several generating functions of combinatorial numbers – including the Bell numbers, Stirling numbers of the second kind, factorials, derangement numbers, involutions, etc. – combinatorially. These are identical to those continued fractions we can get by using the orthogonal polynomial approach (with our a_n and b_n sequences). His approach is purely combinatorial. See the work of P. Barry [48] for some more examples in the intersection of the theory of counting sequences, orthogonal polynomials, and Hankel determinants.

3. Liu and Ma [379] studied the continued fraction representation of the Bell polynomials and some of their generalizations like the ordered Bell polynomials, the r-Bell polynomials (we will discuss these two polynomial sequences in Chapters 6 and 8), the Dowling polynomials, among others.

4. M. Merca [405] discovered the following determinantal identities

$$\det\left(\begin{bmatrix} i+k \\ j+1 \end{bmatrix}\right)_{1 \le i,j \le n} = k!^n \sum_{j=1}^{k} (-1)^{j-1} \frac{1}{j^n} \binom{k}{j},$$

$$\det\left(\begin{Bmatrix} i+k \\ j+1 \end{Bmatrix}\right)_{1 \le i,j \le n} = (n+1)!^{k-1} \sum_{j=1}^{n+1} (-1)^{j-1} \frac{1}{j^{k-1}} \binom{n+1}{j}.$$

5. The Hankel transform of the horizontal sum of Lah numbers is not presented in the On-Line Encyclopedia of Integer Sequences (OEIS). It would therefore be interesting to do a study on this problem. More generally, we can define the *Lah polynomials*

$$L_n(x) = \sum_{k=0}^{n} \left\lfloor \begin{matrix} n \\ k \end{matrix} \right\rfloor x^k,$$

   and then ask for the Hankel determinants of this sequence. The first

Hankel determinants are

$$h_1(x) = 1$$
$$h_2(x) = 2x$$
$$h_3(x) = 4x^2(3 + 4x)$$
$$h_4(x) = 48x^3 \left(16x^3 + 48x^2 + 45x + 12\right)$$
$$h_5(x) = 4608x^4(90 + 648x + 1710x^2 + 2175x^3 + 1440x^4 + 480x^5 + 64x^6)$$

Can one give a description of the coefficient of these polynomials?

(The $L_n(x)$ Lah polynomials will reappear in Section 4.4. They can be related, via their generating function, to the so-called Laguerre polynomials. For Hankel determinantal results for these and other orthogonal polynomials, see, for example, the older [321], and the newer [296].)

6. Interestingly, the Lah numbers appear in the expression of the higher order derivatives of $e^{1/x}$ [183]. For other, combinatorial (concretely Dyck path) interpretations for the Lah numbers, see [109].

7. We defined the Fibonacci numbers among the exercises. A similar sequence, the *Lucas sequence* can be defined as follows: $L_n = L_{n-1} + L_{n-2}$ (this is exactly the same recurrence as for the Fibonacci numbers), but the initial values are $L_0 = 2$, $L_1 = 1$. It can be shown [377] that

$$\sum_{\substack{v_1,\ldots,v_k \geq 1 \\ v_1 + \cdots + v_k = n}} \frac{L_{v_1} \cdots L_{v_k}}{v_1 \cdots v_k} = \frac{k!}{n!} \sum_{j=k}^{n} (-1)^{j-k}(n-j)! \binom{n}{j}\binom{j}{n-j}\left[\genfrac{}{}{0pt}{}{j}{k}\right].$$

8. The Bell and Stirling numbers and Dobiński's formula appear in some interesting contexts. See [467] and its references for probability theory, [192] for a connection with random matrices, and [299, Section 5.5] for representation theory of the symmetric group.

9. The Stirling matrices have been the subject of several studies from a linear algebraic point of view [144, 609].

Chapter 3

The Bell polynomials

3.1 Basic properties of the Bell polynomials

3.1.1 A recursion

The Bell polynomials have many nice and useful properties that are relatively easy to prove. Moreover, we can introduce and practice many important techniques which we can apply later in more general situations. This chapter is not only on the $B_n(x)$ polynomials, but it contains results regarding a generalization of $B_n(x)$ such that we meet some new counting sequences, too.

Let us recall the definition of the Bell polynomials:

$$B_n(x) = \sum_{k=0}^{n} \left\{ {n \atop k} \right\} x^k. \tag{3.1}$$

We already know that $B_n(1) = B_n$, the nth Bell number. By knowing the second kind Stirling numbers, we can easily calculate the Bell polynomials. For example

$$B_4(x) = x^4 + 6x^3 + 7x^2 + x.$$

The recursion for the Stirling numbers can be applied to get a recursion for the Bell polynomials. First of all, we calculate the derivative of $B_n(x)$:

$$B_n'(x) = \sum_{k=0}^{n} \left\{ {n \atop k} \right\} k x^{k-1}.$$

Multiply this by x:

$$x B_n'(x) = \sum_{k=0}^{n} k \left\{ {n \atop k} \right\} x^k.$$

Now apply the recursion of the $\left\{{n \atop k}\right\}$ numbers:

$$
\begin{aligned}
B_n(x) &= \sum_{k=0}^{n} \left\{{n \atop k}\right\} x^k = \sum_{k=0}^{n} \left(\left\{{n-1 \atop k-1}\right\} + k \left\{{n-1 \atop k}\right\} \right) x^k \\
&= \sum_{k=0}^{n} \left\{{n-1 \atop k-1}\right\} x^k + \sum_{k=0}^{n} k \left\{{n-1 \atop k}\right\} x^k \\
&= \sum_{k=0}^{n-1} \left\{{n-1 \atop k}\right\} x^{k+1} + \sum_{k=0}^{n-1} k \left\{{n-1 \atop k}\right\} x x^{k-1} \\
&= x \sum_{k=0}^{n-1} \left\{{n-1 \atop k}\right\} x^k + x \sum_{k=0}^{n-1} k \left\{{n-1 \atop k}\right\} x^{k-1} \\
&= x B_{n-1}(x) + x B'_{n-1}(x).
\end{aligned}
$$

Therefore

$$B_n(x) = x B_{n-1}(x) + x B'_{n-1}(x). \tag{3.2}$$

For instance, we determine $B_4(x)$ from $B_3(x)$ as follows. The Stirling triangle says that

$$B_3(x) = x^3 + 3x^2 + x.$$

The derivative is

$$B'_3(x) = 3x^2 + 6x + 1.$$

By the recursion, $B_4(x)$ must be

$$B_4(x) = x B_3(x) + x B'_3(x) = (x^4 + 3x^3 + x^2) + (3x^3 + 6x^2 + x) = x^4 + 6x^3 + 7x^2 + x.$$

Applying this recurrence, we can determine the Bell polynomials directly, without relying on the Stirling triangle. Indeed, the first polynomial is $B_0(x) = 1$, and then $B_1(x) = x B_0(x) + x B'_0(x) = x$, and so on.

3.1.2   The exponential generating function

If we fix an x, then $B_n(x)$ will be a real sequence for which the generating function is determinable. The resulting function will, of course, depend on x, so it will be a *two-variable* generating function or *bivariate generating function*.

From (3.1) we readily get the exponential generating function after multiplying by $\frac{y^n}{n!}$ and summing over n (in (3.1) the sum can be extended to infinity as if $k > n$ the terms are simply zero):

$$
\sum_{n=0}^{\infty} B_n(x) \frac{y^n}{n!} = \sum_{n=0}^{\infty} \left(\sum_{k=0}^{\infty} \left\{{n \atop k}\right\} x^k \right) \frac{y^n}{n!}
$$

$$
= \sum_{k=0}^{\infty} x^k \left(\sum_{n=0}^{\infty} \left\{{n \atop k}\right\} \frac{y^n}{n!} \right) = \sum_{k=0}^{\infty} x^k \frac{(e^y - 1)^k}{k!}
$$

$$= \sum_{k=0}^{\infty} \frac{[x(e^y - 1)]^k}{k!} = e^{x(e^y - 1)}.$$

That is, the exponential generating function of the Bell polynomials is

$$\sum_{n=0}^{\infty} B_n(x) \frac{y^n}{n!} = e^{x(e^y - 1)}. \tag{3.3}$$

As it is in many cases, this results in a new formula for the Bell polynomials:

$$\sum_{n=0}^{\infty} B_n(x + z) \frac{y^n}{n!} = e^{(x+z)(e^y - 1)} = e^{x(e^y - 1)} e^{z(e^y - 1)}.$$

The right-hand side is a product of two exponential generating functions, one of them belongs to $B_n(x)$, while the other belongs to $B_n(z)$. Hence, the exponential generating function of the product – by (2.7) – is

$$e^{x(e^y - 1)} e^{z(e^y - 1)} = \sum_{n=0}^{\infty} \frac{y^n}{n!} \left(\sum_{k=0}^{n} \binom{n}{k} B_k(x) B_{n-k}(z) \right).$$

Comparing the coefficients, we get a nice formula which we can call *addition theorem* (after changing z to y):

$$B_n(x + y) = \sum_{k=0}^{n} \binom{n}{k} B_k(x) B_{n-k}(y).$$

We had the recursion (1.1) for the Bell numbers. Can we find the polynomial version of it? Recalling (1.1) and the method on how we got the exponential generating function on page 45, the reverse steps yield the wanted result. Taking the derivative on both sides of (3.3), on the left-hand side we get the generating function of $B_{n+1}(x)$, while on the right-hand side we have $xe^y e^{x(e^y - 1)}$. This is equivalent to

$$\sum_{n=0}^{\infty} B_{n+1}(x) \frac{y^n}{n!} = xe^y \sum_{n=0}^{\infty} B_n(x) \frac{y^n}{n!},$$

from which the generalization of (1.1) already follows:

$$B_{n+1}(x) = x \sum_{k=0}^{n} \binom{n}{k} B_k(x).$$

The ordinary generating function of the Bell polynomials is a bit more complicated to deduce. We return to this problem later in a more general situation (see Section 8.4).

3.2   About the zeros of the Bell polynomials

In the following subsections, we present some information about the zeros of the Bell polynomials. Some of them are quite remarkable.

3.2.1   The real zero property

We are going to prove that the Bell polynomials have only real zeros. The importance of this fact will be revealed in subsequent sections.

The recursion

$$B_n(x) = x(B_{n-1}(x) + B'_{n-1}(x))$$

will be extremely useful to prove our claim. A simple corollary of this recursion is the equality

$$B_n(x) = e^{-x} x (e^x B_{n-1}(x))'. \tag{3.4}$$

To see that this equality indeed holds, take the derivative on the right-hand side and note that the exponentials cancel out.

We are going to prove the real zero property by induction. $B_1(x) = x$ and this polynomial satisfies our claim. Now let us suppose that all the zeros of $B_{n-1}(x)$ are real. Then the function $e^x B_{n-1}(x)$ has the same set of zeros, because the exponential function is always positive. Hence, the zeros of the derivatives of $e^x B_{n-1}(x)$ are among the zeros of $e^x B_{n-1}(x)$ by Rolle's[1] theorem[2]. Since $B_{n-1}(x)$ has $n - 1$ zeros, from which $n - 2$ is strictly negative and one is $x = 0$, the derivative must have $n - 2$ negative zeros (left to 0 but right to the greatest negative zero, and among the negative zeros). On the other hand, e^x is asymptotically 0 in $-\infty$; therefore $(e^x B_{n-1}(x))'$ has an additional zero between $-\infty$ and the leftmost zero of $B_{n-1}(x)$. We have found $n - 2 + 1 = n - 1$ negative zeros on the right-hand side of (3.4). There is one zero on the right by the factor x: $x = 0$. Altogether we have found $n - 1$ negative zeros and $x = 0$. This is n in total so we have found that all the zeros of $B_n(x)$ are real and non-positive.

Reading the above proof carefully, we see that we proved more of what we actually stated: the zeros of $B_n(x)$ are not only negative (except $x = 0$), but the negative zeros are *simple* and located among the zeros of $B_{n-1}(x)$ except the leftmost one: this is on the left-hand side of the leftmost zero of $B_{n-1}(x)$. This is an interesting and rather common phenomenon among combinatorially relevant polynomials, and it is called *interlacing property*.

The real-zero property of the Bell polynomials was first noted by Harper in 1967 [279].

---

[1] Michel Rolle (1652-1719), French mathematician.

[2] This theorem roughly says that if a differentiable function has zeros in two different points x_1 and x_2, then it surely has a stationary point between x_1 and x_2 where the derivative is zero. This theorem was proven by Rolle in 1691; however, it was known earlier by the Indian mathematician Bhāshkara (1114-1185).

3.2.2 The sum and product of the zeros of $B_n(x)$

We know from the last section that we can write

$$B_n(x) = (x - t_1^{(n)})(x - t_2^{(n)}) \cdots (x - t_n^{(n)}),$$

where $t_i^{(n)}$ are the zeros of the Bell polynomials ($i = 1, \ldots, n$), and that these numbers are negative, except one zero which is $x = 0$. We choose this to be $t_1^{(n)}$, that is, $t_1^{(n)} = 0$

By *Viète's*[3] *formulas* we can have a bit more knowledge about these zeros. One of these formulas says that if we have a polynomial

$$p(x) = a_n x^n + a_{n-1} x^{n-1} + \cdots + a_1 x + a_0, \qquad (3.5)$$

such that the zeros are $x_1, \ldots, x_n$ (these can be real or complex), that is,

$$p(x) = a_n(x - x_1)(x - x_2) \cdots (x - x_n), \qquad (3.6)$$

then the sum of these zeros is

$$x_1 + x_2 + \cdots + x_n = -\frac{a_{n-1}}{a_n}.$$

Another formula evaluates the product of the zeros:

$$x_1 x_2 \cdots x_n = (-1)^n \frac{a_0}{a_n}. \qquad (3.7)$$

These can easily be seen by comparing the coefficients of x^{n-1} and x^0 on the right-hand sides of (3.5) and (3.6), respectively[4].

These formulas result in the following sum and product formulas with respect to the zeros of the Bell polynomials:

$$t_1^{(n)} + t_2^{(n)} + \cdots + t_n^{(n)} = -\frac{\left\{ {n \atop n-1} \right\}}{\left\{ {n \atop n} \right\}} = -\binom{n}{2} = -\frac{n(n-1)}{2} \quad (n \geq 1),$$

and (taking $B_n(x)/x$ instead)

$$t_2^{(n)} \cdots t_n^{(n)} = -\frac{\left\{ {n \atop 1} \right\}}{\left\{ {n \atop n} \right\}} = -1 \quad (n \geq 2).$$

The sum formula yields that the *average* of the zeros tends to minus infinity:

$$\frac{t_1^{(n)} + t_2^{(n)} + \cdots + t_n^{(n)}}{n} = -\frac{n-1}{2}.$$

The product formula assures that $B_n(x)$ (at least for $n \geq 3$) has always a zero in the interval $] -1, 0[$. (If all the zeros were smaller than minus one, they would not form a product being equal to minus one.)

---

[3]François Viète (1540-1603), French mathematician and lawyer.
[4]More on the history of Viète's formulas can be read in [239].

3.2.3   The irreducibility of $B_n(x)$

Whether the Bell polynomials ever have rational zeros ($n \geq 3$) other than the trivial $t_1^{(n)} = 0$ is unknown. If a non-constant polynomial has no rational zeros, we say that it is *irreducible* over the rationals.

In 1983 J. W. Layman and C. L. Prather [360] gave five different statements equivalent to the irreducibility of the Bell polynomials $B_n(x)/x$. One of these is the non-vanishing property of the Bell polynomials at $x = -1$. This is actually easy to see once we recall the *rational root theorem* [574, p. 109].

The rational root theorem says that if a polynomial

$$p(x) = a_n x^n + a_{n-1} x^{n-1} + \cdots + a_1 x + a_0$$

has a rational zero x^*, i.e., it is of the form $x^* = \frac{a}{b}$ (where a and b have no divisors in common), then a must divide a_0, and b must divide a_n.

Since the constant coefficient and the highest order coefficient in $B_n(x)/x$ are both one, the only possible rational zeros are ± 1. Of course, $B_n(x) > 0$ if $x > 0$, so the only possible rational root is minus one. Thus, we arrive at the following statement: $B_n(x)/x$ is irreducible over the rationals if and only if $B_n(-1) \neq 0$. Since $B_2(x) = 1 + x$, we have that $B_2(-1) = 0$, but this seems to be the only one case. In [360] this was checked up to $n \leq 110$ (up to 900 in [359]), and we checked by computer up to $n = 10\,000$ but none of $B_n(-1)$ numbers vanishes ($n \neq 2$). Layman further studied the arithmetical properties of the $B_n(-1)$ numbers [359].

After these initial studies, many achievements have been made. It is known now, due to T. Amdeberhan, V. De Angelis, and V. H. Moll [18], that there is *at most one* $n > 2$ such that $B_n(-1) = 0$. See [29] for an "economical account" of the proof, and for historical remarks. The paper [589] contains many additional arithmetical properties of this sequence. The conjecture that $B_n(-1) \neq 0$ for all $n > 2$ is attributed to H. Wilf, and it is often referred to as *Wilf's conjecture*. The source of this attribution is seemingly the paper of Y. Yang [608]. In this paper and also in [572], a detailed asymptotic analysis is carried out for this sequence. Among others, it is known that

$$\limsup_{n \to \infty} \frac{\log |B_n(-1)|}{n \log n} = 1.$$

3.2.4   The density of the zeros of $B_n(x)$

Yet another interesting topic is the distribution of the zeros of the Bell polynomials. One question is the following: knowing that the leftmost zero grows with n (see Section 3.2.1), how rapidly does this leftmost zero grow? We study this question later in Section 7.6.5.

Another question is the distribution of the *all* zeros of all the Bell polynomials on the negative real line. This question is hard, like the irreducibility, but in this case a very nice result is known. The statement is described as follows.

Let us construct the set of all the zeros of all the Bell polynomials:

$$\mathcal{T} = \left\{ t_k^{(n)} \mid n = 1, 2, \ldots; \ k = 1, 2, \ldots, n \right\}.$$

We already know that $\mathcal{T} \subset] - \infty, 0]$, but much more is true: in any neighbor of any negative number there is an element of $\mathcal{T}$. In other words, $\mathcal{T}$ is *dense* in $] - \infty, 0]$.

The proof of this density statement uses some heavy machinery from complex function theory (Nevanlinna's theory and the saddle point method), and therefore it is out of the scope of this book, but we give some insight.

First a statement comes, which is interesting in itself.

A derivation formula

Let f be a function, so that we can take its successive derivatives at any real or complex point except maybe zero. Then

$$\left(f(e^{iz}) \right)^{(n)} = i^n \sum_{k=0}^{n} \left\{ {n \atop k} \right\} f^{(k)}(e^{iz}) e^{ikz}.$$

The proof of this statement hinges on the recursion of the Stirling numbers of the second kind. It is true for $n = 1$:

$$\left(f(e^{iz}) \right)' = i \left\{ {1 \atop 1} \right\} f'(e^{iz}) e^{i1z}.$$

We proceed by induction. Supposing that the statement holds up to $n - 1$, we prove it for n:

$$\left(f(e^{iz}) \right)^{(n)} = \left[\left(f(e^{iz}) \right)^{(n-1)} \right]' = \left[i^{n-1} \sum_{k=0}^{n-1} \left\{ {n - 1 \atop k} \right\} f^{(k)}(e^{iz}) e^{ikz} \right]' =$$

$$i^{n-1} \sum_{k=0}^{n-1} \left\{ {n - 1 \atop k} \right\} \left(f^{(k+1)}(e^{iz}) i e^{iz} e^{ikz} + f^{(k)}(e^{iz}) i k e^{ikz} \right) =$$

$$i^n \sum_{k=0}^{n-1} \left\{ {n - 1 \atop k} \right\} \left(f^{(k+1)}(e^{iz}) e^{i(k+1)z} + k f^{(k)}(e^{iz}) e^{ikz} \right) =$$

$$i^n \sum_{k=1}^{n} \left\{ {n - 1 \atop k - 1} \right\} f^{(k)}(e^{iz}) e^{ikz} + i^n \sum_{k=0}^{n-1} \left\{ {n - 1 \atop k} \right\} f^{(k)}(e^{iz}) e^{ikz} =$$

$$i^n \sum_{k=0}^{n} \left\{ {n \atop k} \right\} f^{(k)}(e^{iz}) e^{ikz}.$$

the statement is proved.

Setting $f = \exp$, we know that $\exp^{(k)} = \exp$, and thus

$$\left(e^{e^{iz}}\right)^{(n)} = i^n e^{e^{iz}} \sum_{k=0}^{n} \left\{ {n \atop k} \right\} (e^{iz})^k = i^n e^{e^{iz}} B_n(e^{iz}). \tag{3.8}$$

This derivative formula is interesting in itself, let alone its consequence: the zeros of the derivatives of the double exponential $e^{e^{iz}}$ are the same as the zeros of the Bell polynomials composed by the exponential function (indeed, $e^{e^{iz}}$ has no zeros at all on the right-hand side, so all the vanishing points come from the zeros of $B_n(e^{iz})$).

The zeros of the derivatives of $e^{e^{iz}}$

These points are determined by writing

$$e^{iz} = t_k^{(n)}$$

for appropriate k and n. Taking logarithm carefully ($t_k^{(n)}$ is a negative real number, so we must use the complex logarithm with complex branches) we get that

$$z = -i \log |t_k^{(n)}| + (2l + 1)\pi \quad (l = 0, \pm 1, \pm 2, \dots). \tag{3.9}$$

Let us introduce now a notion which originates from the work of Pólya[5] [468]. The *final set* of a (complex, entire[6]) function f is constituted by those points z which, in any neighbor of themselves contain zeros of infinitely many derivatives of f.

It is, in general, very hard to determine the final set of an entire function. A result from 1980 [210] is the following: the final set of the entire function e^{-e^z} consists of the horizontal lines $y = 2l\pi$ (l runs through the integers). One can see that $e^{e^{iz}} = e^{-i(z+\pi)}$, thus the final set of $e^{e^{iz}}$ is the collection of the vertical lines with abscissa $(2l + 1)\pi$. Now recall (3.9). We see that the abscissas agree, so the logarithms of the absolute values of the elements of $\mathcal{T}$ must be dense on the real line. Since the logarithm is a continuous function, it must be true that $\mathcal{T}$ is dense in $] - \infty, 0[$.

3.2.5   Summation relations for the zeros of $B_n(x)$

Although the zeros of the Bell polynomials are hard to determine one by one[7], several interesting summation relations for the zeros can be found.

---

[5]György Pólya (1887-1985), Hungarian mathematician. He worked on many branches of mathematics, most notably in function theory, probability, combinatorics, and number theory.

[6]A function mapping the complex plane to itself is entire if its Taylor series has infinite radius of convergence.

[7]Numerical approximations can be applied, however.

The sum of the reciprocals of the zeros

We deduce the formulas in this section for general polynomials. Let

$$p(x) = a_n x^n + a_{n-1} x^{n-1} + \cdots + a_1 x + a_0$$

be a polynomial, and let $\alpha_1, \ldots, \alpha_n$ be its zeros (multiple zeros appear in the list as many times as their multiplicity). Then it is possible to write $p_n(x)$ as

$$p_n(x) = (\alpha_1 - x) \cdots (\alpha_2 - x) \cdots (\alpha_n - x) =$$

$$\alpha_1 \alpha_2 \cdots \alpha_n \left(1 - \frac{x}{\alpha_1}\right) \left(1 - \frac{x}{\alpha_2}\right) \cdots \left(1 - \frac{x}{\alpha_n}\right).$$

Let us denote the product of the zeros by α:

$$\alpha = \alpha_1 \alpha_2 \cdots \alpha_n.$$

(Remember that this product is easy to determine by (3.7).) Thus,

$$\frac{1}{\alpha} p_n(x) = \left(1 - \frac{x}{\alpha_1}\right) \left(1 - \frac{x}{\alpha_2}\right) \cdots \left(1 - \frac{x}{\alpha_n}\right). \tag{3.10}$$

The coefficient of x on the right-hand side is

$$-\frac{1}{\alpha_1} - \frac{1}{\alpha_2} - \cdots - \frac{1}{\alpha_n},$$

while on the left-hand side it is a_1/α. Thus,

$$\sum_{k=1}^{n} \frac{1}{\alpha_k} = -\frac{a_1}{\alpha}.$$

If we want to apply this to the Bell polynomials, we should be careful a bit, because one zero of the Bell polynomials is $\alpha_1 = 0$. Therefore, in place of $B_n(x)$, we apply the above to $B_n(x)/x$, and we finally get that

$$\sum_{k=2}^{n} \frac{1}{t_k^{(n)}} = -\left\{ {n \atop 2} \right\} = 1 - 2^{n-1}.$$

The quadratic sum of the reciprocals of the zeros

If we take (3.10) again and apply it for $p_n(-x)$, we get

$$\frac{1}{\alpha^2} p_n(x) p_n(-x) = \left(1 - \frac{x^2}{\alpha_1^2}\right) \left(1 - \frac{x}{\alpha_2^2}\right) \cdots \left(1 - \frac{x}{\alpha_n^2}\right).$$

Comparing the coefficients of x^2 on both sides, we arrive at the nice quadratic summation formula for the zeros of a polynomial:

$$\sum_{k=1}^{n} \frac{1}{\alpha_k^2} = a_1^2 - 2a_2a_0.$$

For the Bell polynomials (analyzing again $B_n(x)/x$) this gives

$$\sum_{k=2}^{n} \frac{1}{(t_k^{(n)})^2} = \left\{ {n \atop 2} \right\}^2 - 2 \left\{ {n \atop 3} \right\}.$$

The general power sum of the reciprocals of the zeros

Taking the logarithm of both sides of (3.10), we have

$$-\frac{p_n'(x)}{p_n(x)} = \sum_{k=1}^{n} \frac{1}{\alpha_k - x} = \sum_{k=1}^{n} \frac{1}{\alpha_k} \frac{1}{1 - \frac{x}{\alpha_k}}. \tag{3.11}$$

The innermost expression, $\frac{1}{1 - \frac{x}{\alpha_k}}$, is a geometric sum, and (supposing that $|x| < |\alpha_k|$ for all k) we have that

$$-\frac{p_n'(x)}{p_n(x)} = \sum_{k=1}^{n} \frac{1}{\alpha_k} \sum_{l=0}^{\infty} \frac{x^l}{\alpha_k^l} = \sum_{l=0}^{\infty} x^l \sum_{k=1}^{n} \frac{1}{\alpha_k^{l+1}}.$$

Let us introduce the following notation for the reciprocal power sum of the zeros:

$$Z_n(l) = \sum_{k=1}^{n} \frac{1}{\alpha_k^l}.$$

Then we almost have our final result:

$$-\frac{p_n'(x)}{p_n(x)} = \sum_{l=0}^{\infty} Z_n(l+1)x^l.$$

Multiplying both sides by x and re-indexing on the right-hand side, we get the below generating function for the reciprocal sums of the zeros of a polynomial:

$$\sum_{l=1}^{\infty} Z_n(l)x^l = -x\frac{p_n'(x)}{p_n(x)}. \tag{3.12}$$

Note that the initial formula (3.11) is not without its own interest. If we apply it for the Bell polynomials, we get that

$$\frac{1}{1 - t_1^{(n)}} + \frac{1}{1 - t_2^{(n)}} + \cdots + \frac{1}{1 - t_n^{(n)}} = \frac{B_{n+1}}{B_n} - 1.$$

(Here we used that $B_n'(1) = B_{n+1} - B_n$. See the Exercises.)

3.3 Generalized Bell polynomials

Interestingly enough, there is a function which incorporates all of the numbers and polynomials we have studied so far and many others. It is worth to study this function in more detail.

Definition 3.3.1. *Let a sequence $z_1, z_2, \ldots$ be given. The function of infinitely many variables defined by the exponential generating function*

$$\sum_{n=0}^{\infty} \left(\sum_{k=0}^{n} B_{n,k}(z_1, z_2, \ldots) x^k \right) \frac{y^n}{n!} = \exp\left[x \left(\sum_{n=1}^{\infty} z_n \frac{y^n}{n!} \right) \right] \qquad (3.13)$$

is called generalized Bell polynomial.

Some examples enlighten how this function incorporates our counting sequences. First let $z_n = 1$ for all n. Then

$$\sum_{n=0}^{\infty} \left(\sum_{k=0}^{n} B_{n,k}(1, 1, \ldots) x^k \right) \frac{y^n}{n!} = \exp\left[x \left(\sum_{n=1}^{\infty} \frac{y^n}{n!} \right) \right] = \exp(x(e^y - 1)),$$

and on the right-hand side we have the exponential generating function of the Bell numbers. Hence,

$$\sum_{k=0}^{n} B_{n,k}(1, 1, \ldots) x^k = B_n(x).$$

By the same reason the coefficients of the powers of x are the Stirling numbers of the second kind:

$$B_{n,k}(1, 1, \ldots) = \left\{ {n \atop k} \right\}. \qquad (3.14)$$

These considerations show that the generalized Bell polynomials cover the theory of the second kind Stirling numbers. And they also show where the name "generalized Bell polynomial" comes.

Now let $z_n = (n-1)!$. What we get is

$$\sum_{n=0}^{\infty} \left(\sum_{k=0}^{n} B_{n,k}(0!, 1!, \ldots) x^k \right) \frac{y^n}{n!} = \exp\left[x \left(\sum_{n=1}^{\infty} \frac{y^n}{n} \right) \right] =$$

$$\exp(x(-\ln(1 - y))) = \left(\frac{1}{1 - y} \right)^x.$$

The reader might remember an Exercise 23 of the previous chapter. In this exercise we saw that $\left(\frac{1}{1-y} \right)^x$ is the exponential generating function of the rising factorials $x^{\overline{n}}$. Since this is a polynomial (with first kind Stirling number coefficients) we get that the generalized Bell polynomials incorporate the theory of the first kind Stirling numbers (see (2.44)), too:

$$\sum_{k=0}^{n} B_{n,k}(0!, 1!, \dots)x^k \;=\; x^{\overline{n}},$$

$$B_{n,k}(0!, 1!, \dots) \;=\; \left[{n \atop k} \right]. \tag{3.15}$$

3.4   Idempotent numbers and involutions

We introduce two more combinatorially interesting sequences which also appear as particular cases of the generalized Bell polynomials.

Idempotent numbers

One can verify that

$$B_{n,k}(1, 2, 3, \dots) = \binom{n}{k} k^{n-k}. \tag{3.16}$$

These numbers will appear in the following combinatorial problem. Let $\iota_n{}^8$ be the nth *idempotent number* which gives that how many functions f are there on an n element set (say, on $\{1, 2, \dots, n\}$) such that

$$f : \{1, 2, \dots, n\} \to \{1, 2, \dots, n\}, \quad \text{and} \quad f(f(x)) = f(x) \text{ for any } x.$$

We will prove that

$$\iota_n = \sum_{k=1}^{n} \binom{n}{k} k^{n-k}. \tag{3.17}$$

Let us see in more detail what the assumption $f(f(x)) = f(x)$ means. If

$$f(x) = y, \text{ then } y = f(x) = f(f(x)) = f(y).$$

Hence, we can prescribe the function in some points and then the images of these points as elements *in the domain* will be fixed points. Therefore, a function f with the above property can be given if we choose some, say k, elements from n and we prescribe the function on these points arbitrarily and on the images of these points the function f is prescribed to be identical. (3.17) follows by summing on the possible values of k.

---

[8]This is the Greek letter *iota*.

The proof of (3.16) is left to the reader.

Involutions

We take one more interesting particular case of the generalized Bell polynomials. A permutation π is called *involution* if applying π two times, the permuted elements are sent back to their original positions. In formula,

$$\pi^2 = \varepsilon,$$

where ε denotes the identity permutation. Note that the cycle decomposition of an involution must not contain cycles of length 3 or more. In other words, an involution contains only fixed points and transpositions (see page 11). Let the number of involutions on n letters be I_n.

An involution can recursively be constructed. Let us suppose first that we have an involution on $n-1$ elements and we add n to this involution as a fixed point. This offers I_{n-1} possibilities. Next, having an involution on $n-1$ elements we can put n into a 1 length cycle. But how many 1-length cycles are there in an involution of $n-1$ elements? To answer the question, we must remember the permutation patterns on page 56. Recalling formula (2.32) we now have a pattern of the form (j_1, j_2), where $j_1 + 2j_2 = n$. Hence, the total number of involutions on $n-1$ elements is

$$I_{n-1} = \sum_{j_1 + 2j_2 = n-1} \frac{(n-1)!}{j_1! 1^{j_1} j_2! 2^{j_2}}.$$

We have to insert n beside a fixed point. As there are j_1 fixed points, we have j_1 possibilities (j_1 varies, of course):

$$\sum_{j_1 + 2j_2 = n-1} \frac{(n-1)!}{j_1! j_2! 2^{j_2}} j_1 = \sum_{j_1 + 2j_2 = n-1} \frac{(n-1)!}{(j_1 - 1)! j_2! 2^{j_2}} =$$

$$\sum_{j_1 + 1 + 2j_2 = n-1} \frac{(n-1)!}{j_1! j_2! 2^{j_2}} = \sum_{j_1 + 2j_2 = n-2} \frac{(n-1)!}{j_1! j_2! 2^{j_2}} =$$

$$(n-1) \sum_{j_1 + 2j_2 = n-2} \frac{(n-2)!}{j_1! j_2! 2^{j_2}} = (n-1) I_{n-2}.$$

In total – considering the first case above – we have that on n elements there are

$$I_n = I_{n-1} + (n-1) I_{n-2} \tag{3.18}$$

involutions.

After introducing the sequence I_n, we show how it is related to the generalized Bell polynomials. This relation is easy to see via the generating functions, so we now determine the exponential generating function of I_n. Invoking the above recursion,

$$\sum_{n=0}^{\infty} I_n \frac{x^n}{n!} = e^{x+\frac{x^2}{2}}. \tag{3.19}$$

From this, it is already obvious how the generalized Bell polynomials cover the involutions:

$$I_n = \sum_{k=0}^{n} B_{n,k}(1,1,0,\dots). \tag{3.20}$$

Another proof of (3.18)

It is known that the set of permutations on n elements can be identified by those $n \times n$ matrices which have exactly one 1 in each row and column, and zeros elsewhere. In fact, such matrices are called *permutation matrices*. For instance, the permutation

$$\begin{pmatrix} 1 & 2 & 3 & 4 & 5 & 6 \\ 3 & 5 & 1 & 6 & 4 & 2 \end{pmatrix}$$

corresponds to the matrix

$$\begin{pmatrix} 0 & 0 & 1 & 0 & 0 & 0 \\ 0 & 0 & 0 & 0 & 1 & 0 \\ 1 & 0 & 0 & 0 & 0 & 0 \\ 0 & 0 & 0 & 0 & 0 & 1 \\ 0 & 0 & 0 & 1 & 0 & 0 \\ 0 & 1 & 0 & 0 & 0 & 0 \end{pmatrix}.$$

One can try to act this matrix on a vector $\begin{pmatrix} a_1 & a_2 & a_3 & a_4 & a_5 & a_6 \end{pmatrix}$ to see that the resulting vector is $\begin{pmatrix} a_3 & a_5 & a_1 & a_6 & a_4 & a_2 \end{pmatrix}$.

To see the validity of (3.18), we first observe that the inverse of a permutation matrix is its transpose (this corresponds to the fact that the inverse of a permutation is the permutation in which we swap the first and second line, and reorder the columns so that the first line elements are in increasing order). Therefore, there are as many involutions as many *self-adjoint* matrices. We proceed by induction. It is easy to see that the unique 1×1 and 2×2 self-adjoint permutation matrices are the identity matrices and $\begin{pmatrix} 0 & 1 \\ 1 & 0 \end{pmatrix}$, reflecting the fact that $I_1 = 1$ and $I_2 = 2$. Now let $n > 2$. If 1 happens to be in the top left corner, the $(n-1) \times (n-1)$ submatrix not including the first row and column must be self-adjoint, and there are I_{n-1} of these by the induction hypothesis. Now let us suppose that 1 appears somewhere in the first row but not in the first column. By symmetry, this 1 appears in the transposed position, too, so after deleting the corresponding rows and columns we have an $(n-2) \times (n-2)$ matrix which must be self-adjoint; there are I_{n-2} of these. Since the 1 in the first row can appear in $n-1$ positions, we have that in this case the total number of matrices is $(n-1)I_{n-2}$. Summing the two disjoint cases, we are done. This proof comes from [151].

3.5 A summation formula for the Bell polynomials

In this short section, we prove a summation formula for the Bell polynomials which we will use later. By (3.13),

$$\exp x \left(\sum_{m=1}^{\infty} z_m \frac{y^m}{m!} \right) = \sum_{k=0}^{\infty} \frac{x^k}{k!} \left(\sum_{m=1}^{\infty} z_m \frac{y^m}{m!} \right)^k.$$

The polynomial theorem helps to expand the inner power:

$$\sum_{k=0}^{\infty} \frac{x^k}{k!} \left[\sum_{j_1+j_2+\cdots=k} \frac{k!}{j_1! j_2! \cdots} \left(\frac{z_1 y^1}{1!} \right)^{j_1} \left(\frac{z_2 y^2}{2!} \right)^{j_2} \cdots \right].$$

If we would like to get the Bell polynomials, we have to look for the coefficient of $\frac{y^n}{n!}$ inside the sum. The power of y is $1j_1 + 2j_2 + \cdots$, the coefficient in question is

$$\sum_{\substack{j_1+j_2+\cdots=k \\ 1j_1+2j_2+\cdots=n}} \frac{n!}{j_1! j_2! \cdots} \left(\frac{z_1}{1!} \right)^{j_1} \left(\frac{z_2}{2!} \right)^{j_2} \cdots$$

This is nothing else but $B_{n,k}(z_1, z_2, \ldots)$:

$$B_{n,k}(z_1, z_2, \ldots) = \sum_{\substack{j_1+j_2+\cdots=k \\ 1j_1+2j_2+\cdots=n}} \frac{n!}{j_1! j_2! \cdots} \left(\frac{z_1}{1!} \right)^{j_1} \left(\frac{z_2}{2!} \right)^{j_2} \cdots \qquad (3.21)$$

Note that this representation gives the already known formulas (2.28) and (2.29) for the Stirling numbers if we choose z_i, as in (3.14) or in (3.15).

3.6 The Faà di Bruno formula

The generalized Bell polynomials have applications outside of combinatorics. Now we show such an application.

If we have two exponential generating functions corresponding to a_n and b_n (we suppose that $b_0 = 0$),

$$f(x) = \sum_{n=0}^{\infty} a_n \frac{x^n}{n!}, \quad g(x) = \sum_{n=1}^{\infty} b_n \frac{x^n}{n!},$$

then it is interesting to ask what is the exponential generating function of $f(g(x))$.

Let us denote $f(g(x))$ by $h(x)$ and its coefficients by c_n. Then

$$f(g(x)) = \sum_{k=0}^{\infty} a_k \frac{g(x)^k}{k!} = h(x) = \sum_{n=0}^{\infty} c_n \frac{x^n}{n!}.$$

Applying the polynomial theorem

$$\sum_{k=0}^{\infty} a_k \frac{g(x)^k}{k!} = \sum_{k=0}^{\infty} \frac{a_k}{k!} \left(\sum_{i=1}^{\infty} b_i \frac{x^i}{i!} \right)^k =$$

$$\sum_{k=0}^{\infty} \frac{a_k}{k!} \left(\sum_{j_1+j_2+\cdots=k} \frac{k!}{j_1!j_2!\cdots} \left(\frac{b_1 x^1}{1!} \right)^{j_1} \left(\frac{b_2 x^2}{2!} \right)^{j_2} \cdots \right).$$

By our assumption this must be equal to $h(x)$; therefore, the coefficients of x^n must be equal to $\frac{c_n}{n!}$ for all k. This holds only if $j_1 + 2j_2 + \cdots = n$, that is,

$$c_n = \sum_{k=1}^{n} a_k \sum_{\substack{j_1+j_2+\cdots=k \\ j_1+2j_2+\cdots=n}} \frac{n!}{j_1!j_2!\cdots} \left(\frac{b_1}{1!} \right)^{j_1} \left(\frac{b_2}{2!} \right)^{j_2} \cdots.$$

These are the coefficients of the composition $f(g(x))$. On the other hand, comparing this with (3.21), we arrived at the fact that c_n equals to the Bell polynomials:

$$c_n = \sum_{k=1}^{n} a_k B_{n,k}(b_1, b_2, \dots).$$

This is the *formula of Faà di Bruno*[9].

Faà di Bruno's formula tells us how to calculate the coefficients in the composition of two functions: first, put the coefficients of the inner function (here $g(x)$) into the generalized Bell polynomial with parameters n, k; then multiply this with the kth coefficient a_k of the outer function (here $f(x)$); finally sum on $k = 1, 2, \dots, n$.

---

[9] Francesco Faà di Bruno (1825-1888), Italian mathematician, catholic priest, and musician.

Exercises

1. Calculate $B_5(x)$ using $B_4(x)$ and the recursion ($B_4(x)$ can be found on p. 85).

2. Prove the polynomial version of the Dobiński formula:

$$B_n(x) = \frac{1}{e^x} \sum_{k=0}^{\infty} \frac{k^n x^k}{k!}.$$

3. Prove that

$$\sum_{k=0}^{n} k \left\{ {n \atop k} \right\} = B_{n+1} - B_n,$$

$$\sum_{k=0}^{n} k^2 \left\{ {n \atop k} \right\} = B_{n+2} - 2B_{n+1}.$$

Hint: use the (3.2) recursion for the Bell polynomials.

4. The previous exercise can be phrased in terms of the Bell polynomials and their derivatives:

$$B_n'(1) = B_{n+1} - B_n,$$
$$B_n''(1) = B_{n+2} - 3B_{n+1} + B_n.$$

Prove these statements.

5. Let ξ be a random variable taking on the values $k = 1, 2, \ldots, n$; such that $P(\xi = k)$ is the probability that a randomly and uniformly chosen partition on n elements contain exactly k blocks. Determine the expected value and variance of ξ.

6. Give a proof for the general derivation formula for the Bell polynomials:

$$B_n^{(k)}(x) = k! \sum_{i=k}^{n} \binom{n}{i} \left\{ {i \atop k} \right\} B_{n-i}(x).$$

(Hint: The exponential generating function of the $B_n(x)$ polynomials is $e^{x(e^y - 1)}$, whence the exponential generating function of $B_n^{(k)}(x)$ is $\frac{\partial^k}{\partial x^k} e^{x(e^y - 1)}$, which has a simple form. Then use (2.19).)

7. Prove the (3.2) recursion by combinatorial means. Hint: fix x to be a positive integer and consider it as a number of colors. Then the term $\left\{ {n \atop k} \right\} x^k$ is the number of partitions of n elements into k blocks such that each block is colored with one of the x colors independently.

8. Calculate the first five idempotent numbers.

9. Show that the exponential generating function of the idempotent numbers is

$$\sum_{n=0}^{\infty} \mathrm{i}_n \frac{x^n}{n!} = e^{x e^x}.$$

10. Verify the following representation of the e^x exponential function:

$$e^x = \sum_{n=0}^{\infty} \mathrm{i}_n \frac{W^n(x)}{n!}.$$

Here $W(x)$ is the Lambert W function, see on p. 110.

11. Show that

$$\sum_{n=0}^{\infty} \mathrm{i}_n x^n = \sum_{n=0}^{\infty} \frac{x^n}{(1 - nx)^{n+1}}.$$

(This example is taken from the *On-Line Encyclopedia of Integer Sequences* (OEIS) and was given by Vladeta Jovović. See http://oeis.org/A000248.)

12. Show that

$$\mathrm{i}_n = \sum_{k=1}^{n} \binom{n-1}{k-1} k \mathrm{i}_{n-k}.$$

(This example is taken from The *On-Line Encyclopedia of Integer Sequences* (OEIS) and was given by James East. See http://oeis.org/A000248.)

13. What is the value of $B_{n,k}(1, 0, 0, \dots)$?

14. Prove that the $\left\lfloor \begin{smallmatrix} n \\ k \end{smallmatrix} \right\rfloor$ Lah numbers (see (2.53)) belong to the generalized Bell polynomial class with arguments $1!, 2!, 3!, \dots$:

$$\left\lfloor \begin{matrix} n \\ k \end{matrix} \right\rfloor = B_{n,k}(1!, 2!, 3!, \dots).$$

15. Prove (3.19) and prove that from (3.19), (3.20) follows.

16. Calculate the fifth involution number I_5 by the recursion given for these numbers. (The initial values are $I_0 = I_1 = 1$.)

17. Prove that the Hankel transform of the involutions are the superfactorials. (Use the "polynomial method" introduced in Section 2.10. To do this, you might consider the generating function

$$\sum_{n=0}^{\infty} I_n(y) \frac{x^n}{n!} = e^{y(x + x^2/2)}.$$

Here $I_n(y)$ can be considered as the involution polynomial.)

18. Determine the coefficient of x^n in $\sin(e^x-1)$ by the Faà di Bruno formula! (It is interesting to note that the second kind Stirling numbers appear in the coefficients.)

Outlook

1. The problem of generalized involutions, when a permutation π satisfies $\pi^k = \varepsilon$ is discussed in [158, p. 257], see also the references therein. The problem was considered originally in [447].

2. Involutions and even more general constructions were studied in great detail in [17] from number theoretical and combinatorial perspective, and detailed asymptotic analysis is also given for I_n in this paper.

3. The Hilbert series of a flag variety can be related to the Stirling numbers of the second kind and the Bell polynomials [303].

4. The reciprocal sum formula (3.12) for the roots of polynomials can be extended to polynomials of infinite degree, and thus getting formula for reciprocal sums of zeros of transcendental entire functions. See the paper of Spiegel [529].

Chapter 4

Unimodality, log-concavity, and log-convexity

4.1 "Global behavior" of combinatorial sequences

It is often useful and interesting to know how a combinatorial sequence looks like "globally." How rapidly does it grow/decrease? Can we find simple uniform bounds for it? Where are its maxima and minima? The first two of these questions will be studied in Chapter 7. Regarding the third question, it turns out that many two-parameter counting sequences grow, reach an index, then decrease, so the localization of the maxima is a well-motivated question.

The most simple example is of the Pascal triangle. We fix the upper parameter n and let the lower parameter k run over the set $\{1, \ldots, n\}$. For instance, if $n = 6$, then $\binom{n}{1}, \ldots, \binom{n}{6}$ is

$$6, \quad 15, \quad 20, \quad 15, \quad 6, \quad 1.$$

It can also be seen by the symmetry of the Pascal triangle that the sequence

$$\binom{n}{k}_{k=0}^{n}$$

grows, reaches its maximum/maxima at the center and then decreases. The index/indices of the maximum/maxima is/are $n/2$ or $(n \pm 1)/2$, depending on the parity of n.

Considering the Stirling triangles (see the tables at the end of the book), we see that they behave similarly. For example $\begin{bmatrix} 6 \\ k \end{bmatrix}$:

$$120, \quad 274, \quad 225, \quad 85, \quad 15, \quad 1,$$

and $\begin{Bmatrix} 6 \\ k \end{Bmatrix}$:

$$1, \quad 31, \quad 90, \quad 65, \quad 15, \quad 1;$$

both grow, reach a maximum, then decrease. Where are the maxima exactly for all n? This is a very hard problem; even today we do not know the answer in the second kind case. To be able to put our question in an exact form, we need some definitions.

4.2 Unimodality and log-concavity

Unimodality, log-concavity, peak and plateau

Definition 4.2.1. *A not necessarily finite sequence $a_0, a_1, \ldots$ is* unimodal[1] *if it increases, reaches its maximum (or maxima), and then decreases. More precisely, $a_0, a_1, \ldots$ is unimodal if there is an index n and a non-negative integer k such that*

$$a_0 \leq a_1 \leq \cdots \leq a_n = a_{n+1} = \cdots = a_{n+k} \geq a_{n+k+1} \geq \cdots$$

For example, the following sequences are all unimodal.

$$4, 4, 4, 4; \quad 1, 2, 4, 8, 10, 10, 10, 10, 6, 3, -2; \quad 7, 9, 9, 9; \quad -3, -1, 5, 7, 6, 5, 4, 3.$$

If there are more than one maximal members in the unimodal sequence (i.e., $k \geq 1$), then these members are called *plateau*. If the maximal element is unique ($k = 0$), it is called *peak*.

There is a stronger property than unimodality, called log-concavity.

Definition 4.2.2. *The sequence $a_0, a_1, \ldots$ is* log-concave *if for all $k \geq 1$*

$$a_k^2 \geq a_{k-1} a_{k+1} \tag{4.1}$$

If it is also true that

$$a_k^2 > a_{k-1} a_{k+1} \tag{4.2}$$

then the sequence is called strictly log-concave.

Before we proceed, we have to remark that we will be covering in this chapter only the basics of the theory of log-concavity and log-convexity of counting sequences. This particular area of combinatorics could itself fill a book. Please see the Outlook at the end of this chapter for further references.

Where does the term "log-concavity" come?

In *convex analysis*, a positive function f is called log-concave if $\log f$ is concave[2]. This means that

$$\log f(tx + (1-t)y) \geq t \log f(x) + (1-t) \log f(y) \quad (0 < t < 1).$$

---

[1]This expression comes from probability theory and statistics. The noun *mode* in statistics is the most often appearing element of a statistical data set. In probability theory, mode is a point where the density function attains its maximum value.

[2]Positivity is necessary to claim for sequences, too. In the opposite case, the statement would not be true: the sequence $-1, -2, -1$ is log-concave ($(-2)^2 \geq (-1)(-1)$) but not unimodal.

In particular, for $t = 1/2$:

$$\log f\left(\frac{x+y}{2}\right) \geq \frac{\log f(x) + \log f(y)}{2}.$$

Taking exponential, and then taking square, we have the equivalent inequality

$$f^2\left(\frac{x+y}{2}\right) \geq f(x)f(y).$$

Now we see the similarity between this and (4.1).

Log-concavity is stronger than unimodality

Why log-concavity is stronger than unimodality? To see this, let us rewrite (4.1):

$$\frac{a_k}{a_{k-1}} \geq \frac{a_{k+1}}{a_k}.$$

This means that the fraction $\frac{a_k}{a_{k-1}}$ decreases as k grows. Therefore, it is first greater than one, reaches or jumps over 1, and then it is smaller than one. In these cases our sequence strictly grows, has a peak or a plateau and then strictly decreases.

If we have strict log-concavity, then the fraction $\frac{a_k}{a_{k-1}}$ strictly decreases. Hence, for at most one index k, the fraction can be 1. This yields that the plateau of the sequence $a_0, a_1, \ldots$ contains *maximum two members*.

A way of testing log-concavity

There is a theorem which is often useful to check log-concavity. Since we work with combinatorial sequences, we suppose that all the a_k coefficients are positive. The statement[3] is as follows. If the polynomial

$$p(x) = a_0 + a_1 x + a_2 x^2 + \cdots + a_n x^n \tag{4.3}$$

has only real zeros (and these zeros are then non-positive by the positivity of the coefficients), then

$$a_k^2 \geq a_{k-1}a_{k+1} \frac{k}{k-1} \frac{n-k+1}{n-k} \quad (1 \leq k \leq n-1). \tag{4.4}$$

---

[3]Inequality (4.4) is also called as *Newton's inequality* after Isaac Newton who intended to solve the problem of counting the number of imaginary roots of the algebraic equation $a_0 x^n + a_1 x^{n-1} + \cdots + a_1 x + a_0$. During this attempt, he claimed (but gave no proof) that the number of imaginary roots cannot be less than the sign changes in the sequence

$$a_0^2, \left(\frac{a_1}{\binom{n}{1}}\right)^2 - \frac{a_2}{\binom{n}{2}}\frac{a_0}{\binom{n}{0}}, \ldots, \left(\frac{a_{n-1}}{\binom{n}{n-1}}\right)^2 - \frac{a_n}{\binom{n}{n}}\frac{a_{n-2}}{\binom{n}{n-2}}, a_n^2.$$ If all the solutions of the above equation are real, then all the entries in the above sequence must be non-negative. From here, (4.4) follows. More details can be found in [453].

Since $\frac{k}{k-1}\frac{n-k+1}{n-k} > 1$, it follows that $a_0, \ldots, a_n$ is strictly log-concave. Before proving this theorem, we see some examples on how it works – but even before that, we give a short and interesting historical remark.

The log-concavity of the binomial coefficients

Let us set $a_k = \binom{n}{k}$ (n is fixed). Then by the binomial theorem

$$p(x) = \binom{n}{0} + \binom{n}{1}x + \cdots + \binom{n}{n}x^n = (x+1)^n.$$

This polynomial has only real zeros (-1 with multiplicity n), so the assumption of the theorem holds. Thus, a_k is strictly log-concave:

$$\binom{n}{k}^2 \geq \binom{n}{k-1}\binom{n}{k+1}\frac{k}{k-1}\frac{n-k+1}{n-k} \geq \binom{n}{k-1}\binom{n}{k+1}. \tag{4.5}$$

The log-concavity of the Stirling numbers of the first kind

Turning to the Stirling numbers of the first kind, set $a_k = \begin{bmatrix} n \\ k \end{bmatrix}$. By (2.41)

$$p(x) = \sum_{k=0}^{n} \begin{bmatrix} n \\ k \end{bmatrix} x^k = x(x+1)(x+2)\cdots(x+n-1),$$

which has only real zeros. This results in the important observation that the sequence $\begin{bmatrix} n \\ k \end{bmatrix}_{k=0}^{n}$ is strictly log-concave: it strictly increases, has at most two maxima and then strictly decreases. While (4.5) can easily be proven by simple algebra without relying on (4.5), the log-concavity of $\begin{bmatrix} n \\ k \end{bmatrix}$ cannot be proven in a so simple way – the real zero test is a useful tool.

In fact, more than log-concavity is true: $\left(\begin{bmatrix} n \\ k \end{bmatrix}\right)_{k=0}^{n}$ has a *unique peak* if $n > 2$ as was shown by Paul Erdős[4] [218]. For a given n we denote the index of this unique peak by K_n. It can be proven that

$$\left[\ln n - \frac{1}{2}\right] < K_n < [\ln n] \quad (n > 188),$$

where $[\cdot]$ is the integer part. The proof can be found in the cited paper of Erdős. The maximizing index is therefore located close to $\ln n$. This is smaller than $n/2$ so the maximum of the first kind Stirling number sequence is not around the center but occurs much earlier.

The proof of (4.4)

---

[4]Paul Erdős (1913-1996), Hungarian mathematician. He was one of the most productive mathematicians in history who published approximately 1500 articles with 231 co-authors!

Now let us prove (4.4). In place of the polynomial (4.3), we consider the two-variable polynomial

$$q(x, y) = a_0 y^n + a_1 x y^{n-1} + \cdots + a_{n-1} x^{n-1} y + a_n x^n.$$

Note that

$$p\left(\frac{x}{y}\right) = y^n q(x, y),$$

hence the zeros of $q(x, y)$ are the same as that of $p(x)$, apart from the 0 of multiplicity n which comes from the factor y^n. If now we take the derivative of $q(x, y)$ $k-1$ times with respect to x and we take the derivative of the result $n - k - 1$ times with respect to the variable y, then only three terms survive for any k. We verify this in an example. Let $k = 6$ and $n = 10$. Then a part of $q(x, y)$ is

$$\cdots + a_4 x^4 y^6 + a_5 x^5 y^5 + a_6 x^6 y^4 + a_7 x^7 y^3 + a_8 x^8 y^2 + \cdots .$$

Differentiating this as we described before, only the three central terms survive:

$$5! \cdot 5 \cdot 4 \cdot 3 a_5 y^2 + 6!4! a_6 xy + 7 \cdot 6 \cdot 5 \cdot 4 \cdot 3 \cdot 3! a_7 x^2.$$

Factoring out y^2 we get

$$y^2 \left(\frac{5!5!}{2} a_5 + 6!4! a_6 \frac{x}{y} + \frac{7!3!}{2} a_7 \frac{x^2}{y^2} \right).$$

In the parentheses, there is a second-degree polynomial of $\frac{x}{y}$. The zeros of this polynomial are real by the reasons mentioned above. Hence, the discriminant is surely non-negative:

$$(6!4! a_6)^2 - 4 \frac{7!3!}{2} a_7 \frac{5!5!}{2} a_5 \geq 0.$$

Rearranging and simplifying:

$$a_6^2 \geq a_5 a_7 \frac{7}{6} \frac{5}{4}.$$

This is exactly what we wanted to get. The general case works similarly and leads directly to (4.4).

4.3 Log-concavity of the Stirling numbers of the second kind

The log-concavity via the Bell polynomials

By the application of the theorem of the last section, it is not hard to show that the second kind Stirling numbers form a log-concave sequence. To do this, we only need to know that the

$$B_n(x) = \sum_{k=0}^{n} \left\{ {n \atop k} \right\} x^k$$

Bell polynomials have only real zeros. This statement and its proof were the content of Section 3.2. It is interesting to know that the shape of the Stirling number triangle depends on the real zero property of the Bell polynomials.

The consequence of this real zero property is that the Stirling numbers of the second kind with fixed upper parameter form log-concave sequences:

$$\left\{ {n \atop k} \right\}^2 \geq \left\{ {n \atop k-1} \right\} \left\{ {n \atop k+1} \right\}.$$

The real zero property of the Bell polynomials was proven by Harper in 1967 [279], the log-concavity independently by Dobson in 1968 [200] (where it is remarked that the log-concavity inequality was presented also by D. Klarner in an unpublished paper), see also [374]. Estimation on the maximizing index was carried out in [42, 115, 277, 317, 318, 319, 403, 491, 593] and in a paper of Yu [612] by probabilistic methods. About the uniqueness of the maximizing index, see [119, 330, 593].

Estimations for the maximizing index

There is a long standing conjecture about the second kind Stirling numbers: it is conjectured to be true that the sequences $\left\{ {n \atop k} \right\}_{k=0}^{n}$ always have a peak if $n > 2$. This conjecture is verified up to $n = 10^6$ [119].

For the maximizing index (or indices), there are some estimations. For example, it is true [592] that if the maximizing index is K_n, then

$$\frac{n}{\ln n} < K_n < \frac{n}{\ln n - \ln \ln n} \quad (n \geq 13).$$

It was shown by statistical tools [279] that for big n the maximizing index K_n asymptotically equals $\frac{n}{\ln n}$. We shall prove this later in Section 7.3.

Another interesting fact about the index K_n was given by R. Canfield and C. Pomerance [119] in 2002. Let $W(n)$ be the (unique) solution of the transcendental equation

$$xe^x = n. \tag{4.6}$$

Then for sufficiently large n, and also for $2 \leq n \leq 1200$

$$K_n \in \{ \lfloor e^{W(n)} - 1 \rfloor, \lceil e^{W(n)} - 1 \rceil \}$$

(here $\lfloor x \rfloor$ is the greatest integer not exceeding x, $\lceil x \rceil$ is the smallest integer which is not smaller than x).

Solving (4.6) is not an easy task at all[5]. The function that describes the solutions is called *Lambert*[6] *W function* (see the Appendix).

Even more recently, in 2009 Yu [612] found a similar bound for K_n which is not asymptotic but valid for all $n \geq 2$:

$$\lfloor e^{W(n)} \rfloor - 2 \leq K_n \leq \lfloor e^{W(n)} \rfloor + 1.$$

A statement about the growth of K_n

We give a statement about K_n. Remember: we do not know whether K_n is unique or not, so now we choose the smallest maximizing index to be K_n (by log-concavity, there are at most two consecutive maxima). We state that

$$K_{n+1} = K_n \quad \text{or} \quad K_{n+1} = K_n + 1 \quad (n \geq 1).$$

Therefore the maximizing index grows at most by one in each line. This result is due to Dobson [200].

We prove the statement by induction. For $n = 1, 2$ the statement is true. Now let us suppose that up to a given n it is true that $K_i \leq K_n$ if $i \leq n$. To prove the statement for $n + 1$, we use the recursion:

$$\begin{Bmatrix} n+1 \\ k \end{Bmatrix} = \begin{Bmatrix} n \\ k-1 \end{Bmatrix} + k \begin{Bmatrix} n \\ k \end{Bmatrix}.$$

For parameters $n + 1, k - 1$ this becomes

$$\begin{Bmatrix} n+1 \\ k-1 \end{Bmatrix} = \begin{Bmatrix} n \\ k-2 \end{Bmatrix} + (k-1) \begin{Bmatrix} n \\ k-1 \end{Bmatrix}.$$

Subtracting this from the previous equation, we have

$$\begin{Bmatrix} n+1 \\ k \end{Bmatrix} - \begin{Bmatrix} n+1 \\ k-1 \end{Bmatrix} =$$

$$\begin{Bmatrix} n \\ k-1 \end{Bmatrix} - \begin{Bmatrix} n \\ k-2 \end{Bmatrix} + k \left(\begin{Bmatrix} n \\ k \end{Bmatrix} - \begin{Bmatrix} n \\ k-1 \end{Bmatrix} \right) + \begin{Bmatrix} n \\ k-1 \end{Bmatrix}.$$

If $2 \leq k \leq K_n$ (remember: $\begin{Bmatrix} n \\ 0 \end{Bmatrix} = 0$) then both differences are positive which means that

$$\begin{Bmatrix} n+1 \\ k \end{Bmatrix} - \begin{Bmatrix} n+1 \\ k-1 \end{Bmatrix} \geq 0.$$

So, for $2 \leq k \leq K_n$ the sequence $\begin{Bmatrix} n \\ k \end{Bmatrix}$ grows and

$$K_{n+1} \geq K_n. \tag{4.7}$$

---

[5]This equation has infinitely many solutions for any n if we enable x to be complex.

[6]Johann Heinrich Lambert (1728-1777), Swiss mathematician, physicist and astronomer. He proved first that π is irrational.

Now let $K_n + 2 \le k \le n + 1$. Then we use the recursion (1.5). We rewrite this recursion by invoking the symmetry $\binom{n}{m} = \binom{n}{n-m}$:

$$\left\{ {n+1 \atop k} \right\} = \sum_{m=1}^{n+1} \binom{n}{m-1} \left\{ {n+1-m \atop k-1} \right\} =$$

$$\sum_{m=1}^{n+1} \binom{n}{n-m+1} \left\{ {n-m+1 \atop k-1} \right\} = \sum_{m=0}^{n} \binom{n}{m} \left\{ {m \atop k-1} \right\}.$$

Writing down this again for the parameters $n+1, k-1$ and subtracting from the above we have

$$\left\{ {n+1 \atop k} \right\} - \left\{ {n+1 \atop k-1} \right\} = \sum_{m=0}^{n} \binom{n}{m} \left(\left\{ {m \atop k-1} \right\} - \left\{ {m \atop k-2} \right\} \right).$$

Since $K_n + 2 \le k$ it follows that $K_n + 1 \le k - 1$, and that the quantity in the parenthesis is negative:

$$\left\{ {n+1 \atop k} \right\} - \left\{ {n+1 \atop k-1} \right\} < 0.$$

This results that K_n must be smaller than any possible k. In other words, $K_{n+1} < k$ for any k for which $K_n + 2 \le k \le n + 1$. Then it follows that

$$K_{n+1} < K_n + 2.$$

This, together with (4.7), is what we stated before.

4.4   The log-concavity of the Lah numbers

Thanks to the simple formula

$$\left\lfloor {n \atop k} \right\rfloor = \frac{n!}{k!} \binom{n-1}{k-1} \tag{4.8}$$

given for the Lah numbers, it costs nothing to show that these numbers are strictly log-concave, i.e.,

$$\left\lfloor {n \atop k-1} \right\rfloor \left\lfloor {n \atop k+1} \right\rfloor < \left\lfloor {n \atop k} \right\rfloor^2.$$

By (4.8), this inequality holds if and only if

$$k(k-1)(n-k) < (k+1)k(n-k+1).$$

And this inequality is clearly true.

We now locate the maxima of the Lah numbers. The relations

$$\left\lfloor {n \atop k-1} \right\rfloor \underset{>}{\overset{\leq}{\gtreqless}} \left\lfloor {n \atop k} \right\rfloor$$

are equivalent to

$$k^2 \underset{>}{\overset{\leq}{\gtreqless}} n+1,$$

i.e., $k = \sqrt{n+1}$. Consequently, if $n+1$ is a square number, then there are two maximizing indices (the Lah numbers have a plateau)

$$K_{n,1} = \sqrt{n+1} - 1, \quad \text{and} \quad K_{n,2} = \sqrt{n+1}.$$

If, in turn, $n+1$ is not a square, then the maximizing index is unique (we have a peak), and

$$K_n = \lfloor \sqrt{n+1} \rfloor.$$

These results are valid for all $n \geq 1$.

The above statements can be found in [456] where the results are given in a more general form (with respect to the r-Lah numbers).

The zeros of the Lah polynomials

We have learned that the real zero property of a polynomial with positive coefficients implies the log-concavity of the coefficient sequence. Although we have just seen the strict log-concavity of $\left\lfloor {n \atop k} \right\rfloor$, it is not without interest to know that the *Lah polynomials*

$$L_n(x) = \sum_{k=0}^{n} \left\lfloor {n \atop k} \right\rfloor x^k$$

have the real zero property (note also that log-concavity does not imply the real zero property, as one of the exercises shows).

Our approach is similar to what we have done for the Bell polynomials. Remembering (2.54) we can easily give a recursive formula for the Lah polynomials:

$$L_{n+1}(x) = nL_n(x) + xL_n(x) + xL_n'(x).$$

This is equivalent to

$$L_{n+1}(x) = \frac{e^{-x}}{x^{n-1}} \left(x^n e^x L_n(x) \right)'. \tag{4.9}$$

(Just perform the derivation and simplify.)

We show that all the polynomials $L_n(x)$ have only real zeros; they have $n-1$ negative zeros and one zero at $x=0$. This statement is true for $n \geq 1$.

We proceed by induction. The $L_1(x) = x$ and $L_2(x) = x^2 + 2x$ polynomials certainly satisfy the statement. Let us suppose that we have proven the statement up to a given n. Now invoke (4.9). By our assumption, the function $x^n e^x L_n(x)$ has a zero at $x = 0$ of multiplicity $n + 1$, and it has $n - 1$ negative zeros as well. By Rolle's theorem, we infer that $(x^n e^x L_n(x))'$ has zeros among the zeros of $L_n(x)$ (there are $n - 2$), it has a zero between the rightmost negative zero of $L_n(x)$ and $x = 0$. And, because e^x tends to zero as $x \to -\infty$, there must be another extremum of $x^n e^x L_n(x)$ somewhere on the left-hand side of the leftmost zero of $L_n(x)$. This gives that the derivative has a zero in this place. These are n negative real zeros. That L_{n+1} has a zero at $x = 0$ also can be seen: the already mentioned zero of multiplicity $n + 1$ of $x^n e^x L_n(x)$ at $x = 0$ reduces to a zero of multiplicity n after taking the derivative, and the x^{n-1} in the denominator cancels out all of these except one. This will be the zero of $L_{n+1}(x)$ at $x = 0$. The proof is complete.

4.5 Log-convexity

4.5.1 The log-convexity of the Bell numbers

It was observed by K. Engel that the Bell numbers satisfy the *reverse* of the log-concavity inequality [216].

Definition 4.5.1. *The sequence* $a_0, a_1, \ldots$ *is* log-convex *if for all* $k \geq 1$

$$a_k^2 \leq a_{k-1} a_{k+1}. \tag{4.10}$$

Engel's observation can therefore be phrased as follows: the Bell number sequence forms a log-convex sequence:

$$B_n^2 \leq B_{n-1} B_{n+1} \quad (n \geq 1).$$

See also [116]. The proof of Engel depends on some probabilistic results due to Harper [279].

The proof of log-convexity of B_n

The proof we are going to show is due to Spivey [532].
The *total* number of blocks in *all* the partitions of an n-element set is

$$\sum_{k=1}^{n} k \begin{Bmatrix} n \\ k \end{Bmatrix}.$$

In Exercise 3 of Chapter 3, the reader was asked to prove that

$$S_n := \sum_{k=1}^{n} k \begin{Bmatrix} n \\ k \end{Bmatrix} = B_{n+1} - B_n.$$

To show this, separate the partitions counted by B_{n+1} into (1) those which have a set consisting of the single element $n+1$ and (2) those which do not. It should be clear that there are B_n of the former. Also, there are S_n of the latter because each partition in group (2) is formed by adding $n+1$ to a set in a partition of $\{1, 2, \ldots, n\}$. Thus, $B_{n+1} = B_n + S_n$.

The *average* number of blocks is the total number of blocks divided by the number of partitions. Hence, this average (for a fixed n) is equal

$$A_n = \frac{1}{B_n} \sum_{k=1}^{n} k \left\{ {n \atop k} \right\} = \frac{B_{n+1}}{B_n} - 1.$$

Clearly, it is enough to show that A_n is an increasing sequence[7]. Each partition of $\{1, 2, \ldots, n+1\}$ can be associated with a partition of $\{1, 2, \ldots, n\}$ by removing the element $n+1$ from the set containing it. Under the inverse of this mapping, each partition of $\{1, 2, \ldots, n\}$ consisting of k sets maps to k partitions consisting of k sets (if $n+1$ is placed in an already-existing set) and one partition consisting of $k+1$ sets (if $n+1$ is placed in a set by itself) out of the partitions of $\{1, 2, \ldots, n+1\}$. Thus, partitions of $\{1, 2, \ldots, n\}$ with more sets map to more partitions of $\{1, 2, \ldots, n+1\}$ containing the same number of sets as well as one partition with one more set. This raises the average number of sets as we move from n elements to $n+1$ elements, and thus A_n is increasing. The proof is therefore done.

See other proofs in [36] (in this paper the Bell numbers agree with the sequence $b_2(n)$), where it is also proved that

$$1 \leq \frac{B_{n-1}B_{n+1}}{B_n^2} \leq 1 + \frac{1}{n},$$

and even the fact that $B_n/n!$ is log-concave (for this statement, see the below subsection). Another (probabilistic and more general) proof is presented in [92].

4.5.2 The Bender-Canfield theorem

There is a condition for real sequences which is often useful to prove log-convexity or log-concavity. The statement is the following. Let $1, t_1, t_2, \ldots$ be a log-concave sequence of real numbers, and define a_n via the generating function

$$\sum_{n=0}^{\infty} a_n x^n = \exp\left(\frac{t_1}{1} x^1 + \frac{t_2}{2} x^2 + \frac{t_3}{3} x^3 + \cdots \right). \tag{4.11}$$

Then a_n is log-concave and $n! a_n$ is log-convex.

This theorem was proved in [59] by E. A. Bender and R. Canfield. Let us see how to use it.

---

[7] $A_n \geq A_{n-1}$ if and only if $\frac{B_{n+1}}{B_n} - 1 \geq \frac{B_n}{B_{n-1}} - 1$, and this holds if and only if $B_n^2 \leq B_{n-1}B_{n+1}$.

For instance, let us re-prove the log-convexity of the Bell numbers. Knowing that

$$\sum_{n=0}^{\infty} \frac{B_n}{n!} u^n = \exp(\exp x - 1) = \exp\left(\frac{1}{1!}x^1 + \frac{1}{2!}x^2 + \frac{1}{3!}x^3 + \cdots\right).$$

Thus, we choose $t_i = \frac{1}{(i-1)!}$ for all $i \geq 1$. Since

$$\frac{1}{(i-2)!}\frac{1}{i!} \leq \frac{1}{(i-1)!},$$

the sequence t_i is log-concave. Then the Bender-Canfield theorem results that $B_n/n!$ is log-concave, and $n!B_n/n! = B_n$ is log-convex.

The log-convexity of the involutions

We met the sequence I_n counting involutions in Section 3.4. We saw that

$$\sum_{n=0}^{\infty} I_n \frac{x^n}{n!} = e^{x + \frac{x^2}{2}}.$$

Putting $x_1 = x_2 = 1$, and $x_i = 0$ for $i > 2$ in (4.11), we see that we just get the result for the involutions. Since this x_i sequence is log-concave, I_n is log-convex, while $I_n/n!$ is log-concave.

The log-convexity of the L_n numbers

The horizontal sum L_n of the Lah numbers is studied in Section 2.9. In Exercise 28 of Chapter 2, the reader was asked to prove that

$$\sum_{n=0}^{\infty} L_n \frac{x^n}{n!} = e^{\frac{x}{1-x}}.$$

This result can easily be used to prove the log-convexity of the sequence L_n. Indeed, since

$$e^{\frac{x}{1-x}} = \exp(x + x^2 + x^3 + \cdots),$$

by virtue of the geometric series, we have to choose $x_i = i$ in (4.11). This sequence is log-concave, and thus L_n is log-convex, while $L_n/n!$ is log-concave.

Exercises

1. Prove that the ratio of similarly indexed second kind Stirling numbers and binomial coefficients is log-concave:

$$\left(\frac{\left\{{n\atop k}\right\}}{\binom{n}{k}}\right)^2 \geq \frac{\left\{{n\atop k-1}\right\} \left\{{n\atop k+1}\right\}}{\binom{n}{k-1} \binom{n}{k+1}}$$

   for all $n \geq 1$. (See [423] for the details.)

2. According to the previous example, there is a number Ω_n (the smallest of maximum two indices) such that

$$\frac{\left\{{n\atop \Omega_n}\right\}}{\binom{n}{\Omega_n}} \geq \frac{\left\{{n\atop k}\right\}}{\binom{n}{k}} \quad (k = 1, \ldots, n).$$

   Determine the asymptotic behaviour of Ω_n. Yous should get, as a first approximation, that

$$\Omega_n \sim \frac{n}{\log(n)}.$$

   A bit more careful analysis could lead to

$$\Omega_n \sim \frac{n}{\log\left(n + 1 - \frac{n}{\log(n+1)}\right)}.$$

   (See again [423].)

3. Show that if the coefficients of $p(x)$ and $q(x)$ form log-concave sequences, then the same statement holds true for the coefficients of $p(x)q(x)$.

4. Show that if the polynomial $\sum_{i=0}^{n} a_i x^i$ has only real zeros, then the polynomial $\sum_{i=0}^{n} \frac{a_i}{i!} x^i$ has the same property [96, Theorem 2.4.1].

5. Give an example of a short sequence which is log-concave but the attached polynomial $p(x)$ has complex zeros. (One is $3x^3 + 5x^2 + 7x + 3$.)

6. Show that an a_n sequence of positive numbers is log-convex if and only if a_{n+1}/a_n is increasing.

7. The counterpart of the previous exercise is as follows: show that an a_n sequence of positive numbers is log-concave if and only if a_{n+1}/a_n is decreasing.

8. Show that the solution of Equation (4.6) is unique if n is a positive integer and x is real.

9. Let x be a positive number and let a_n be recursively defined as

$$a_1 = x, \quad a_n = x^{a_{n-1}} \ (n > 1).$$

Look for the maximal number x for which the sequence a_n is still convergent. (The limit of this sequence for a given x (if it exists) is called *power tower*.)

10. Show that if $h(x)$ is the limit of the above-defined power tower, then

$$h(x) = -\frac{W(-\ln x)}{\ln x},$$

where W is the Lambert function defined by (4.6).

11. Show that the sequence of the Bell numbers is convex, i.e.,

$$B_n \le \frac{1}{2}(B_{n-1} + B_{n+1}).$$

(Hint: use the arithmetic-geometric mean inequality.)

12. Prove that any log-convex sequence is convex.

13. Show that the E_n Euler (zigzag) number sequence is log-convex. (See [380, Example 2.2].)

14. Show that the $a_n = n!$ sequence is log-convex.

15. Verify the following derivation formulas with respect to the Lah polynomials for all non-negative integer n:

$$L_n'(1) = L_{n+1} - (n+1)L_n,$$
$$L_n''(1) = (2n+1)L_n - L_{n+1}.$$

Outlook

1. For more on the real zero property of the sequences appearing in this book, see the Outlook of Chapter 10, and Subsection 7.6.5.

2. Log-concavity results can often be proved by injectivity. If we can exhibit an injective function from pairs counted by $a_{k-1}a_{k+1}$ to those corresponding to $a_k \cdot a_k = a_k^2$, then we clearly have that $a_{k-1}a_{k+1} \leq a_k^2$. This approach is described in details in [505].

3. An interesting question is about the unimodality in the Stirling triangle not in the horizontal direction but in other directions, for example through the diagonal formed by $\left\{{n-k \atop k}\right\}$, where k runs. See the thesis [559] for examples on non-horizontal unimodality in the second kind Stirling triangle; and [543] for the log-concavity of $\left\{{n-ak \atop c+bk}\right\}$, where k runs. For similar studies in the Pascal triangle, see [542], and the next point. This question with respect to the multinomial coefficients was studied in [543].

4. B.-X. Zhu [614] studied log-concavity in triangular arrays which are rather general. They include the Stirling numbers, the so-called Jacobi-Stirling and Legendre-Stirling numbers, and central factorials (see the citations in [614] for further information on these sequences which were attracting a great deal of attention of combinatorialists recently). Zhu's triangle is defined by the recurrence

$$T_{n,k} = (a_1 k^2 + a_2 k + a3)T_{n-1,k} + (b_1 k^2 + b_2 k + b_3)T_{n-1,k-1},$$

together with the initial condition $T_{n,k} = 0$ unless $0 \leq k \leq n$, and $T_{0,0} = 1$.

5. Log-concavity is preserved by the binomial transform. This fact is useful in the theory of stochastic processes [320]. Standard reference for this, related, and more general results is the book of Brenti [96]. It was relatively recently proven [583] by Yi Wang that the more general transform

$$b_n = \sum_{k=0}^{n} \binom{n+a}{k+b} a_k$$

still preserves the log-concavity of the sequence a_n for arbitrary non-negative integer a and b.

These can be extended from binomial transform to *binomial convolution*. Having two sequences a_k and b_n, then their binomial convolution is

$$c_n = \sum_{k=0}^{n} \binom{n}{k} a_k b_{n-k}.$$

If both a_n and b_n are log-concave, then so is the sequence c_n. Moreover, Yi Wang's result extends to binomial convolution: the sequence

$$c_n = \sum_{k=0}^{n} \binom{n+a}{k+b} a_i b_{n-k}$$

is log-concave if both a_n and b_n are log-concave. For these results, see [584, Corollary 3.4-3.5].

6. Log-convexity is studied in great detail in the paper [380]. See also [535]. In [380] it was shown that if a_n and b_n are log-convex sequences, then so is $a_n + b_n$. Moreover, the log-convexity is preserved under binomial convolution, i.e., if a_n and b_n are log-convex sequences, then

$$c_n = \sum_{k=0}^{n} \binom{n}{k} a_k b_{n-k}$$

is also log-convex. It is also preserved by the binomial and Stirling transformations of both kinds: the

$$b_n = \sum_{k=0}^{n} \binom{n}{k} a_k,$$

$$c_n = \sum_{k=0}^{n} \left[\begin{matrix} n \\ k \end{matrix} \right] a_k,$$

and

$$d_n = \sum_{k=0}^{n} \left\{ \begin{matrix} n \\ k \end{matrix} \right\} a_k$$

sequences are all log-convex given that a_n is itself log-convex.

For deeper results on this topic, see Chapter 7 in [84]. This book is worth to be studied not only if one is interested in log-cancavity of sequences, but also because it basically covers all the basics of enumerative combinatorics. [84] could actually be cited in each Outlook in this book.

7. By many transformations, the real zero property of polynomials is preserved. More on this topic can be found in Brenti's book [96].

8. Log-convexity can be generalized to polynomials. A polynomial $f(q)$ is said to be *q-log-convex*, if

$$f_{n+1}(q)f_{n-1}(q) \geq_q f_n^2(q),$$

where $f(q) \geq_q g(q)$ means that the difference $f(q) - g(q)$ has only non-negative coefficients. Log-concavity is defined similarly. Maybe the most elementary treatment is of Sagan [504]. See also [140, 378, 614, 615] on q-log-convexity.

9. The *binomial coefficients bisection problem* (BCBP) is the following. Find all the vectors $\delta_0, \delta_1, \ldots, \delta_n \in \{-1, 1\}$ such that

$$\sum_{k=0}^{n} \delta_k \binom{n}{k} = 0.$$

This problem was studied in [180, 246, 294, 295]. By the symmetry of the Pascal triangle, there are always solutions, but the question of finding all the solutions is a harder problem.

Chapter 5

The Bernoulli and Cauchy numbers

5.1 Power sums

The Stirling numbers can be used to calculate sums of powers of consecutive integer numbers. Let us first recall the well-known formulas from elementary mathematics:

$$1 + 2 + 3 + \cdots + n = \frac{1}{2}(n^2 + n).$$

The sum of the squares of the first n numbers is

$$1^2 + 2^2 + 3^2 + \cdots + n^2 = \frac{1}{6}(2n^3 + 3n^2 + n).$$

For the third powers we have that

$$1^3 + 2^3 + 3^3 + \cdots + n^3 = \frac{1}{4}(n^4 + 2n^3 + n^2).$$

We can easily recognize a rule: the sums of the powers of the first n positive integers can be expressed as polynomials of n. It would be interesting to know what are the coefficients of these polynomials in general. The sums on the left-hand sides will be called *power sums*.

The Stirling numbers pop up in this problem via the relation (2.46):

$$x^p = \sum_{k=0}^{p} \left\{ {p \atop k} \right\} x^{\underline{k}}. \tag{5.1}$$

On the left-hand side there is the pth power of a real number x. Sum on x from 1 to $n-1$. (We shall see later that summing up to $n-1$ in place of n makes the resulting expressions somewhat simpler looking.)

$$1^p + 2^p + \cdots + (n-1)^p = \sum_{x=1}^{n-1} \sum_{k=0}^{p} \left\{ {p \atop k} \right\} x^{\underline{k}}.$$

If we could find a simple polynomial expression for the right-hand side we

123

would be ready. What we immediately see is that we should know the sums of the falling factorials, since

$$\sum_{x=1}^{n-1}\sum_{k=0}^{p}\left\{{p\atop k}\right\}x^{\underline{k}} = \sum_{k=0}^{p}\left\{{p\atop k}\right\}\sum_{x=1}^{n-1}x^{\underline{k}}.$$

In one of the exercises at the end of this chapter the reader is asked to prove[1] that

$$\sum_{x=1}^{n-1}x^{\underline{k}} = \frac{n^{\underline{k+1}}}{k+1}. \tag{5.2}$$

Applying this identity, we can have the intermediate formula

$$1^p + 2^p + \cdots + (n-1)^p = \sum_{k=0}^{p}\left\{{p\atop k}\right\}\sum_{x=1}^{n}x^{\underline{k}} = \sum_{k=0}^{p}\left\{{p\atop k}\right\}\frac{n^{\underline{k+1}}}{k+1},$$

or its equivalent, the rather simple

$$1^p + 2^p + \cdots + (n-1)^p = \sum_{k=0}^{p}k!\left\{{p\atop k}\right\}\binom{n}{k+1}. \tag{5.3}$$

The falling factorials can be expressed by the signed Stirling numbers as we have seen in (2.45):

$$n^{\underline{k+1}} = \sum_{m=0}^{k+1}\left[{k+1\atop m}\right]n^m.$$

Substituting this into the above, we get that

$$\sum_{k=0}^{p}\left\{{p\atop k}\right\}\frac{n^{\underline{k+1}}}{k+1} = \sum_{k=0}^{p}\left\{{p\atop k}\right\}\frac{1}{k+1}\sum_{m=0}^{k+1}\left[{k+1\atop m}\right]n^m$$

This is already a polynomial of n! To get the final form, we sum inside up to $p+1$ in place of $k+1$ (we can do this, because the new terms are all zero). Why $p+1$? Because m can be at most $k+1$, and k can be at most p. Hence,

$$\sum_{k=0}^{p}\left\{{p\atop k}\right\}\frac{1}{k+1}\left[{k+1\atop m}\right] = \sum_{m=0}^{p+1}n^m\left(\sum_{k=0}^{p}\left\{{p\atop k}\right\}\frac{1}{k+1}\left[{k+1\atop m}\right]\right).$$

Or, together with the initial-left hand side

$$1^p + 2^p + \cdots + (n-1)^p = \sum_{m=0}^{p+1}n^m\left(\sum_{k=0}^{p}\left\{{p\atop k}\right\}\frac{1}{k+1}\left[{k+1\atop m}\right]\right). \tag{5.4}$$

---

[1]The most simple proof uses mathematical induction. A more sophisticated and very elegant way to prove (5.2) is using finite differences – a tool which is similar in many ways to analysis but it is on discrete sets.

We see now that the power sums are indeed polynomials, and the coefficients of these polynomials are expressible in terms of the Stirling numbers of both kinds.

Two other rapid facts are (1) The power sum of the first $n - 1$ positive integers with power p is a polynomial of n (this is why we sum up to $n - 1$ instead of n); and of degree $p + 1$. (2) The constant term of these power sum polynomials is always zero (because the term $m = 0$ in (5.4) is simply 0).

The expression (5.3) asks for a combinatorial proof. The following one was found by D. Callan [113].

Combinatorial proof of the power sum formula (5.3)

For convenience, on both sides we write n in place of $n - 1$, and on the right-hand side we run the sum from $k = 1$:

$$\sum_{i=1}^{n} i^p = \sum_{k=1}^{p+1} (k - 1)! \left\{ {p + 1 \atop k} \right\} \binom{n}{k}. \tag{5.5}$$

We will consider sequences of $p + 1$ positive integers such that all the elements of the sequences are at most n; moreover, the last entry of the sequence is also the largest. Repetitions are allowed. Thus, for example, when $n = 7$ and $p = 8$, the nine-element sequence $6, 7, 4, 7, 6, 2, 4, 7, 7$ is a possible one. We will enumerate such sequences in two ways.

First, we count them by the last entry i. Let i be fixed between one and n. Then for each entry of the sequence we can choose from i numbers, so altogether we have i^p choices for the sequence. The last, $(p + 1)$th element is always fixed, as it must be the greatest. Summing over i we have $\sum_{i=1}^{n} i^k$, the left-hand side of our formula.

Second, we count such sequences by the number k of distinct integers occurring among the entries. Thus, k can vary between one and $p + 1$. First, choose those k distinct integers from that of n in $\binom{n}{k}$ ways. We permute these elements such that the last element is the greatest. There are $(k - 1)!$ such permutations. Next, choose a partition for the $p + 1$ entries into k non-empty blocks (in $\left\{ {p+1 \atop k} \right\}$ ways), and order the blocks in increasing order according to the blocks' maximal element. We have $(k - 1)! \left\{ {p+1 \atop k} \right\} \binom{n}{k}$ choices for such permutation-partition pairs. To match our sequences and the permutation-block pairs, we place the jth entry of the permutation into the positions shown by the jth block. For example, let the set of different elements be $\{2, 4, 6, 7\}$ (so $k = 4$), and choose the permutation 6247 on these elements. Then choose the partition $\{1, 5\}$, $\{6\}$, $\{3, 7\}$, $\{2, 4, 8, 9\}$ of the $p+1 = 9$ elements. Now put 6 into the first and fifth positions, put 2 into the sixth position, put 4 into the third and seventh position, etc. We arrive at the sequence $6, 7, 4, 7, 6, 2, 4, 7, 7$. It can be seen that the sequences and the partition-permutation pairs have a bijective correspondence.

5.1.1   Power sums of arithmetic progressions

The power sum formula (5.5) can easily be generalized so that we are summing *arithmetic progressions'* powers. Arithmetic sequences are of the form $r, r + m, r + 2m, \ldots, r + nm$. We will prove very briefly that

$$\sum_{i=1}^{n}(r + im)^p = m^p \sum_{k=1}^{p+1}(k-1)!\left\{\begin{matrix}p+1\\k\end{matrix}\right\}\left[\binom{n+r/m}{k} - \binom{r/m}{k}\right],$$

given that m is not zero[2].

Applying (5.5) with $n = n + t$ and $n = t$, and subtracting the second expression from the first, it comes that

$$\sum_{i=1}^{n}(t + i)^p = \sum_{k=1}^{p+1}(k-1)!\left\{\begin{matrix}p+1\\k\end{matrix}\right\}\left[\binom{n+t}{k} - \binom{t}{k}\right].$$

As this formula is valid for all positive integer t, and on both sides we have polynomials of t, it follows that the formula is valid for all *real* t. Let us write t as a rational number $t = r/m$. Then multiply both sides of the last formula with m^p. We get that

$$\sum_{i=1}^{n}(r + im)^p = m^p \sum_{k=1}^{p+1}(k-1)!\left\{\begin{matrix}p+1\\k\end{matrix}\right\}\left[\binom{n+r/m}{k} - \binom{r/m}{k}\right].$$

This is exactly the formula that we wanted to verify.

5.2   The Bernoulli numbers

In the previous section, we have reached our aim and have found an expression for the power sum polynomials. We can go even further because it is possible to express the coefficients of the power sum polynomials in another way using the Bernoulli numbers. These numbers are important in many areas of mathematics and even in physics.

The B_n *Bernoulli*[3] *numbers* are usually defined by the exponential generating function

$$\sum_{n=0}^{\infty} B_n \frac{x^n}{n!} = \frac{x}{e^x - 1}. \tag{5.6}$$

---

[2] This result was presented in [353], too.

[3] Jacob Bernoulli (1654-1705), Swiss mathematician and member of the famous Bernoulli family. The Bernoulli numbers appear in the book *Ars Conjectandi*. This book was posthumously published in 1713.

The Bell numbers are also denoted by B_n, but the Bell numbers and Bernoulli numbers will not appear together in this book so there will be no risk of confusion during using the notation B_n.

We are going to connect our power sum formula to the Bernoulli numbers. We begin with the exponential generating function of

$$\sum_{k=0}^{p} \left\{ {p \atop k} \right\} \frac{1}{k+1} \left[{k+1 \atop m} \right]. \tag{5.7}$$

First determine the generating function of

$$\frac{1}{k+1} \left[{k+1 \atop m} \right]:$$

$$x \sum_{k=0}^{\infty} \frac{1}{k+1} \left[{k+1 \atop m} \right] \frac{x^k}{k!} = \sum_{k=0}^{\infty} \left[{k+1 \atop m} \right] \frac{x^{k+1}}{(k+1)!} = \sum_{k=1}^{\infty} \left[{k \atop m} \right] \frac{x^k}{k!} = \frac{\ln^m(1+x)}{m!},$$

from which one can infer

$$\sum_{k=0}^{\infty} \frac{1}{k+1} \left[{k+1 \atop m} \right] \frac{x^k}{k!} = \frac{1}{x} \frac{\ln^m(1+x)}{m!}. \tag{5.8}$$

We consider the exponential generating function of (5.7). This is an often used shortcut: if we are lucky, the generating function of the sequence we are studying has a simple form and we can connect it to other, already known generating functions. Multiplying (5.7) with $\frac{x^p}{p!}$ and summing over p, we have

$$\sum_{p=0}^{\infty} \frac{x^p}{p!} \sum_{k=0}^{p} \left\{ {p \atop k} \right\} \frac{1}{k+1} \left[{k+1 \atop m} \right] = \sum_{k=0}^{\infty} \frac{1}{k+1} \left[{k+1 \atop m} \right] \sum_{p=0}^{\infty} \frac{x^p}{p!} \left\{ {p \atop k} \right\} =$$

$$\sum_{k=0}^{\infty} \frac{1}{k+1} \left[{k+1 \atop m} \right] \frac{(e^x-1)^k}{k!} = \frac{1}{e^x-1} \frac{\ln^m(1+e^x-1)}{m!} = \frac{1}{e^x-1} \frac{x^m}{m!}. \tag{5.9}$$

In the last step, we used the exponential generating function in (5.8) and that of the second kind Stirling numbers.

The

$$\frac{1}{e^x-1} \frac{x^m}{m!}$$

generating function is not equal to (5.6) but we can rewrite it:

$$\frac{1}{e^x-1} \frac{x^m}{m!} = \frac{x^{m-1}}{m!} \frac{x}{e^x-1} = \frac{x^{m-1}}{m!} \sum_{n=0}^{\infty} B_n \frac{x^n}{n!} = \frac{1}{m!} \sum_{n=0}^{\infty} B_n \frac{x^{n+m-1}}{n!} =$$

$$\frac{1}{m!} \sum_{n=m-1}^{\infty} B_{n-m+1} \frac{x^n}{(n-m+1)!}.$$

Since

$$\frac{1}{n+1}\binom{n+1}{m} = \frac{n!}{m!(n-m+1)!},$$

$$B_{n-m+1}\frac{1}{n!}\frac{1}{n+1}\binom{n+1}{m} = B_{n-m+1}\frac{1}{m!(n-m+1)!},$$

we see that

$$\frac{1}{m!}\sum_{n=0}^{\infty}B_n\frac{x^{n+m-1}}{n!} = \sum_{n=0}^{\infty}B_{n-m+1}\frac{x^n}{n!}\frac{1}{n+1}\binom{n+1}{m}.$$

Comparing this with the coefficients of (5.9), we get the wanted relation between the coefficients of the power sum polynomials and Bernoulli numbers (if $m > 0$):

$$\sum_{k=0}^{n}\left\{\begin{matrix}n\\k\end{matrix}\right\}\frac{1}{k+1}\left\lceil\begin{matrix}k+1\\m\end{matrix}\right\rceil = \frac{1}{n+1}\binom{n+1}{m}B_{n-m+1}. \qquad (5.10)$$

With this information we can rewrite (5.4) as follows:

$$1^p + 2^p + \cdots + (n-1)^p = \frac{1}{p+1}\sum_{m=1}^{p+1}n^m\binom{p+1}{m}B_{p+1-m}. \qquad (5.11)$$

5.2.1   The Bernoulli polynomials

Expression of power sums via the Bernoulli polynomials

The *Bernoulli polynomials* are defined as

$$B_n(x) = \sum_{m=0}^{n}\binom{n}{m}x^m B_{n-m}.$$

Clearly, $B_n = B_n(0)$. The Bernoulli polynomials are of high importance in the analytic theory of numbers[4].

From our present viewpoint, the Bernoulli polynomials are useful, because (5.11) can be expressed in a very compact form with them. Note that the right-hand side of (5.11) is almost a Bernoulli polynomial:

$$\sum_{m=1}^{p+1}n^m\binom{p+1}{m}B_{p+1-m} = B_{p+1}(n) - B_{p+1}.$$

---

[4]Analytic number theory studies number theoretical questions via the methods of complex analysis. The Bernoulli numbers and polynomials appear as some special values of the Riemann zeta function and its relatives.

Thus, we get yet another expression (and the simplest looking one) for our power sums:

$$1^p + 2^p + \cdots + (n-1)^p = \frac{1}{p+1}(B_{p+1}(n) - B_{p+1}). \qquad (5.12)$$

This gives another reason why it is better to sum up to $n-1$: the argument of the Bernoulli polynomial is the simplest one in this case.

Power sums of arithmetic progressions and the Bernoulli polynomials

The initial point in the proof that the power sums can be expressed by the Stirling numbers was (5.1). When we consider power sums of arithmetic progressions, we should consider expressions like

$$(mx + r)^p = \sum_{k=0}^{p} m^k W_{m,r}(p,k) x^{\underline{k}}, \qquad (5.13)$$

where $W_{m,r}(p,k)$ are some appropriate coefficients, similar to the Stirling numbers (m^k could have been put into the $W_{m,r}(p,k)$ coefficients). Such coefficients indeed exist, and they are called *r-Whitney numbers of the second kind*[5]. We do not delve into the theory of these numbers; for the details, see [418, 429]. It is enough to know that the whole process can be done, mutatis mutandis, in the case of the arithmetic progressions. Summing $(mx + r)^p$ we get the above expression with the r-Whitney numbers, and, on the other hand, the r-Whitney sum can be transformed by expanding the falling factorial (and using the first kind Stirlings) similarly as we did in (5.4). This latter step was analyzed in [50, Theorem 1.], and we get that

$$r^p + (r+m)^p + \cdots + (r+(n-1)m)^p = \sum_{i=1}^{p+1} n^i \left(\sum_{k=0}^{p} m^k W_{m,r}(p,k) \frac{1}{k+1} \left[\begin{matrix} k+1 \\ i \end{matrix} \right] \right).$$

One sees that this indeed looks like (5.4).

Then we invoke formula (6) in [418],

$$\sum_{k=0}^{p} m^k W_{m,r}(p,k) \frac{1}{k+1} \left[\begin{matrix} k+1 \\ i \end{matrix} \right] = \frac{m^p}{p+1} \binom{p+1}{i} B_{p+1-i}\left(\frac{r}{m}\right),$$

to express that result in terms of the Bernoulli polynomials:

$$r^p + (r+m)^p + \cdots + (r+(n-1)m)^p = \frac{m^p}{p+1} \sum_{i=1}^{p+1} n^i \binom{p+1}{i} B_{p+1-i}\left(\frac{r}{m}\right).$$

---

[5] These numbers were discovered by several authors independently, see [167, 418].

A bit of transformation on the right-hand side is as follows:

$$\frac{m^p}{p+1} \sum_{i=1}^{p+1} n^i \binom{p+1}{i} B_{p+1-i} \left(\frac{r}{m}\right) =$$

$$\frac{m^p}{p+1} \sum_{i=0}^{p+1} n^{p+1-i} \binom{p+1}{i} B_i \left(\frac{r}{m}\right) - B_{p+1} \left(\frac{r}{m}\right).$$

We need the *translation property* of the Bernoulli polynomials:

$$B_n(x+y) = \sum_{k=0}^{n} \binom{n}{k} B_k(x) y^{n-k}$$

to realize that the first sum in the second line above equals $B_{p+1} \left(n + \frac{r}{m}\right)$. Hence, we arrive at the pretty expression

$$r^p + (r+m)^p + \cdots + (r+(n-1)m)^p = \frac{m^p}{p+1} \left[B_{p+1} \left(n + \frac{r}{m}\right) - B_{p+1} \left(\frac{r}{m}\right)\right].$$

$$(5.14)$$

This gives back (5.12) as a particular case when $r = 0$ and $m = 1$ (similarly as (5.13) goes into the relation involving the Stirling numbers; and in this case $W_{1,0}(n, k) = \{{n \atop k}\}$).

Formula (5.14) was given in [52, 290], and [133, 283] contain evaluations in terms of Bernoulli numbers.

5.3   The Cauchy numbers and Riordan arrays

There are two other interesting sequences we would like to study in this chapter. These are the Cauchy numbers of the first and second kind.

Also, we present an extremely powerful tool to prove combinatorial identities – the Riordan array technique.

5.3.1   The Cauchy numbers of the first and second kind

Definition 5.3.1. *The* Cauchy numbers of the first and second kind[6] *are defined, respectively, by the definite integrals*

$$c_n = \int_0^1 x(x-1)(x-2) \cdots (x-n+1) dx,$$

$$C_n = \int_0^1 x(x+1)(x+2) \cdots (x+n-1) dx.$$

---

[6]The numbers $c_n/n!$ are sometimes called *Laplace numbers*.

Both kinds of Cauchy numbers are rational numbers (they are definite integrals of polynomials), and C_n is always positive.

Note that we can rewrite the definition of the Cauchy numbers in a shorter form if we take the binomial coefficients into account:

$$c_n = \int_0^1 x^{\underline{n}} dx = n! \int_0^1 \binom{x}{n} dx,$$

$$C_n = \int_0^1 x^{\overline{n}} dx = (-1)^n n! \int_0^1 \binom{-x}{n} dx = n! \int_0^1 \binom{x+n-1}{n} dx.$$

(For the definition of the binomial coefficients for real upper parameters see (2.51).)

These numbers are connected to the Stirling numbers. This connection immediately comes if we expand the falling and rising factorials by using the Stirling numbers:

$$c_n = \int_0^1 x(x-1)(x-2)\cdots(x-n+1)dx = \sum_{k=0}^n \left[{n \atop k}\right] \frac{1}{k+1},$$

$$C_n = \int_0^1 x(x+1)(x+2)\cdots(x+n-1)dx = \sum_{k=0}^n \left[{n \atop k}\right] \frac{1}{k+1}.$$

The generating function of c_n and C_n comes easily:

$$\sum_{n=0}^\infty c_n \frac{x^n}{n!} = \frac{x}{\log(1+x)}, \tag{5.15}$$

$$\sum_{n=0}^\infty C_n \frac{x^n}{n!} = \frac{x}{(x-1)\log(1-x)}. \tag{5.16}$$

We prove the first; the second one is left as an exercise. Recalling the Stirling number expression for c_n we have that

$$\sum_{n=0}^\infty c_n \frac{x^n}{n!} = \sum_{n=0}^\infty \frac{x^n}{n!} \sum_{k=0}^n \left[{n \atop k}\right] \frac{1}{k+1}.$$

We can extend the inner sum for $k > n$, because these terms are zero but with this step we make the two sums interchangeable:

$$\sum_{n=0}^\infty \frac{x^n}{n!} \sum_{k=0}^n \left[{n \atop k}\right] \frac{1}{k+1} = \sum_{k=0}^\infty \frac{1}{k+1} \sum_{n=0}^\infty \left[{n \atop k}\right] \frac{x^n}{n!}.$$

Applying the (2.36) exponential generating function of the signed Stirlings we can finalize the proof:

$$\sum_{k=0}^\infty \frac{1}{k+1} \sum_{n=0}^\infty \left[{n \atop k}\right] \frac{x^n}{n!} = \sum_{k=0}^\infty \frac{1}{k+1} \frac{1}{k!} \log^k(1+x) = \frac{x}{\log(1+x)},$$

as we stated.

Now we deduce additional identities for c_n and C_n by using the Riordan array method.

5.3.2   Riordan arrays

The Riordan arrays are extremely useful if we would like to calculate sums of the form

$$\sum_{k=0}^{n} d_{n,k} a_k, \tag{5.17}$$

where a_k is a given sequence and $d_{n,k}$ is a triangular sequence as, for example, the binomial coefficients or the Stirling numbers. By definition, the *Riordan*[7] *array* is the lower triangular infinite matrix

$$d_{n,k} = [x^n] d(x)(x \cdot h(x))^k \quad (n \geq 1), \tag{5.18}$$

where d and h are two almost arbitrary generating functions. It must hold true that $d(0) \neq 0$. Also, if we want our array to be invertible, which equivalently means that $d_{n,n} \neq 0$ for all n, then we must suppose that $h(0) \neq 0$. In this case, the Riordan array is *proper*. Above $[x^n] f(x)$ denotes the coefficient of x^n in the generating function $f(x)$. Thus, for example,

$$[x^n] e^x = \frac{1}{n!}.$$

If we write down $d_{n,k}$ in matrix form where n is the row index and k is the column index, then we see that the resulting matrix will be a lower triangular matrix as we said. This is so, because $[x^n] d(x)(x \cdot h(x))^k$ will be obviously zero if $n < k$.

In symbols, we write that

$$\mathcal{R}(d_{n,k}) = (d(x), h(x)) = (d, h),$$

and we say that $d_{n,k}$ is a Riordan array or *Riordan matrix*. It is determined by d and h via (5.18).

---

[7] John Riordan (1903-1988), American mathematician.

The following examples are easy to establish:

$$\mathcal{R}\left(\frac{k!}{n!}\begin{bmatrix} n \\ k \end{bmatrix}\right) = \left(1, \frac{1}{x}\log\left(\frac{1}{1-x}\right)\right),$$

$$\mathcal{R}\left(\frac{k!}{n!}\overline{\begin{bmatrix} n \\ k \end{bmatrix}}\right) = \left(1, \frac{1}{x}\log(1+x)\right),$$

$$\mathcal{R}\left(\frac{k!}{n!}\begin{Bmatrix} n \\ k \end{Bmatrix}\right) = \left(1, \frac{e^x - 1}{x}\right), \tag{5.19}$$

$$\mathcal{R}\left(\frac{k!}{n!}\left|\begin{matrix} n \\ k \end{matrix}\right|\right) = \left(1, \frac{1}{1-x}\right),$$

$$\mathcal{R}\left(\frac{k!}{n!}\binom{n}{k}\right) = \left(\frac{1}{1-x}, \frac{1}{1-x}\right). \tag{5.20}$$

For example, let us take the first Riordan array $\mathcal{R}\left(\frac{k!}{n!}\begin{bmatrix} n \\ k \end{bmatrix}\right)$. We know that

$$\sum_{n=k}^{\infty} \begin{bmatrix} n \\ k \end{bmatrix}\frac{x^n}{n!} = \frac{1}{k!}\log^k\left(\frac{1}{1-x}\right);$$

therefore

$$[x^n]\frac{k!}{n!}\begin{bmatrix} n \\ k \end{bmatrix} = [x^n]\log^k\left(\frac{1}{1-x}\right).$$

This must be rewritten in the form (5.18). Choosing $d(x) = 1$ and $h(x) = \frac{1}{x}\log\left(\frac{1}{1-x}\right)$ we are done, because

$$[x^n]\frac{k!}{n!}\begin{bmatrix} n \\ k \end{bmatrix} = [x^n]d(x)(x \cdot h(x))^k = [x^n]1 \cdot \left(\frac{x}{x}\log^k\left(\frac{1}{1-x}\right)\right) =$$

$$[x^n]\log^k\left(\frac{1}{1-x}\right).$$

As we mentioned, the Riordan array method is useful if we would like to find sums of the form (5.17). To demonstrate this statement, we are going to prove the following identity (which is due to R. Sprugnoli [533]):

$$\sum_{k=0}^{n} d_{n,k}a_k = [x^n]d(x)f(x \cdot h(x)), \tag{5.21}$$

where $f(x)$ is the generating function of the sequence a_n.

To prove this identity, we need only to find the generating function of the left-hand side. It will be $d(x)f(x \cdot h(x))$ as our identity suggests.

$$\sum_{n=0}^{\infty} x^n \left(\sum_{k=0}^{n} d_{n,k}a_k\right) = \sum_{k=0}^{\infty} a_k \left(\sum_{n=0}^{\infty} x^n d_{n,k}\right) =$$

$$\sum_{k=0}^{\infty} a_k d(x)(x \cdot h(x))^k = d(x)f(x \cdot h(x)),$$

as we stated.

5.3.3   Some identities for the Cauchy numbers

Three more identities for the Cauchy numbers are added here. Two of them are applications of the Riordan method. This short subsection is added to show additional applications of this useful tool. The results of this section are taken from [406].

The first identity that we are going to prove is

$$\sum_{k=0}^{n}\left\{{n \atop k}\right\}c_k = \frac{1}{n+1} \quad (n \geq 0). \tag{5.22}$$

A trivial transformation first:

$$\sum_{k=0}^{n}\left\{{n \atop k}\right\}c_k = n!\sum_{k=0}^{n}\frac{k!}{n!}\left\{{n \atop k}\right\}\frac{c_k}{k!}.$$

Then, by (5.21),

$$\sum_{k=0}^{n}\left\{{n \atop k}\right\}c_k = n![x^n]d(x)f(x \cdot h(x)),$$

where

$$(d(x), h(x)) = \left(1, \frac{e^x - 1}{x}\right), \quad \text{and} \quad f(x) = \frac{x}{\log(1 + x)}$$

by (5.19) and (5.15). Therefore,

$$\sum_{k=0}^{n}\left\{{n \atop k}\right\}c_k = n![x^n]\frac{x\left(\frac{e^x-1}{x}\right)}{\log\left(1 + x\left(e^x - 1x\right)\right)} = n![x^n]\frac{e^x - 1}{x}.$$

Since, as it is easy to see, $\frac{e^x-1}{x}$ is the generating function of $\frac{1}{(n+1)!}$,

$$n![x^n]\frac{e^x - 1}{x} = \frac{1}{n+1},$$

and (5.22) follows.

The dual of this identity is

$$\sum_{k=0}^{n}\left\{{n \atop k}\right\}(-1)^k C_k = \frac{(-1)^n}{n+1}.$$

The reader is called to prove this result.

Recurrence for the Cauchy numbers

A recurrence can also be established for the first kind Cauchy numbers:

$$c_n = n!\sum_{k=0}^{n-1}\frac{c_k}{k!}\frac{(-1)^{n-k+1}}{n-k+1} \quad (n \geq 1). \tag{5.23}$$

(Note that this is indeed a recurrence, because it expresses c_n by c_k where $k < n$. This recurrence, in fact, uses *all* the former members up to c_{n-1}.)

To prove the recurrence, we use a very simple trick. We sum on $k = 0, \ldots, n$ and then we separate the term $k = n$. Thus, rewriting the recurrence, we see that it is equivalent to

$$n! \sum_{k=0}^{n} \frac{c_k}{k!} \frac{(-1)^{n-k+1}}{n-k+1} = 0,$$

unless $n = 0$. Hence, the generating function of this sequence must be $f(x) = 1$, and so

$$1 = \frac{x}{\log(1+x)} \frac{\log(1+x)}{x}$$

with

$$\frac{x}{\log(1+x)} = \sum_{n=0}^{\infty} \frac{c_n}{n!} x^n, \quad \text{and} \quad \frac{\log(1+x)}{x} = \sum_{n=0}^{\infty} \frac{(-1)^n}{n+1} x^n,$$

whence Cauchy's product leads to the recursion. The dual of this recurrence for the second kind Cauchy numbers can be found in the Exercises.

A relation between the Cauchy and Bernoulli numbers

Finally, we give a relation between the Cauchy and Bernoulli numbers:

$$-\frac{B_n}{n} = \sum_{k=1}^{n} (-1)^k \left\{ {n \atop k} \right\} \frac{C_k}{k} \quad (n \geq 1). \tag{5.24}$$

To see this relation, we must apply the Riordan method with $\mathcal{R}\left(\frac{k!}{n!}\left\{{n \atop k}\right\}\right)$ and with $a_k = (-1)^k C_k/k$. To this end, we only have to find the generating function of $(-1)^k C_k/k$. If we divide (5.16) by x we get that

$$\sum_{n=0}^{\infty} (-1)^n C_n \frac{x^{n-1}}{n!} = \frac{1}{(x+1)\log(1+x)},$$

and then we integrate. The integral will not be convergent because of the presence of $1/x$ on the left-hand side. Therefore, we subtract $1/x$ from both sides:

$$\sum_{n=1}^{\infty} (-1)^n C_n \frac{x^{n-1}}{n!} = \frac{1}{(x+1)\log(1+x)} - \frac{1}{x},$$

whence

$$\sum_{n=1}^{\infty} (-1)^n \frac{C_n}{n} \frac{x^n}{n!} = \int_0^x \left(\frac{1}{(t+1)\log(1+t)} - \frac{1}{t} \right) dt =$$

$$\log \frac{\log(1+x)}{x}.$$

This means that

$$f(x) = \log \frac{\log(1+x)}{x}.$$

Remember that the attached Riordan array is

$$\mathcal{R}\left(\frac{k!}{n!}\left\{{n \atop k}\right\}\right) = \left(1, \frac{e^x - 1}{x}\right),$$

so

$$\sum_{k=1}^{n}(-1)^k\left\{{n \atop k}\right\}\frac{C_k}{k} = n!\sum_{k=1}^{n}\frac{k!}{n!}\left\{{n \atop k}\right\}(-1)^k\frac{C_k}{kk!} =$$

$$n![x^n]d(x)f(x \cdot h(x)) = n![x^n]1 \cdot \log\frac{\log(1 + e^x - 1)}{e^x - 1} = n![x^n]\log\frac{x}{e^x - 1}.$$

We would be ready if we could show that

$$[x^n]\log\frac{x}{e^x - 1} = -\frac{B_n}{nn!} \quad (n \geq 1).$$

This is left to the reader. For the dual of (5.24), see the Exercises.

Exercises

1. Prove by induction (without using the Stirling numbers) the three power sum identities at the beginning of the chapter.

2. Now solve the previous problem by using (5.4).

3. Prove (5.2) by induction.

4. Calculate the first six Bernoulli numbers.

5. Show that $B_{2n+1} = 0$ for all $n > 1$.

6. Verify the identity

$$B_n = \sum_{k=0}^{n} (-1)^k \frac{k!}{k+1} \left\{ {n \atop k} \right\} \quad (n \geq 0).$$

(See p. 232, Exercise 24 for a generalization to polynomials. Hint: specialize (5.10).)

7. Verify the identity

$$B_n = \sum_{k=0}^{n} (-1)^k k! \left\{ {n+1 \atop k+1} \right\} H_{k+1}$$

which connects the Bernoulli, Stirling and harmonic numbers.

8. Applying the orthogonality, prove the dual of the previous formula:

$$H_{n+1} - \frac{1}{n!} \sum_{k=0}^{n} (-1)^k \left[{n+1 \atop k+1} \right] B_k.$$

9. Prove (5.16).

10. Present and prove the binomial theorem using the language of Riordan arrays (see (5.20)).

11. Show that the generating function of the binomial transform of the sequence a_n with generating function $f(x)$ is

$$\frac{1}{1-x} f \left(\frac{x}{1-x} \right),$$

which is just Euler's transformation formula, see (2.14). (Euler's formula was established with special tricks on p. 45. This result shows in itself why the general framework of the Riordan method is so useful.)

12. By using the Riordan method, show that the exponential generating function of the Bell polynomials $B_n(x)$ is $e^{x(e^y-1)}$.

13. Prove the identity of Wang [582]:

$$\sum_{k=0}^{n} \binom{n}{k}(H_n - H_k)\frac{c_k}{k!} = n.$$

14. Show that the following recursion connecting the first and second kind Cauchy numbers is valid:

$$(-1)^n c_n = C_n - nC_{n-1} \quad (n \geq 1).$$

(Hint: use the trivial identity

$$\frac{x}{\log(1+x)} = (1+x)\frac{x}{(1+x)\log(1+x)}.)$$

15. Verify the dual of (5.22):

$$\sum_{k=0}^{n} \left\{ {n \atop k} \right\}(-1)^k C_k = \frac{(-1)^n}{n+1}.$$

16. Prove that

$$C_n = n!\sum_{k=0}^{n-1} \frac{C_k}{k!}\frac{1}{(n-k)(n-k+1)}$$

holds for $n \geq 1$. This identity is the dual of (5.23).

17. Prove the dual formula of (5.24):

$$\sum_{k=1}^{n} \left[{n \atop k} \right]\frac{B_k}{k} = -\frac{C_n}{n} \quad (n \geq 1).$$

18. If, in the previous example, there is no denominator, then the next formula arises:

$$\sum_{k=1}^{n} \left[{n \atop k} \right]B_k = -\frac{(n-1)!}{n+1} \quad (n \geq 1).$$

Prove this by using the Riordan technique (see [533, p. 288]).

19. Prove the orthogonality property of the signed first kind and second kind Stirling numbers discussed in Section 2.6 via the Riordan array method. (See p. 285 in [533].)

20. Prove the following interesting relation between the Bernoulli numbers and Euler numbers:

$$B_{2n} = (-1)^{n-1}\frac{2n}{4^{2n} - 2^{2n}}E_{2n-1} \quad (n > 0).$$

(Hint: use the exponential generating function of the Bernoulli numbers and André's generating function result on p. 80.)

Outlook

1. There are studies of power sums when the consecutive members of the arithmetic progression are of opposite sign (these are called *alternating power sums*). See [431] and the citations in it.

2. In [501] the weighted sum

$$\sum_{j=0}^{m-1} (aj + b)^2 z^j$$

of squares was studied in detail.

3. The finite sum of consecutive powers of a fixed positive integer m were studied by B. Sury [553], who found that

$$1 + m + m^2 + \cdots + m^n = \sum_{j \geq 0} (-1)^j \binom{n-j}{j} m^j (1+m)^{n-2j} \quad (m > 0, n \geq 0).$$

In [396] this is proved combinatorially, using tiling with dominoes.

4. A good reference for the Riordan array method is a paper by R. Sprugnoli [533] where a large number of examples is available. The Riordan arrays first appeared in [514]. A basic survey was written by P. Barry [47]. In [141] one can encounter nice formulas with Stirling numbers, like

$$\sum_{k=0}^{n} \binom{n}{k} k^r = \sum_{k=0}^{r} \left\{ {r \atop k} \right\} n^{\underline{k}} 2^{n-k},$$

and many others.

5. The history of the formula in Exercise 6 can be found in [259]. More connections between Stirling and Bernoulli numbers is contained in [517]. Both of these sources contain, for example, the relation

$$B_n = \sum_{k=1}^{n} \frac{(-1)^k \binom{n+1}{k+1}}{\binom{n+k}{k}} \left\{ {n+k \atop k} \right\}.$$

See also [270].

6. The following sequence of polynomials, here we defined them by their exponential generating function, often appears in the literature:

$$\sum_{n=0}^{\infty} E_n(x) \frac{t^n}{n!} = \frac{2e^{xt}}{e^t + 1}.$$

Unfortunately, here we have a clash: these polynomials are called Euler polynomials, and the numbers $E_n = 2^n E_n \left(\frac{1}{2}\right)$ are Euler numbers, which are completely different from "our" Euler numbers on Section 1.8. Still, these numbers are related to the Stirling numbers of the second kind, and the coefficients of $E_n(x)$ are determined by $\left\{{n \atop k}\right\}$. See [271] for the details.

7. One sees from the very first lines of this chapter that the first power sums can be described as a polynomial with integer coefficients times a rational number. This is not a difficult statement: we know that the power sums are polynomials, and each polynomial can be multiplied by a rational number such that the coefficients become integers. The reciprocal of the smallest such number is what we are talking about, carried from the right to the left. This number sequence has been the subject of many studies. See [329] for a recent source of results and citations, and also A064538 in The On-Line Encyclopedia of Integer Sequences.

8. The Cauchy numbers have an interesting relationship to the Riemann zeta function, see [79] and the references therein.

9. The Bernoulli numbers, their connection to the Riemann zeta function, their generalizations, and their importance in analytic number theory are studied in detail in [34].

Chapter 6

Ordered partitions

6.1 Ordered partitions and the Fubini numbers

6.1.1 The definition of the Fubini numbers

The number of partitions of an n element set is given by the nth Bell number. If the elements in the block are ordered, we get the L_n numbers – the horizontal sums of the Lah numbers. If, in turn, the order of the blocks is taken into account, we arrive at another counting sequence: that of the Fubini numbers. In this chapter we study this sequence; it will turn out that it is connected also to some interesting counting problems regarding permutations.

Without taking the order of the blocks into account, the partitions

$$1, 5|2, 3|4, 6$$

and

$$2, 3|1, 5|4, 6$$

are identical. In this chapter we do distinguish them and make the following definition.

Definition 6.1.1. *Let F_n denote the number of all the partitions of an n-set such that we take the order of the blocks in the individual partitions into account. F_n is the nth ordered partition number, or nth Fubini[1] number. The Fubini numbers are also called* ordered Bell numbers[2].

Going back to our very first example in the book, we saw that four elements have 15 partitions:

$$1|2|3|4$$

$$1|2|3, 4, \quad 1|3|2, 4, \quad 1|4|2, 3, \quad 2|3|1, 4, \quad 2|4|1, 3, \quad 3|4|1, 2$$

$$1|2, 3, 4, \quad 2|1, 2, 4, \quad 3|1, 2, 4, \quad 4|1, 2, 3, \quad 1, 2|3, 4, \quad 1, 3|2, 4, \quad 1, 4|2, 3$$

$$1, 2, 3, 4$$

---

[1] Guido Fubini (1879-1943), Italian mathematician.
[2] Sometimes *preferential arrangement* is also used.

Since the first line contains four blocks, we have $4! = 24$ different ordered partitions on this level. In the second line, there are six 3-partitions; therefore, there are $6 \cdot 3!$ ordered 3-partitions, and so on. Altogether

$$F_4 = 4! \begin{Bmatrix} 4 \\ 1 \end{Bmatrix} + 3! \begin{Bmatrix} 4 \\ 3 \end{Bmatrix} + 2! \begin{Bmatrix} 4 \\ 2 \end{Bmatrix} + 1! \begin{Bmatrix} 4 \\ 1 \end{Bmatrix} = 75.$$

This argument is valid for any finite size of sets, so

$$F_n = \sum_{k=1}^{n} k! \begin{Bmatrix} n \\ k \end{Bmatrix}. \tag{6.1}$$

We set $F_0 = 1$.

6.1.2   Two more interpretations of the Fubini numbers

There are other ways we can look at the ordered partitions. We give here some examples.

Surjective functions

Going back to p. 17, we see there that the number of surjective functions from an n-set to a k-set is $k! \begin{Bmatrix} n \\ k \end{Bmatrix}$. Summing over k, we infer that the Fubini numbers count *all* the surjective functions on an n-set onto some set (of cardinality necessarily between one and n).

Competitions with ties

Another interpretation comes if we consider a competition of n people where *draws* (or *ties*) are allowed. The nth Fubini number F_n gives the number of the possible outputs in such a competition. This is so because an ordered k-partition on n people can be considered as an output of a competition where the n people (better to say, their indices) in the same blocks are classified equal (draws).

This interpretation offers a recursive formula for the Fubini numbers. To enumerate all the outputs of our competition, we can choose k competitors from n, and these go to the first position. This can be done in $\binom{n}{k}$ ways. Then the remaining $n - k$ competitors can be considered as competitors of a new competition for the second, third,... position. They can be classified in F_{n-k} ways. Summing over k, put these into a formula:

$$F_n = \sum_{k=1}^{n} \binom{n}{k} F_{n-k}. \tag{6.2}$$

Applying the $\binom{n}{k} = \binom{n}{n-k}$ symmetry,

$$F_n = \sum_{k=0}^{n-1} \binom{n}{k} F_k. \tag{6.3}$$

It is worth it to compare this with the (1.1) recursion for the Bell numbers.

6.1.3 The Fubini numbers count chains of subsets

Let us fix a set S of n elements. We say that a sequence

$$U_1 \subset U_2 \subset \cdots \subset U_k \tag{6.4}$$

of subsets of S is a *chain* if $U_i \subset U_{i+1}$ and $U_i \neq U_{i+1}$ for all possible is. The length of the chain (6.4) is defined to be k.

Let a chain be called *full* if $U_1 = \emptyset$ and $U_k = S$. We are going to see that the number of all the full chains on an n element set S is the n-th Fubini number.

Indeed, $U_1 = \emptyset$, and we would like to choose an U_2, as the next element of the chain. If the chain is of length two, we choose U_2 to be S. If it is longer, then we have to choose a non-empty proper subset of S. If the chain is of length three, we must choose $U_3 = S$. Otherwise, we have to fix the proper subset U_3 of S such that it is disjoint from U_2. We continue the process in this manner until eventually we reach $U_k = S$. Obviously, $U_2 \setminus U_1 (= U_1), U_3 \setminus U_2, \ldots, U_k \setminus U_{k-1}$ form a partition of S. In the case of chains, the order in which we have chosen our chain elements (blocks) matters, since, for example, the two chains

$$\emptyset \subset \{1,2\} \subset \{1,2,3,4\} = S,$$

and

$$\emptyset \subset \{3,4\} \subset \{1,2,3,4\} = S$$

are different (even if the corresponding partitions, $1,2|3,4$ and $3,4|1,2$, are the same). For this reason, the number of chains of length k for a set of n elements is $k! \left\{ {n \atop k} \right\}$. Summing over k, we get the number of all the possible full chains for an n element set, and this is the number F_n.

6.1.4 The generating function of the Fubini numbers

In order to find the exponential generating function of the Fubini numbers, we use (6.1). Summing over n after multiplying with $\frac{x^n}{n!}$ we get that

$$\sum_{n=0}^{\infty} F_n \frac{x^n}{n!} = \sum_{n=0}^{\infty} \left(\sum_{k=1}^{n} k! \left\{ {n \atop k} \right\} \right) \frac{x^n}{n!}.$$

Here k can run from 0 to infinity, so that we can interchange the sums:

$$\sum_{n=0}^{\infty}\left(\sum_{k=1}^{n}k!\left\{{n\atop k}\right\}\right)\frac{x^n}{n!} = \sum_{k=0}^{\infty}k!\left(\sum_{n=0}^{\infty}\left\{{n\atop k}\right\}\frac{x^n}{n!}\right) = \sum_{k=0}^{\infty}k!\frac{(e^x-1)^k}{k!} =$$

$$\sum_{k=0}^{\infty}(e^x-1)^k = \frac{1}{1-(e^x-1)}.$$

The last sum is a simple geometric series (see (2.2)). The exponential generating function is therefore

$$\sum_{n=0}^{\infty}F_n\frac{x^n}{n!} = \frac{1}{2-e^x}. \tag{6.5}$$

This results in a nice infinite sum representation for the Fubini numbers, which can be considered as the analogue of the Dobiński formula (2.17). Rewriting the generating function:

$$\frac{1}{2-e^x} = \frac{1}{2}\frac{1}{1-\frac{e^x}{2}} = \frac{1}{2}\sum_{k=0}^{\infty}\left(\frac{e^x}{2}\right)^k = \frac{1}{2}\sum_{k=0}^{\infty}\frac{e^{kx}}{2^k} =$$

$$\sum_{k=0}^{\infty}\frac{1}{2^{k+1}}\sum_{n=0}^{\infty}\frac{k^n x^n}{n!} = \sum_{n=0}^{\infty}\frac{x^n}{n!}\sum_{k=0}^{\infty}\frac{k^n}{2^{k+1}}.$$

Comparing the coefficients,

$$F_n = \sum_{k=0}^{\infty}\frac{k^n}{2^{k+1}}. \tag{6.6}$$

As the Fubini numbers are also called ordered Bell numbers, we may call this nice result an *ordered Dobiński formula*.

6.1.5   The Hankel determinants of the Fubini numbers

To determine the Hankel determinants of the Fubini numbers, we use the just proven ordered Dobiński formula and the orthogonal polynomial theory tool we got to know in Subsection 2.10.3.

We are looking for an orthogonal polynomial sequence such that the attached functional L is such that

$$L(x^n) = F_n.$$

The *Meixner polynomials* are the corresponding polynomials we are looking for. In Section 1.9 of [338], we see that for the $M_n(x;\beta,c)$ polynomials

$$\sum_{x=0}^{\infty}\frac{(\beta)_x}{x!}c^x M_n(x;\beta,c)M_m(x;\beta,c) = \frac{c^{-n}n!}{(\beta)_n(1-c)^\beta}\delta_{nm}$$

for $\beta > 0$ and $0 < c < 1$. Here

$$\delta_{nm} = \begin{cases} 1, & \text{if } n = m; \\ 0, & \text{otherwise} \end{cases}$$

is the *Kronecker delta symbol*.

We read out the functional L which is[3]

$$L(f(x)) = \sum_{t=0}^{\infty} \frac{(\beta)_t}{t!} c^t f(t).$$

Setting $\beta = 1$ and $c = \frac{1}{2}$ we have that

$$L(x^n) = \sum_{t=0}^{\infty} \frac{1}{2^t} t^n = 2F_n,$$

by the ordered Dobiński formula. The factor two is irrelevant here because it can be incorporated into L and the orthogonality relation. The (1.9.4) recurrence in [338] for the normalized Meixner polynomials

$$p_n(x) = \left(\frac{c}{c-1}\right)^n (\beta)_n M_n(x; \beta, c)$$

is, in our particular choice of parameters,

$$p_{n+1}(x) = (x - (3n+1))p_n(x) - 2n^2 p_{n-1}(x),$$

meaning that

$$a_n = -(3n+1), \quad \text{and} \quad b_n = 2n^2.$$

Substituting this into (2.64), we get that

$$\begin{vmatrix} F_0 & F_1 & F_2 & \cdots & F_n \\ F_1 & F_2 & F_3 & \cdots & F_{n+1} \\ \vdots & & & & \\ F_n & F_{n+1} & F_{n+2} & \cdots & F_{2n} \end{vmatrix} = 2^{\frac{n(n+1)}{2}} \prod_{k=1}^{n} k!^2 \quad (n = 1, 2, \dots).$$

This result can be found in [233], see the table at the end of the paper.

6.2 Fubini polynomials

It is worth it to introduce the Fubini polynomials because they will have interesting connections to permutations. The *Fubini polynomials* are defined

---

[3]The functional L is not an integral as in our example in Subsection 2.10.3 but an infinite sum. Infinite sums can be considered as special integrals when, in place of summing over areas, we sum over numbers where the "area" covered by the numbers is one. This is why the authors of [338] kept the summation index being x.

as

$$F_n(x) = \sum_{k=0}^{n} k! \left\{ {n \atop k} \right\} x^k$$

for all $n \geq 0$.

The fourth Fubini polynomial, for instance, is

$$F_4(x) = 24x^4 + 36x^3 + 14x^2 + x.$$

In this section we study the main properties of these polynomials. First, we note that the exponential generating function of $F_n(x)$ can be deduced similarly as we did for (6.5):

$$\sum_{n=0}^{\infty} F_n(y) \frac{x^n}{n!} = \frac{1}{1 - y(e^x - 1)}. \tag{6.7}$$

A recurrence for $F_n(x)$

The $F_n(x)$ polynomials satisfy a recurrence that can be proven by the recursion of the Stirling numbers (similarly as we did with the Bell polynomials). Since

$$\left\{ {n \atop k} \right\} = k \left\{ {n-1 \atop k} \right\} + \left\{ {n-1 \atop k-1} \right\},$$

we have that

$$F_n(x) = \sum_{k=1}^{n} k! k \left\{ {n-1 \atop k} \right\} x^k + \sum_{k=1}^{n} k! \left\{ {n-1 \atop k-1} \right\} x^k.$$

For the first sum

$$\sum_{k=1}^{n} k! k \left\{ {n-1 \atop k} \right\} x^k = x \left(\sum_{k=1}^{n} k! \left\{ {n-1 \atop k} \right\} x^k \right)' = x F'_{n-1}(x),$$

while for the second

$$\sum_{k=1}^{n} k! \left\{ {n-1 \atop k-1} \right\} x^k = x \left(\sum_{k=1}^{n} (k-1)! \left\{ {n-1 \atop k-1} \right\} x^k \right)' = x(x F_{n-1}(x))' =$$

$$x F_{n-1}(x) + x^2 F'_{n-1}(x).$$

Finally,

$$F_n(x) = x[F_{n-1}(x) + (1 + x) F'_{n-1}(x)].$$

Note that this yields the relation

$$F_n(x) = x((1 + x) F_{n-1}(x))'. \tag{6.8}$$

The real zero property of $F_n(x)$

Based on this relation, we can study the zero structure of the Fubini polynomials and use this knowledge to prove the log-concavity of the coefficients. As we shall see, the zeros of $F_n(x)$ are all real, non-positive and greater than -1. Therefore, the coefficients $k!\left\{{n \atop k}\right\}$ form a log-concave sequence in k (this notion was introduced in Chapter 4):

$$\left(k!\left\{{n \atop k}\right\}\right)^2 \geq (k-1)!\left\{{n \atop k-1}\right\}(k+1)!\left\{{n \atop k+1}\right\}, \tag{6.9}$$

whence the nice inequality

$$\left\{{n \atop k}\right\}^2 \geq \frac{k}{k+1}\left\{{n \atop k-1}\right\}\left\{{n \atop k+1}\right\} \tag{6.10}$$

follows.

Now let us see why the zeros of $F_n(x)$ are real and fall into $]-1,0]$. We proceed by induction. For $F_1(x) = x$ the statement is true. The non-zero root of $F_2(x) = x + 2x^2$ is negative and greater than -1. The other is $x = 0$. Note that $F_n(x)$ does not contain a constant term if $n > 0$, so $x = 0$ is always a zero with multiplicity one.

Now let us prove our statement for general n. Recall (6.8). The right-hand side already contains the $x = 0$ zero with multiplicity one, so we can concentrate on the negative zeros. We will show that the derivative on the right-hand side has zeros only in the interval $]-1,0[$. The function

$$(1+x)F_{n-1}(x)$$

has a zero at $x = -1$ because of the factor $1 + x$. The other zeros come from the equation $F_{n-1}(x) = 0$. The $x = 0$ is a solution, and by the induction hypothesis $F_{n-1}(x)$ has $n-2$ roots in $]-1,0[$. Hence, by Rolle's theorem, the derivative has a zero on the right-hand side of -1 but on the left of all the zeros of $F_{n-1}(x)$. Moreover, the derivative has zeros among the zeros of $F_{n-1}(x)$ and, in addition, between the rightmost non-zero root of $F_{n-1}(x)$ and $x = 0$. Altogether, we counted $n-1$ negative zeros, all of them in $]-1,0[$. Hence, we have found all the zeros of the n degree polynomial $F_n(x)$ and these possess the properties we wanted to demonstrate.

6.3 Permutations, ascents, and the Eulerian numbers

In this section we study ascents, descents, and runs of permutations. Later we shall see that these are connected to ordered partitions.

6.3.1   Ascents, descents, and runs

Definition 6.3.1. *Let a permutation*

$$\pi = \begin{pmatrix} 1 & 2 & 3 & \cdots & n \\ i_1 & i_2 & i_3 & \cdots & i_n \end{pmatrix}$$

be given. The position j in (the bottom line of) π is called ascent *if $i_j < i_{j+1}$. If $i_j > i_{j+1}$ we say that the position j is a* descent.

For example, in

$$\begin{pmatrix} 1 & 2 & 3 & 4 & 5 & 6 \\ 3 & 1 & 4 & 6 & 2 & 5 \end{pmatrix}$$

the positions $2, 3, 5$ are ascents, while $1, 4$ are descents.

The following definition is strongly related.

Definition 6.3.2. *The consecutive maximal increasing subsequences of permutations are called* run*s.*

In the above example, there are three runs: 3; $1, 4, 6$; and $2, 5$.

Next we study how the ascents, descents and runs are related. If we want to count the ascents and descents in a permutation, we go from left to the right and compare two consecutive elements: the first with the second, then the second with the third, and so on, and finally we compare the penultimate with the last element. Altogether, we have $n - 1$ comparisons and in each position we have two alternatives: we can have an ascent or a descent. Hence,

$$\text{ascents} + \text{descents} = n - 1. \tag{6.11}$$

On the other hand, a permutation always begins with a run and this run ends at the first descent. Then a new run follows until the next descent, and so on. This means that two neighboring runs are separated by a descent, so

$$\text{runs} = \text{descents} + 1.$$

The above two results together yield that

$$\text{runs} = n - \text{ascents}. \tag{6.12}$$

Having the fact that the number of ascents determines the number of runs and descents, it is enough to deal with ascents. We therefore introduce the notion of Eulerian numbers[4] which count permutations with a given number of ascents.

---

[4]Not to be confused with the Euler numbers E_n which count zigzag permutations!

6.3.2 The definition of the Eulerian numbers

Definition 6.3.3. *The number of n-permutations which contain exactly k ascents is given by the* Eulerian number *with parameters n and k, and it is denoted by $\left\langle {n \atop k} \right\rangle$*[5].

There is a table for the first Eulerian numbers at the end of the book. For now, let us see a simple example. All the permutations on the 3-set $\{1, 2, 3\}$ are listed here:

$$\begin{pmatrix} 1 & 2 & 3 \\ 1 & 2 & 3 \end{pmatrix}, \quad \begin{pmatrix} 1 & 2 & 3 \\ 2 & 1 & 3 \end{pmatrix}, \quad \begin{pmatrix} 1 & 2 & 3 \\ 3 & 1 & 2 \end{pmatrix},$$

$$\begin{pmatrix} 1 & 2 & 3 \\ 1 & 3 & 2 \end{pmatrix}, \quad \begin{pmatrix} 1 & 2 & 3 \\ 2 & 3 & 1 \end{pmatrix}, \quad \begin{pmatrix} 1 & 2 & 3 \\ 3 & 2 & 1 \end{pmatrix}.$$

As it is easy to see, the numbers of ascents in these permutations are $2, 1, 1, 1, 1, 0$, respectively. Therefore

$$\left\langle {3 \atop 0} \right\rangle = 1, \quad \left\langle {3 \atop 1} \right\rangle = 4, \quad \left\langle {3 \atop 2} \right\rangle = 1.$$

(The numbers of runs are $1, 2, 2, 2, 2, 3$, respectively, while the numbers of descents are $0, 1, 1, 1, 1, 2$.)

If a permutation

$$\begin{pmatrix} 1 & 2 & 3 & \cdots & n \\ i_1 & i_2 & i_3 & \cdots & i_n \end{pmatrix}$$

contains k ascents, then its reverse

$$\begin{pmatrix} 1 & 2 & 3 & \cdots & n \\ i_n & i_{n-1} & i_{n-2} & \cdots & i_1 \end{pmatrix}$$

contains $n - k - 1$ ascents, so

$$\left\langle {n \atop k} \right\rangle = \left\langle {n \atop n - k - 1} \right\rangle \tag{6.13}$$

holds, which is similar to the symmetry of the binomial coefficients.

Since any n-permutation contains at least 0 and at most $n - 1$ ascents, it is also obvious that

$$\sum_{k=0}^{n-1} \left\langle {n \atop k} \right\rangle = n!.$$

---

[5]This notation is not so common as the notation for the Stirling numbers. It was introduced by Donald E. Knuth, American mathematician, in 1973 [335, Section 5.1.3]; but $\left\langle {n \atop k} \right\rangle$ counted there the permutations with k runs. The signs $\langle$ and $\rangle$ refer to the relation signs, and the relation signs refer to the comparisons when we are looking for ascents and descents. It is also important to note that the number of permutations with $k - 1$ descents (or, equivalently, k runs) is also often called Eulerian number, and for these numbers the notation $A_{n,k}$ is common. By the relation (6.12) between runs and ascents we have that $A_{n,k} = \left\langle {n \atop n-k} \right\rangle$.

6.3.3 The basic recursion for the Eulerian numbers

If we introduce a new counting sequence, one of the first tasks is to find a recursion for the sequence if possible[6].

To find a recursion for $\left\langle {n \atop k} \right\rangle$, which is the number of n-permutations with k ascents, we can insert the last element n into an $(n-1)$-permutation. We then have two possibilities:

(1) The existing $(n-1)$-permutation already has k ascents. Then, we must insert our new element n to keep the number of ascents unchanged. So we can insert n into the first position, or right after an ascent position. We had k ascents and these, together with the first position, give $k+1$ possibilities, $(k+1)\left\langle {n-1 \atop k} \right\rangle$ in total.

(2) Once we insert n, it is possible that we get a new ascent. In this case, the initial permutation must have $k-1$ ascents (we cannot have more than one new ascent inserting one element). Now we can insert n right after any descent position, or in the last position. The number of descents on $n-1$ elements is $n-2-(k-1) = n-k-1$, plus the last position: $(n-k)\left\langle {n-1 \atop k-1} \right\rangle$. Hence,

$$\left\langle {n \atop k} \right\rangle = (k+1)\left\langle {n-1 \atop k} \right\rangle + (n-k)\left\langle {n-1 \atop k-1} \right\rangle. \tag{6.14}$$

We set $\left\langle {0 \atop 0} \right\rangle = 1$ and $\left\langle {n \atop 0} \right\rangle = 0$ if $n \geq 1$.

The Eulerian numbers are connected to the binomial coefficients, Stirling numbers and Fubini numbers. Now we describe these relations.

6.3.4 Worpitzky's identity

The *Worpitzky*[7] *identity*[8] is a combinatorial equality which connects the Eulerian numbers and binomial coefficients:

$$m^n = \sum_{k=0}^{n} \left\langle {n \atop k} \right\rangle \binom{m+k}{n}. \tag{6.15}$$

A proof is as follows. We take all the functions of the form

$$f : N \to M,$$

where $|N| = n$ and $|M| = m$. There are m^n such functions. We show that the right-hand side equals this number.

---

[6]The sequences appearing in this book are "easy enough" to find recursions for. But this is not always the case. A famous example is that of $L(n)$: these count the number of $n \times n$ *Latin squares*. There is no known recursion for this sequence. Even worse, there is not any effective formula to calculate $L(n)$ in general.

[7]Julius Daniel Theodor Worpitzky (1835-1895), German mathematician.

[8]According to our present knowledge, this identity first appeared in a Chinese mathematical article by Li Shan-Lan in 1867. For $n \leq 5$ it was known to the Japanese Yoshiuke Matsunaga, who died in 1744. See [459, p. 14.].

Such a function f can be represented by a list: we prescribe the images of $1, 2, \ldots, n$ by f and put the images into a list

$$(a_1, a_2, \ldots, a_n),$$

where $a_i \in \{1, 2, \ldots, m\}$. Put these elements in increasing order and make a permutation as follows. If the increasing order of $a_1, \ldots, a_n$ is $a_{i_1}, a_{i_2}, \ldots, a_{i_n}$, then let the permutation be

$$\begin{pmatrix} 1 & 2 & \cdots & n \\ i_1 & i_2 & \cdots & i_n \end{pmatrix}.$$

To see how this works in practice, take $m = 3$ and $n = 6$, and let f be given by the list

$$(2, 3, 3, 1, 2, 2).$$

(Hence, $f(1) = 2$, $f(2) = 3$, and so on.) In increasing order

$$(1, 2, 2, 2, 3, 3). \tag{6.16}$$

The corresponding permutation reads as

$$\begin{pmatrix} 1 & 2 & 3 & 4 & 5 & 6 \\ 4 & 1 & 5 & 6 & 2 & 3 \end{pmatrix}.$$

(This is so because the first and unique 1 was originally in the fourth position, the first 2 in the first position, the second 2 was in the fifth position, and so on.)

The next observation is crucial. In the arising permutation, there is an ascent if and only if in the ordered list there is an *equality* between two elements. (In our example, the permutation has ascent in the $2, 3, 5$ positions (1-5, 5-6, 2-3), and in (6.16) there is an equality in these positions with the following elements (2-3, 3-4, 5-6). These agree with the given positions in the first line of the permutation.)

In general, if $a_{i_j} = a_{i_{j+1}}$ then $i_j < i_{j+1}$ (ascent). Now we make the monotonically increasing sequence

$$1 \leq a_{i_1} \leq a_{i_2} \leq \cdots \leq a_{i_n} \leq m \tag{6.17}$$

to be strictly increasing. For the above reason, equality appears only at ascents, we have inequality in descent positions. Supposing that we have k ascents in the permutations (there are $\left\langle {n \atop k} \right\rangle$ such permutations), the strictly increasing modification will be

$$1 \leq a_{i_1} < a_{i_2}(+1) < a_{i_3}(+2) < \cdots < a_{i_n}(+k-1) \leq m+k.$$

Here the terms in parenthesis are added only when there is an equality (k cases). How many such strictly increasing sequences are there? We choose

n elements from $m + k$ (and then put them in increasing order, but this is uniquely determined), so the answer is $\binom{m+k}{n}$.

Summing over k, we are done. All the possible functions can be given in such a way, so Worpitzky's identity is proved[9].

6.3.5 A relation between Eulerian numbers and Stirling numbers

There are several relations connecting the Eulerian and Stirling numbers (see also Section 6.5), but the following is the simplest one.

$$m! \left\{ {n \atop m} \right\} = \sum_{k=0}^{m} \left\langle {n \atop k} \right\rangle \binom{k}{n-m}. \tag{6.18}$$

The left-hand side counts the number of ways we can group n elements into m blocks such that the order of the blocks counts. We have to show that the right-hand side counts the same grouping. First note that $\binom{k}{n-m} = \binom{k}{m-n+k}$ and so

$$\sum_{k=0}^{m} \left\langle {n \atop k} \right\rangle \binom{k}{n-m} = \sum_{k=0}^{m} \left\langle {n \atop n-k} \right\rangle \binom{n-k}{m-k}.$$

Let an n-permutation be given with k runs. The runs – whose positions obviously matter in a permutation – give k blocks. If $k = m$, we are ready because there is a one-to-one correspondence between the blocks and runs. Hence, let us suppose that $k < m$. Now we have to split some runs to get new blocks. More concretely, we need $m - k$ new runs to get $m - k$ new blocks and hence m blocks in total. We know that k runs mean $n - k = n - 1 - (k-1)$ ascending positions, and from these we have to choose $m - k$: $\binom{n-k}{m-k}$ possibilities (in a descent position the run terminates, so we cannot get new runs splitting here). Since there are $\left\langle {n \atop n-k} \right\rangle$ n-permutations with $n - k$ ascents, altogether we have $\binom{n-k}{m-k} \left\langle {n \atop n-k} \right\rangle$ different possibilities for a fixed k. Summing over $k = 0, 1, \ldots, m$ we get the assertion.

The inverse of (6.18) can also be proven. This expresses the Eulerian numbers by the Stirling numbers:

$$\left\langle {n \atop k} \right\rangle = \sum_{m=0}^{n} m! \left\{ {n \atop m} \right\} \binom{n-m}{k} (-1)^{n-m-k}.$$

We give a formal proof. Multiply both sides of (6.18) by x^{n-m} and sum over the possible values of m:

$$\sum_{m=0}^{n} m! \left\{ {n \atop m} \right\} x^{n-m} = \sum_{k=0}^{n} \left\langle {n \atop k} \right\rangle \sum_{m=0}^{n} \binom{k}{n-m} x^{n-m} = \sum_{k=0}^{n} \left\langle {n \atop k} \right\rangle (x+1)^{k},$$

---

[9]This proof was presented to the author by Gábor Nyul.

where in the last equality we used the binomial theorem. Substitute $x - 1$ in place of x and apply the binomial theorem once again:

$$\sum_{m=0}^{n} m! \left\{ {n \atop m} \right\} (x-1)^{n-m} = \sum_{m=0}^{n} m! \left\{ {n \atop m} \right\} \sum_{k=0}^{n-m} \binom{n-m}{k} (-1)^{n-m-k} x^k =$$

$$\sum_{k=0}^{n} x^k \left(\sum_{m=0}^{n} m! \left\{ {n \atop m} \right\} \binom{n-m}{k} (-1)^{n-m-k} \right) = \sum_{k=0}^{n} \left\langle {n \atop k} \right\rangle x^k. \tag{6.19}$$

On both sides, we have polynomials of x. These are equal if and only if all the coefficients are equal. The statement is therefore proved.

From this statement we immediately get a relation to the Fubini numbers. Substitute $x = 2$ in (6.19):

$$F_n = \sum_{k=0}^{n} \left\langle {n \atop k} \right\rangle 2^k. \tag{6.20}$$

In the following section, we prove this identity combinatorially.

6.4 The combination lock game

In this section we present a special construction which helps to prove the combinatorial relation (6.20) between the Eulerian numbers and Fubini numbers. The section is based on a paper of D. J. Velleman and G. S. Call [571].

Let us consider a combination lock with n numbered buttons. To open the lock, we have to push the buttons in a *given* order. Of course, if there is no simultaneous button pressing allowed, there are $n!$ possibilities. Therefore, to make the situation more interesting, we permit *simultaneous* button pressings. Then, for example, a possible combination on 8 elements is

$$3 - 7 - 8, \ 4 - 6, \ 1 - 2, \ 5$$

(which means that first we press the third, seventh and eighth buttons in one step simultaneously and then the fourth and sixth buttons, and so on). The order of the different button presses counts, but the order of the simultaneous button presses does not count. Hence, for example

$$3 - 7 - 8, \ 4 - 6, \ 1 - 2, \ 5$$

and

$$3 - 7 - 8, \ 4 - 6, \ 5, \ 1 - 2$$

are different but

$$3 - 7 - 8, \; 4 - 6, \; 1 - 2, \; 5$$

and

$$3 - 8 - 7, \; 6 - 4, \; 1 - 2, \; 5$$

are the same. We suppose that in a given step the simultaneously pressed buttons are listed in increasing order. How many combinations are there? Let us denote this number by F_n for a fixed number n of buttons. It is obvious that $F_1 = 1$. To open the lock we push some buttons first, simultaneously. Let the number of the firstly pushed buttons be k. We can choose these buttons on $\binom{n}{k}$ ways. Then the remaining $n - k$ buttons can be handled as buttons of a new combination lock. The number of the possibilities for this auxiliary lock is F_{n-k}. Summing over k, we get the total number of different lock opening combinations:

$$F_n = \sum_{k=1}^{n} \binom{n}{k} F_{n-k}.$$

This is the same recursion as (6.2). Since the initial values are also the same, the two sequences must be equal (one certainly anticipated this by the notation). F_n is the sequence of the Fubini numbers. (The combination lock approach and the competition approach are equivalent, as one can see easily.)

After this simple discovery about the connection between combination locks and Fubini numbers, we are going to see how to prove (6.20) combinatorially.

A combination lock can be transformed into a permutation by simply writing down the list of the button pushes. For example, from the above 4 step combinations

$$3 - 7 - 8, \; 4 - 6, \; 1 - 2, \; 5$$

we have that the attached permutation is

$$\begin{pmatrix} 1 & 2 & 3 & 4 & 5 & 6 & 7 & 8 \\ 3 & 7 & 8 & 4 & 6 & 1 & 2 & 5 \end{pmatrix}.$$

Such a permutation will contain at most as many runs as many simultaneous button pushes we had[10] (4 in the example). At most, because the simultaneously pushed buttons are in increasing order and they correspond to runs in the permutation. But it can happen that two (or more) consecutive simultaneous button pushes correspond to one common run. This happens in our example: 1-2-5 gives one run; however, they belong to two button pushes (1-2 and 5). Another example

$$5; 3, 4; 1, 2$$

contains three steps and three runs:

$$1, 2; 5; 3, 4$$

---

[10] Remember that we agreed to list the buttons in increasing order in the simultaneous button pushes.

contains three steps but only two runs; while in

$$1, 2; 3, 4; 5$$

the three steps result only in one run.

The runs and steps can be related in the reverse way. Having an n-permutation with k runs, our goal is to find a unique process to map this permutation to a combination lock. We can form new combination locks from a permutation of k runs. This combination lock will contain at least k steps, by the above argument. Therefore, we have to separate the k runs into more parts if the combination contains more steps. We perform this in the following way. We put marks among the elements of the runs which indicate the end of the sequence of simultaneously pushed buttons[11]. Among n elements there are $n - 1$ places to put the mark, but we do not put marks at the end of the k runs. So from $n - 1$ we lose $k - 1$ places, remains $n - 1 - (k - 1) = n - k$ places. In each place, we can choose whether we put a mark or not. This gives 2^{n-k} possibilities.

Finally, remembering that k runs give $n - k$ ascents, we have

$$F_n = \sum_{k=1}^{n} \left\langle {n \atop n - k} \right\rangle 2^{n-k},$$

which (after reversing the order of the summation) is nothing else but (6.20).

A final observation. A k step combination lock can be interpreted as a k-partition on n elements, where the blocks are formed by the simultaneously pushed buttons. There are $\left\{ {n \atop k} \right\}$ k-partitions but – as the order of the steps counts – we have to multiply this with $k!$. Summing over k, we get all the combination locks:

$$F_n = \sum_{k=1}^{n} k! \left\{ {n \atop k} \right\}.$$

This is just the (6.1) definition of the Fubini numbers!

6.5 Relations between the Eulerian and Fubini polynomials

Identity (6.20) is a particular case of the important result (6.19). To put (6.19) into a short form, we introduce the *Eulerian polynomials*:

$$E_n(x) = \sum_{k=0}^{n} \left\langle {n \atop k} \right\rangle x^k.$$

---

[11]For instance, to get $1, 2; 3, 4; 5$ from $\begin{pmatrix} 1 & 2 & 3 & 4 & 5 \\ 1 & 2 & 3 & 4 & 5 \end{pmatrix}$, we put marks between two and three, and four and five.

Note that $E_n(x)$ is a polynomial of degree $n-1$ and not of degree n.
We immediately have a relation for $E_n(x)$ by (6.19):

$$E_n(x) = \sum_{k=0}^{n} k! \begin{Bmatrix} n \\ k \end{Bmatrix} (x-1)^{n-k}. \tag{6.21}$$

This is a theorem of *Frobenius*[12]. Substituting $x = 2$, we get (6.20). On the
other hand, taking the Fubini polynomials into account, (6.21) is equivalent
to

$$(x-1)^n F_n \left(\frac{1}{x-1} \right) = E_n(x). \tag{6.22}$$

The inverse of this relation is

$$F_n(x) = x^n E_n \left(\frac{x+1}{x} \right).$$

The exponential generating function of $E_n(x)$

These transformation formulas directly lead us to the exponential generating function of the Eulerian polynomials:

$$\sum_{n=0}^{\infty} E_n(x) \frac{y^n}{n!} = \frac{x-1}{x - e^{(x-1)y}}.$$

Indeed, knowing the exponential generating function of the Fubini polynomials (see (6.7)) and (6.22) we have that

$$\sum_{n=0}^{\infty} E_n(x) \frac{y^n}{n!} = \sum_{n=0}^{\infty} F_n \left(\frac{1}{x-1} \right) \frac{[(x-1)y]^n}{n!} =$$

$$\frac{1}{1 - \frac{1}{x-1}(e^{(x-1)y} - 1)} = \frac{x-1}{x - e^{(x-1)y}}.$$

The exponential generating function of $E_n(x)$

The following recursion holds true for the Eulerian polynomials:

$$E_n(x) = (1 + (n-1)x)E_{n-1}(x) + (x - x^2)E'_{n-1}(x). \tag{6.23}$$

The proof is left to the reader.

It is not a novelty that recursions sometimes help us find some information
on the root structure of polynomials (see pages 147 and 88). We could proceed

---

[12] Ferdinand Georg Frobenius (1849-1917), German mathematician.

now as before; however, (6.22) makes the situation easier. The root structure of the Fubini polynomials is known to us, so we see directly that

$$0 = E_n(x) = (x-1)^n F_n\left(\frac{1}{x-1}\right) \tag{6.24}$$

is possible only if $x = 1$ or

$$F_n\left(\frac{1}{x-1}\right) = 0.$$

Since the Eulerian polynomials are positive when $x > 0$, $x = 1$ is not a solution, so only the second case remains. The zeros of the Fubini polynomials are real and belong to $]-1,0]$, hence $x \in]-\infty, 0[$. Thus, all the zeros of the Eulerian polynomials are real and negative.

It follows from this property that the Eulerian numbers are strictly log-concave (see Section 4.2). For a fixed n, the Eulerian numbers increase, attain at most two maxima and then decrease. By the symmetry (6.13) there are two maxima *in the middle* of the sequence when n is even (we have a plateau here), and there is one maximum if n is odd (this is called peak).

Moreover, the following inequality holds by the strict log-concavity (see (4.2)):

$$\left\langle{n \atop k}\right\rangle^2 \geq \left\langle{n \atop k-1}\right\rangle\left\langle{n \atop k+1}\right\rangle.$$

Combinatorial proofs of this log-concavity property were presented in [86, 244]. See also [83, p. 17-21].

6.6 An application of the Eulerian polynomials

In this section we show how to apply the Eulerian numbers to evaluate some interesting infinite sums. The bridge between these infinite sums and the Eulerian numbers is the Worpitzky identity (6.15).

The infinite sums the Eulerian polynomials help to evaluate are of the form

$$\sum_{i=0}^{\infty} i^n x^i,$$

where n is a fixed positive integer. The evaluation reads as

$$\sum_{i=0}^{\infty} i^n x^i = \frac{x^n}{(1-x)^{n+1}} E_n\left(\frac{1}{x}\right). \tag{6.25}$$

To prove this identity consider the left-hand side as a generating function (in fact, this is really a generating function belonging to $a_i = i^n$). If we can show that the right-hand side has the same coefficients as a generating function, we will be ready. Earlier we saw (Exercise 23 in Chapter 2) that

$$\frac{1}{(1-x)^{n+1}} = \sum_{j=0}^{\infty} \binom{n+j}{j} x^j,$$

that is,

$$\frac{x^n}{(1-x)^{n+1}} E_n\left(\frac{1}{x}\right) = \left(\sum_{j=0}^{\infty} \binom{n+j}{j} x^j\right) \sum_{k=0}^{n} \left\langle \begin{matrix} n \\ k \end{matrix} \right\rangle x^{n-k}.$$

We are looking for the ith coefficient in this expression. This belongs to those indices j for which $j + n - k = i$ or $j = i + k - n$. The coefficient in question is

$$\sum_{k=0}^{n} \binom{n+i+k-n}{i+k-n} \left\langle \begin{matrix} n \\ k \end{matrix} \right\rangle = \sum_{k=0}^{n} \binom{i+k}{n} \left\langle \begin{matrix} n \\ k \end{matrix} \right\rangle = i^n.$$

The last equality comes from Worpitzky's identity.

6.7   Differential equation of the Eulerian polynomials

The recursion of the Eulerian numbers results in an interesting differential equation for the exponential generating function.

Denote the two variable exponential generating functions of the Eulerian polynomials by f:

$$f(x, y) := \frac{x-1}{x - e^{(x-1)y}}.$$

Then it can be seen that f satisfies the following partial differential equation:

$$(x - x^2)\frac{\partial f}{\partial x} + (yx - 1)\frac{\partial f}{\partial y} + f = 0. \tag{6.26}$$

Let us analyze the members from left to right. The first term is

$$x\frac{\partial f}{\partial x} = x\sum_{n=0}^{\infty}\left(\sum_{k=0}^{n} \left\langle \begin{matrix} n \\ k \end{matrix} \right\rangle x^k\right)' \frac{y^n}{n!} = \sum_{n=0}^{\infty}\left(\sum_{k=0}^{n} k\left\langle \begin{matrix} n \\ k \end{matrix} \right\rangle x^k\right) \frac{y^n}{n!}.$$

The second term is

$$-x^2\frac{\partial f}{\partial x} = -\sum_{n=0}^{\infty}\left(\sum_{k=0}^{n} k\left\langle \begin{matrix} n \\ k \end{matrix} \right\rangle x^{k+1}\right) \frac{y^n}{n!},$$

the third

$$yx\frac{\partial f}{\partial y} = \sum_{n=0}^{\infty} \left(\sum_{k=0}^{n} \left\langle {n \atop k} \right\rangle x^{k+1} \right) n\frac{y^n}{n!},$$

while the fourth has the form

$$-\frac{\partial f}{\partial y} = -\sum_{n=0}^{\infty} \left(\sum_{k=0}^{n} \left\langle {n \atop k} \right\rangle x^{k} \right) \frac{y^{n-1}}{(n-1)!}.$$

Bring this last term to the right-hand side of the equation, change the indices properly, and finally take everything under a common summation on n:

$$\sum_{n=0}^{\infty} \left[\left(\sum_{k=0}^{n} k \left\langle {n \atop k} \right\rangle x^{k} \right) - \left(\sum_{k=1}^{n+1} (k-1) \left\langle {n \atop k-1} \right\rangle x^{k} \right) + \right.$$

$$\left. \left(\sum_{k=1}^{n+1} n \left\langle {n \atop k-1} \right\rangle x^{k} \right) + \left(\sum_{k=0}^{n} \left\langle {n \atop k} \right\rangle x^{k} \right) \right] \frac{y^n}{n!} = \sum_{n=0}^{\infty} \left(\sum_{k=0}^{n} \left\langle {n+1 \atop k} \right\rangle x^{k} \right) \frac{y^n}{n!}.$$

Comparing the coefficients of y and then of x we get that

$$k\left\langle {n \atop k} \right\rangle - (k-1)\left\langle {n \atop k-1} \right\rangle + n\left\langle {n \atop k-1} \right\rangle + \left\langle {n \atop k} \right\rangle = \left\langle {n+1 \atop k} \right\rangle.$$

This, after collecting the corresponding terms, is nothing else but the (6.14) recursion.

6.7.1 An application of the Fubini polynomials

There is a surprising application of the Fubini polynomials and of the relation (6.25). We are going to prove that the derivatives of the cotangent function are the Fubini polynomials of the cotangent itself:

$$\cot^{(n)}(z) = (2i)^n (\cot(z) - i) F_n \left(\frac{i\cot(z) - 1}{2} \right).$$

Here $i = \sqrt{-1}$. To prove this interesting fact, first note that

$$\cot(z) = i\left(1 + \frac{2}{e^{2iz} - 1} \right). \tag{6.27}$$

Upon the substitution $e^{iz} = y$ we have that the nth derivative of the cotangent equals to

$$\cot^{(n)}(z) = i^{n+1} \left(y\frac{d}{dy} \right)^n \left(1 + \frac{2}{y^2 - 1} \right) = i^{n+1} \left(y\frac{d}{dy} \right)^n \frac{2}{y^2 - 1}.$$

Here $y\frac{d}{dy}$ is an operation, it derives what appears on its right and then multiplies by y. Its nth power means that we perform this operation n times, one after another. In the following step we expand $\frac{2}{y^2-1}$ into a geometric series and then perform the $\left(y\frac{d}{dy}\right)^n$ operation. We arrive at an attractive identity:

$$\cot^{(n)}(z) = -(2i)^{n+1}\sum_{k=1}^{\infty}k^n y^{2k},$$

where, as before, $y = e^{iz}$. The right-hand side is first expressed by the Eulerian polynomials via (6.25):

$$-(2i)^{n+1}\sum_{k=1}^{\infty}k^n y^{2k} = -(2i)^{n+1}\frac{y^{2n}}{(1-y^2)^{n+1}}E_n\left(\frac{1}{y^2}\right),$$

then we convert this into the Fubini polynomials by (6.24):

$$-(2i)^{n+1}\frac{y^{2n}}{(1-y^2)^{n+1}}E_n\left(\frac{1}{y^2}\right) = -(2i)^{n+1}\frac{y^{2n}}{(1-y^2)^{n+1}}\left(\frac{1-y^2}{y^2}\right)^n$$

$$F_n\left(\frac{1}{1/y^2-1}\right) = (2i)^{n+1}\frac{1}{y^2-1}F_n\left(\frac{y^2}{1-y^2}\right).$$

The correspondence $e^{iz} = y$ and (6.27) gives the statement.

By similar methods, it can be shown that

$$\tan^{(n)}(z) = (-2i)^n(\tan(z)-i)F_n\left(\frac{i\tan(z)-1}{2}\right).$$

By changing $z \to iz$ we get the hyperbolic function counterparts:

$$\coth^{(n)}(z) = 2^n(\coth(z)-1)F_n\left(-\frac{\coth(z)+1}{2}\right),$$

$$\tanh^{(n)}(z) = 2^n(\tanh(z)-1)F_n\left(-\frac{\tanh(z)+1}{2}\right).$$

These results appeared in [2], and were further extended to tan, sec, csc functions and their hyperbolic counterparts in [182].

Exercises

1. Calculate the first five Fubini numbers.

2. How many runs, ascents and descents are there in the permutation

$$\begin{pmatrix} 1 & 2 & 3 & 4 & 5 & 6 & 7 & 8 & 9 & 10 \\ 3 & 8 & 6 & 7 & 2 & 5 & 1 & 4 & 10 & 9 \end{pmatrix}?$$

3. Prove that

$$n! = \sum_{k=0}^{n} \left[{n \atop k} \right] F_k.$$

4. Calculate the first five lines of the Eulerian triangle.

5. Let $A(n, k)$ be the number of n-permutations with k *runs*. Prove that

$$A(n, k) = (k + 1)A(n - 1, k + 1) + (n - k)A(n - 1, k).$$

These numbers are also called Eulerian numbers, as we noted in the footnote on p. 149. For more on this kind of definition see the book of M. Bóna [83]. There, on p. 8., one can find that

$$A(n, k) = \sum_{i=0}^{k} (-1)^i \binom{n + 1}{i} (k - i)^n$$

which is a simple closed form for the $A(n, k)$ numbers. Rewriting this formula using ascents in place of runs and re-indexing we get the closed form for the $\left\langle {n \atop k} \right\rangle$ Eulerian numbers, too:

$$\left\langle {n \atop k} \right\rangle = \sum_{i=0}^{k} (-1)^i \binom{n + 1}{i} (k - i + 1)^n.$$

6. Show that

$$\left\langle {n \atop 1} \right\rangle = \left\langle {n \atop n - 2} \right\rangle = 2^n - n - 1,$$

$$\left\langle {n \atop 2} \right\rangle = \left\langle {n \atop n - 3} \right\rangle = 3^n - (n + 1)2^n + \frac{1}{2}n(n + 1).$$

(Hint: use the symmetry relation (6.13) together with the previous exercise.)

7. Express $E_4(x)$ by the help of $F_4(x)$ using (6.22).

8. Calculate the infinite sum

$$\sum_{i=0}^{\infty} \frac{i^4}{2^i}.$$

9. Demonstrate that the following particular values of the derivative of $E_n(x)$ are as follows:

$$E_n'(1) = (n-1)!\binom{n}{2},$$

$$E_n''(1) = 2(n-2)!\left(\binom{n}{3} + 3\binom{n}{4}\right).$$

10. Prove that for the Fubini polynomials

$$F_n'(1) = \frac{F_{n+1} - F_n}{2}.$$

11. Show that the product of the zeros of the nth Eulerian polynomial is $(-1)^{n-1}$. Prove also that their sum is $n + 1 - 2^n$. Note that the first result yields that $E_n(x)$ has zeros between -1 and 0, and, also, some zeros are smaller than -1.

12. Prove that

$$\sum_{k=0}^{n} \begin{bmatrix} n \\ k \end{bmatrix} F_k = 2^{n-1} n!.$$

Use combinatorial arguments or apply the Riordan method. See p. 286-287 in [533].

13. Show that the Fubini numbers form a log-convex sequence. (This is a consequence of (6.1) plus the fact that the Stirling transform preserves log-convexity, together with the log-convexity of the factorial sequence. See the Outlook of Chapter 4, and directly Example 2.6 in [380]. The log-convexity was studied in [616], too.)

14. We have seen in (6.9) that the sequence $\left(k!\left\{{n \atop k}\right\}\right)_{k=0}^{n}$ is log-concave. Thus, it has a peak or a plateau. Let us denote the index of the peak (or the leftmost of the two edges of the plateau) by M_n. Show that

$$M_n \sim \frac{n}{2\log 2}.$$

(The details can be found in [424].)

15. Prove that the $n \times n$ Hankel determinants of the Fubini polynomials is

$$x^{\binom{n}{2}}(1+x)^{\binom{n}{2}} \prod_{k=1}^{n-1} k!^2.$$

16. Prove that the $n \times n$-sized Hankel determinants of the polynomial sequence $\frac{x^k}{(1-x)^{k+1}} E_k\left(\frac{1}{x}\right)$ is

$$(-1)^n \frac{x^{\binom{n}{2}}}{(x-1)^{n^2}} \prod_{k=1}^{n-1} k!^2 \quad (|x| < 1,\, n \geq 1).$$

(Hint: use (6.25) together with the Meixner polynomials mentioned in Subsection 6.1.5.)

17. Let $n \geq 3$ be an odd number. Show that for the maximal Eulerian numbers in two consecutive rows the following ratio is valid:

$$\frac{\max_k \left\langle {n \atop k} \right\rangle}{\max_k \left\langle {n-1 \atop k} \right\rangle} = n + 1.$$

(Hint: by symmetry and log-concavity of the Eulerian numbers we infer that if n is odd, the maximizing index for $\left\langle {n \atop k} \right\rangle$ is $\overline{n} := \frac{n-1}{2}$, while for even n it is $\frac{n}{2} - 1$ *and* $\frac{n}{2}$. Then the exercise is equivalent to

$$\left\langle {n \atop \overline{n}} \right\rangle = (n+1)\left\langle {n-1 \atop \overline{n}-1} \right\rangle = (n+1)\left\langle {n-1 \atop \overline{n}} \right\rangle.$$

Applying the basic recurrence of the Eulerian numbers for $\left\langle {n \atop \overline{n}} \right\rangle$ we can see now that these equations are indeed valid.)

Outlook

1. The Eulerian numbers and first kind Stirling numbers appear in a very interesting context. J. M. Holte [285] studied digit carries during adding some *random* numbers in some base b. The corresponding Markov chain and its transition matrix can be related to $\left\langle {n \atop k} \right\rangle$ and to $\left[{n \atop k} \right]$.

2. The chains that we have mentioned in Subsection 6.1.3 are just the tip of an iceberg. The underlying theory is the theory of *partially ordered sets* and *lattices*. The chains we mentioned in our text belong to the subset lattice. More on this interesting theory can be found in [10] (see also the Outlook of Chapter 8).

3. The series
$$\sum_{i=0}^{\infty} i^{\overline{n}} \frac{x^i}{i!}, \quad \text{and} \quad \sum_{i=0}^{\infty} \frac{i^{\overline{n}}}{i^x}$$
were studied in [460].

4. Ascents, descents, runs can be defined on infinite permutations (bijections of the positive integers), too. Let L_k denote the average length of the kth run in a uniformly taken infinite random permutation. Here "uniformly taken" means that each of the $n!$ possible relative orderings of the first n elements are equally likely. Then, it can be proven that, for all k,
$$L_k = \sum_{n=1}^{\infty} \left\langle {n \atop k-1} \right\rangle \frac{1}{n!}.$$
In particular,
$$L_1 = e - 1 \approx 1.71828,$$
$$L_2 = e^2 - 2e \approx 1.95249,$$
$$L_3 = e^3 - 3e^2 + \frac{3}{2}e \approx 1.99579.$$

So the average length of the first run is slightly smaller than that of the second, and the second one is slightly smaller than the third one. This does not go this way, however. The L_k sequence is not monotone. An exact formula for L_k was given by B. J. Gassner [245], see also [335, Section 5.1.3]. Among others, it is known that $\lim_{k\to\infty} L_k = 2$. These numbers were applied in the determination of average rib lengths of postorder trees [595].

5. An attractive symmetrical identity is known for the Eulerian numbers [152]:

$$\sum_{k=a-1}^{a+b} \binom{a+b}{k} \left\langle \begin{matrix} k \\ a-1 \end{matrix} \right\rangle = \sum_{k=b-1}^{a+b} \binom{a+b}{k} \left\langle \begin{matrix} k \\ b-1 \end{matrix} \right\rangle,$$

which is valid for all positive integers a and b. Here, $\left\langle \begin{smallmatrix} 0 \\ 0 \end{smallmatrix} \right\rangle = 0$, instead of the usual $\left\langle \begin{smallmatrix} 0 \\ 0 \end{smallmatrix} \right\rangle = 1$.

Chapter 7

Asymptotics and inequalities

For very large parameters, it is usually hard to determine the concrete values of combinatorial sequences[1]. If we are ready to some trade-off – not requiring the exact value but a feasible estimation – then the task can be solved.

Also, sometimes we would only like to know "how rapidly" a given sequence grows. If we take a look on the values of superfactorials and the Bell numbers in the tables at the end of the book, we see that in some sense the superfactorial sequence grows faster than the sequence of the Bell numbers. But *how rapidly* do these sequences grow? Usually we consider the answer given if we can approximate our sequence in question with a "simpler" sequence for which the growth is known or obvious. Let us see what this exactly means.

It seems to be acceptable to consider two sequences "equally growing" if the ratio of the members tends to some finite limit as the index tends to infinity[2]. By appropriate rescaling we can suppose that this limit, if it exists, is one. Thus, we say that two sequences a_n and b_n are *asymptotically equal* if

$$\lim_{n \to \infty} \frac{a_n}{b_n} \to 1.$$

Asymptotic equality is shortly denoted by the symbol

$$a_n \sim b_n.$$

For example, if $a_n = (n + 2)^2$, and $b_n = (n + 7)^2$, then $a_n \sim b_n$.

In the second part of the chapter, we provide some inequalities which are often useful in many contexts when some estimation of a given sequence is enough for our purpose. Such inequalities may provide more tractable expressions to work with.

---

[1]What method would the reader use to calculate, say, $\left[\begin{smallmatrix} 10^{12} \\ 43527 \end{smallmatrix} \right]$? Even computer could not calculate this number in a manageable timeframe.

[2]In this section, some notions will appear that we have not considered so far, like convergence or some notions from analysis and set theory. We suppose that the reader is familiar with these.

7.1   The Bonferroni inequality

One way to study how rapidly the second kind Stirling numbers grow is through the *Bonferroni*[3] *inequality* which is a set theoretical relation interesting in itself.

Let Ω be a non-empty finite set, and let $A_1, \ldots, A_n$ some non-empty subsets of Ω. Moreover, let

$$m = \max\left\{ |T| : T \subset \{1, 2, \ldots, n\}, \bigcap_{i \in T} A_i \neq \emptyset \right\}, \tag{7.1}$$

the number of the maximal index set such that the intersection of A_i's is still non-empty ($i \in T$). Then $m \geq 1$, and when $l = 1, 2, \ldots, m$ then

$$\sum_{j=1}^{l} (-1)^{j-1} \sum_{\substack{T \subset \{1,2,\ldots,n\} \\ |T| = j}} \left| \bigcap_{i \in T} A_i \right| \begin{cases} = \left| \bigcup_{i=1}^{n} A_i \right|, & \text{if } l = m, \\[2mm] > \left| \bigcup_{i=1}^{n} A_i \right|, & \text{if } l < m, \text{ and } l \text{ odd}, \\[2mm] < \left| \bigcup_{i=1}^{n} A_i \right|, & \text{if } l < m, \text{ and } l \text{ even}. \end{cases}$$

A modern presentation of this result was published by Horst Wegner German mathematician, see [592] (in one step his argument is simplified in the below text). In the particular case, when $l = m = n$, (in this case there is at least one common element in all the A_i sets), the inequality is actually an equality:

$$\sum_{j=1}^{n} (-1)^{j-1} \sum_{\substack{T \subset \{1,2,\ldots,n\} \\ |T| = j}} \left| \bigcap_{i \in T} A_i \right| = \left| \bigcup_{i=1}^{n} A_i \right|.$$

This is the famous *inclusion-exclusion principle*. In particular, for $n = 2$:

$$|A_1| + |A_2| - |A_1 \cap A_2| = |A_1 \cup A_2|,$$

while for $n = 3$ it is

$$|A_1| + |A_2| + |A_3| - |A_1 \cap A_2| - |A_1 \cap A_3| - |A_2 \cap A_3|$$

$$+ |A_1 \cap A_2 \cap A_3| = |A_1 \cup A_2 \cup A_3|.$$

The proof of the Bonferroni-inequality

To prove Bonferroni's inequality, we need to introduce the *characteristic*

---

[3]Carlo Emilio Bonferroni (1892-1960), Italian probability theorist.

function. The function χ_A takes the value 1 on elements belonging to A, while any other elements not belonging to A are mapped to 0:

$$\chi_a(\omega) = \begin{cases} 1, & \text{if } \omega \in A; \\ 0, & \text{if } \omega \notin A. \end{cases}$$

To prove Bonferroni's inequality in the spirit of Wegner we still need an ingredient in the form of the following lemma.

Let Ω be a set as above, and let $A_1, \ldots, A_n$ be subsets of Ω that their intersection $\bigcap_{i=1}^n A_i$ is non-empty. Moreover, for all $\omega \in \bigcup_{i=1}^n A_i$ let us define $m(\omega) = |\{i : \omega \in A_i\}|$. Then $m(\omega) \geq 1$, and

$$\sum_{j=1}^l (-1)^{j-1} \sum_{\substack{T \subset \{1,2,\ldots,n\} \\ |T|=j}} \chi_{\bigcap_{i \in T} A_i}(\omega) = 1 - (-1)^l \binom{m(\omega) - 1}{l}$$

for all $l = 1, 2, \ldots, n$.

The proof of this statement is as follows. Since $\omega \in \bigcup_{i=1}^n A_i$, ω is contained in some A_i, whence $m(\omega) \geq 1$ is indeed satisfied. If $j = 1, 2, \ldots, n$, then

$$\sum_{\substack{T \subset \{1,2,\ldots,n\} \\ |T|=j}} \chi_{\bigcap_{i \in T} A_i}(\omega) = \sum_{\substack{T \subset \{i : \omega \in A_i\} \\ |T|=j}} \chi_{\bigcap_{i \in T} A_i}(\omega) = \binom{m(\omega)}{j}.$$

Hence, the left-hand side of our lemma can be written as

$$\sum_{j=1}^l (-1)^{j-1} \sum_{\substack{T \subset \{1,2,\ldots,n\} \\ |T|=j}} \chi_{\bigcap_{i \in T} A_i}(\omega) = \sum_{j=1}^l (-1)^{j-1} \binom{m(\omega)}{j} =$$

$$(-1)\sum_{j=1}^l (-1)^j \binom{m(\omega)}{j} = (-1)\left[\sum_{j=0}^l (-1)^j \binom{m(\omega)}{j} - 1\right] =$$

$$1 - \sum_{j=0}^l (-1)^j \binom{m(\omega)}{j} = 1 - (-1)^l \binom{m(\omega) - 1}{l}.$$

In the last step – that was done by Wegner slightly differently – we used the fact that

$$\sum_{j=0}^l (-1)^j \binom{m}{j} = (-1)^l \binom{m - 1}{l}$$

for all positive integer m.

Having the lemma proven, we can proceed to prove the Bonferroni-inequality. For the left-hand side:

$$\sum_{j=1}^l (-1)^{j-1} \sum_{\substack{T \subset \{1,2,\ldots,n\} \\ |T|=j}} \left|\bigcap_{i \in T} A_i\right| = \sum_{j=1}^l (-1)^{j-1} \sum_{\substack{T \subset \{1,2,\ldots,n\} \\ |T|=j}} \sum_{\omega \in \bigcup_{i=1}^n A_i} \chi_{\bigcap_{i \in T} A_i}(\omega)$$

$$= \sum_{\omega \in \bigcup_{i=1}^{n} A_i} \left[1 - (-1)^l \binom{m(\omega) - 1}{l} \right] = \left| \bigcup_{i=1}^{n} A_i \right| - (-1)^l \sum_{\omega \in \bigcup_{i=1}^{n} A_i} \binom{m(\omega) - 1}{l}.$$

Now let us recall the meaning of m. We see that $m = \max_{\omega \in \bigcup_{i=1}^{n} A_i} m(\omega)$. Hence, if $l = m$, then

$$\binom{m(\omega) - 1}{l} = 0,$$

since the upper parameter is smaller than the lower. If $l < m$, then

$$\binom{m(\omega) - 1}{l} > 0$$

for at least one ω. This positive number is added (when l is odd) or subtracted (when l is even) from $|\bigcup_{i=1}^{n} A_i|$. This yields Bonferroni's inequality.

Later in Subsection 10.2.2, we further use these results.

7.2   The asymptotics of the second kind Stirling numbers

7.2.1   First approach

An under- and overestimation for $\left\{ {n \atop k} \right\}$ via the Bonferroni inequality

We have proven already (see (2.20)) that

$$\left\{ {n \atop k} \right\} = \frac{1}{k!} \sum_{j=0}^{k} \binom{k}{j} j^n (-1)^{k-j} = \frac{1}{k!} \sum_{j=0}^{k-1} (-1)^j \binom{k}{j} (k - j)^n \qquad (7.2)$$

A bit more is true. These sums enclose the second kind Stirling numbers in the following sense:

$$\frac{1}{k!} \sum_{j=0}^{l} (-1)^j \binom{k}{j} (k - j)^n \begin{cases} = \left\{ {n \atop k} \right\} & \text{if } l = k - 1, \\[2mm] > \left\{ {n \atop k} \right\} & \text{if } 0 \le l < k - 1, \text{ and } l \text{ is even,} \\[2mm] < \left\{ {n \atop k} \right\} & \text{if } 1 \le l < k - 1, \text{ and } l \text{ is odd.} \end{cases}$$

The proof is as follows. Multiplying with $k!$, $k! \left\{ {n \atop k} \right\}$ is the number of surjective functions mapping an n element set into a k element set (remember Section 1.6). These functions can be counted in the following way. Let the sets X, Y

be of cardinality n and k, respectively. Moreover, let A_y be the set of functions that do *not* have y in their co-domain:

$$A_y = \{f : X \to Y \mid y \notin f(X)\} \quad (y \in Y).$$

By this definition each non-surjective function belongs to some A_y, so

$$k^n - k!\left\{\begin{matrix} n \\ k \end{matrix}\right\} = \left| \bigcup_{y \in Y} A_y \right|.$$

Let $g : X \to Y$ be a function for which $g(x) = y_0$ for all x. By this definition $g \in \bigcap_{y \in Y \setminus \{y_0\}} A_y$. On the other hand, there is no function that it takes no value y^4, hence $\bigcap_{y \in Y} A_y = \emptyset$. This means that we can choose $|Y| - 1$ sets from the A_y sets such that their intersection is non-empty (remember g). That is,

$$m = \max \left\{ |T| : T \subset Y, \bigcap_{y \in Y} A_y \neq \emptyset \right\} = k - 1.$$

By the definition of A_y the $\bigcap_{y \in T} A_y$ intersection contains functions with co-domain $Y \setminus T$, i.e.,

$$\left| \bigcap_{y \in T} A_y \right| = (k - |T|)^n.$$

For all $j = 1, 2, \ldots, k - 1$

$$\sum_{\substack{T \subset Y \\ |T| = j}} \left| \bigcap_{y \in T} A_y \right| = \sum_{\substack{T \subset Y \\ |T| = j}} (k - |T|)^n = \binom{k}{j}(k - j)^n.$$

This, substituted into the Bonferroni-inequality and rearranged, yields the above given statement.

The asymptotics of $\left\{\begin{smallmatrix} n \\ k \end{smallmatrix}\right\}$

The sums presented in the previous statement for certain l are bigger than $\left\{\begin{smallmatrix} n \\ k \end{smallmatrix}\right\}$, for certain l they are smaller. Let us see what these inequalities say in the simplest cases when $l = 0, 1$.

$$\frac{1}{k!}\left(k^n - \binom{k}{1}(k - 1)^n \right) < \left\{\begin{matrix} n \\ k \end{matrix}\right\} < \frac{1}{k!} k^n.$$

---

[4]In fact, the functions $f \subset A \times \emptyset$ would be counterexamples to this statement, but we supposed that $y \neq \emptyset$.

After simplification:

$$\frac{k^n}{k!} - \frac{(k-1)^n}{(k-1)!} < \left\{ {n \atop k} \right\} < \frac{k^n}{k!}. \tag{7.3}$$

(If n is not much larger than k, then the left-hand side can be negative, so the lower estimation is more informative when n is much larger than k.) From here the asymptotic behavior of $\left\{ {n \atop k} \right\}$ can already be seen. If we divide by k^n and multiply by $k!$ then we have

$$1 - k \left(\frac{k-1}{k} \right)^n < \frac{k!}{k^n} \left\{ {n \atop k} \right\} < 1.$$

If k is fixed and n tends to infinity, then the left-hand side tends to one. By the comparison test, we get the asymptotic description of the second kind Stirling numbers when k is fixed and n tends to infinity:

$$\left\{ {n \atop k} \right\} \sim \frac{k^n}{k!}. \tag{7.4}$$

Note that this result readily follows from (7.2) after dividing by k^n.

7.2.2   Second approach

We now show another approach that is simpler but, at the same time, it gives a weaker result. Begin with the (2.43) horizontal generating function:

$$\sum_{m=0}^{n} \left\{ {n \atop m} \right\} x(x-1)(x-2) \cdots (x-m+1) = x^n.$$

Substitute $x = k$ and rewrite the falling factorial so that

$$\sum_{m=0}^{k} \left\{ {n \atop m} \right\} \frac{k!}{(k-m)!} = k^n.$$

From here, by separating the last term, and dividing by $k!$ we get

$$\sum_{m=0}^{k-1} \left\{ {n \atop m} \right\} \frac{1}{(k-m)!} + \left\{ {n \atop k} \right\} = \frac{k^n}{k!}.$$

The right-hand side of (7.3) follows since the sum is positive on the left-hand side is positive. On the other hand, by using this last relation again:

$$\left\{ {n \atop k} \right\} = \frac{k^n}{k!} - \sum_{m=0}^{k-1} \left\{ {n \atop m} \right\} \frac{1}{(k-m)!} > \frac{k^n}{k!} - \frac{1}{(k-1)!} \sum_{m=0}^{k-1} \left\{ {n \atop m} \right\} \frac{(k-1)!}{(k-1-m)!} =$$

$$\frac{k^n}{k!} - \frac{1}{(k-1)!} \sum_{m=0}^{k-1} \left\{ {n \atop m} \right\} (k-1)^m = \frac{k^n}{k!} - \frac{(k-1)^n}{(k-1)!}.$$

This is nothing else but the left-hand side of (7.3).

That this is indeed weaker than the Bonferroni inequality comes from the fact that it is equivalent to (7.3), and this is the $l = 0$ and $l = 1$ particular cases of Bonferroni. Higher values of l give better approximations.

More about the approximation of $\left\{ {n \atop k} \right\}$ can be found in the Outlook of this chapter.

7.3 The asymptotics of the maximizing index of the Stirling numbers of the second kind

Once we know that the $\left\{ {n \atop k} \right\}$ numbers behave like $k^n/k!$ if n is "big", we can run k from 0 to n. It looks that the sequence $k^n/k!$ grows, reaches a maximum then decreases – like the log-concave sequence $\left(\left\{ {n \atop k} \right\} \right)_{k=0}^{n}$ itself! Here is the graph[5] for $n = 10$:

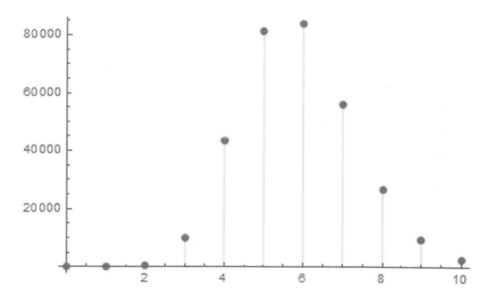

Therefore, if we could find the maximum of the function $x^n/x!$, we could find an approximation for the maximizing index K_n for the Stirling numbers.

---

[5]The graph was drawn in *Mathematica 10*.

To reach this aim, we need some new knowledge from *special function theory*. This part of mathematics studies the properties of functions "useful in some sense"[6] A function is considered to be special if it is relatively widely used and applied. Elementary functions (sin, arctan, log) are, of course, special. The Bessel polynomials we will encounter in Chapter 9 are also special as well as the Lambert W function (see pp. 110–111). We need two other special functions here, the *Euler Gamma function* and the *Digamma function*.

7.3.1   The Euler Gamma function and the Digamma function

The Euler Gamma function

The Euler Gamma function generalizes the $n!$ function to *real* numbers (and even to complex numbers if one wants).

Definition 7.3.1. *The* Euler Gamma function *is defined by the improper integral*

$$\Gamma(x) = \int_0^\infty t^{x-1}e^{-t}dt,$$

where $x > 0$.

It can be seen that $\Gamma(1) = 1 = 0!$, and by partial integration it follows that $\Gamma(2) = 1 = 1!$, $\Gamma(3) = 2 = 2!$, and, in general,

$$\Gamma(n) = (n-1)!.$$

Partial integration says actually much more, it comes easily that Γ satisfies the functional equation

$$\Gamma(x+1) = x\Gamma(x), \tag{7.5}$$

Although the above integral is divergent when $x \le 0$, but the functional equation (7.5) still enables us to extend Γ to negative (non-integer) values. Indeed, for example,

$$-\frac{1}{2}\Gamma\left(-\frac{1}{2}\right) = \Gamma\left(-\frac{1}{2}+1\right),$$

whence

$$\Gamma\left(-\frac{1}{2}\right) = -2\Gamma\left(\frac{1}{2}\right).$$

On the other hand, the integral definition of the Γ function is valid and makes sense not only for integers but for real or even complex numbers when

---

[6]Some mathematicians indeed say that special function theory should be called "useful function theory."

the real part is positive. Therefore, the Γ function (and thus the factorial) is extensible to the whole complex plane (except the non-positive integers)[7].

The Digamma function

Definition 7.3.2. *The* Digamma function, $\psi(x)$, *is the logarithmic derivative of the Euler Gamma function:*

$$\psi(x) = (\log \Gamma(x))' = \frac{\Gamma'(x)}{\Gamma(x)}.$$

A straight consequence of (7.5) is the functional equation for the Digamma function:

$$\psi(x+1) = \frac{1}{x} + \psi(x).$$

We will not prove it, but it is interesting to know that $\psi(1)$ is a well-known constant:

$$\psi(1) = -\gamma,$$

where γ is the so-called *Euler-Mascheroni*[8] *constant*, or simply Euler constant. Its numerical value is

$$\gamma = 0,5772156649\ldots.$$

There is a whole book dedicated to the constant γ [281]. See also Section 1.5 in the excellent book [230].

What we will particularly need from the theory of the Digamma function is its asymptotics for large positive real x:

$$\psi(x) \sim \log x - \frac{1}{2x} + \cdots. \tag{7.6}$$

Here we use the $\sim$ symbol in an extended sense: the ratio of the two sides tends to one as x tends to ∞.

7.3.2 The asymptotics of K_n

We are now ready to derive the asymptotics

$$K_n \sim \frac{n}{\log n}$$

---

[7]There are different functions that are extensions of the factorial, in the sense that they satisfy the functional equation $f(x+1) = xf(x)$ (together with the necessary normalization $f(1) = 1$), but Γ is considered to be the most useful one. What is more important, it is known that if, in addition, f is log-convex, then there is *only one* such f and it is *the* Gamma function. This characterization is the content of the famous *Bohr-Mollerup theorem*. See [35] for a good treatise and a classical on the Gamma function, and [89] for a modern treatment.

[8]Lorenzo Mascheroni (1750-1800), Italian mathematician.

for the maximizing index of $\left\{{n \atop k}\right\}$. Actually, we are going to prove the slightly better approximation

$$K_n \sim \frac{n}{\log n - \log\log n}.$$

Knowing that for large n and fixed k

$$\left\{{n \atop k}\right\} \sim \frac{k^n}{k!},$$

we suppose that K_n approximately equals to the maximum of the function[9] $\frac{x^n}{x!}$. Rewrite this function by using the Γ function:

$$\frac{x^n}{x!} = \frac{x^n}{\Gamma(x+1)} = \frac{x^{n-1}}{\Gamma(x)}.$$

Taking the derivative and applying the definition of the Digamma function, we can see after some algebra that the maximum must satisfy the simple equation

$$x\psi(x) = n - 1. \tag{7.7}$$

Substitute now the (7.6) approximation of the Digamma function. It follows that the maximum must satisfy the *approximative* equation

$$n - \frac{1}{2} = x\log x. \tag{7.8}$$

We know better the log function than the ψ but it still does not lead to a solvable equation like the quadratics, for example. But not everything is lost. Take the logarithm of both sides:

$$\log\left(n - \frac{1}{2}\right) = \log x + \log\log x.$$

Substituting $\log x = t$ (so that $x = e^t$), we have the equivalent equation

$$\log\left(n - \frac{1}{2}\right) = t + \log t. \tag{7.9}$$

Now recall the definition of the Lambert W function on pp. 110–111. Taking the logarithm of that equation, we have that the solution of (7.9) is simply $W\left(n - \frac{1}{2}\right)$. That is,

$$x = \exp t = \exp^{W\left(n - \frac{1}{2}\right)}.$$

Now, again by (4.6),

$$W(a)e^{W(a)} = a$$

---

[9]It seems to be suspicious to use the only asymptotically valid $\left\{{n \atop k}\right\} \sim \frac{k^n}{k!}$ relation. Fortunately, $\frac{k^n}{k!}$ is close to $\left\{{n \atop k}\right\}$ for moderately large n, too.

for any a, thus $e^{W(a)} = \frac{a}{W(a)}$ and the exponential function can be eliminated:

$$x = \exp^{W\left(n - \frac{1}{2}\right)} = \frac{n - \frac{1}{2}}{W\left(n - \frac{1}{2}\right)}.$$

This is the approximate maximum of the function $x^n/x!$ what we were looking for. Hence,

$$K_n \sim \frac{n - \frac{1}{2}}{W\left(n - \frac{1}{2}\right)}.$$

If we do not like the presence of the Lambert function, we can approximate it by a better known function, exactly as we did with the Digamma passing from (7.7) to (7.8). However, to be able to do this, we would need how rapidly $W(n)$ grows as $n \to \infty$. Recalling again the defining equation $xe^x = n$ of $W(n)$, we see that if n is large, x must be large as well. If, on the other hand, x is large, then e^x is exponentially large and x can be neglected in xe^x. Hence, in the first approximation, $e^x = n$, and this equation is solvable. It comes that

$$W(n) \sim \log n$$

for large n. This is already enough for us to see that

$$K_n \sim \frac{n - \frac{1}{2}}{W\left(n - \frac{1}{2}\right)} \sim \frac{n - \frac{1}{2}}{\log\left(n - \frac{1}{2}\right)}.$$

If n is large, the term $-1/2$ can be neglected and we get

$$K_n \sim \frac{n}{\log n}.$$

This is what we wanted to prove.

If we would use a second approximation in the defining equation of W, we could get that

$$W(n) \sim \log n - \log \log n.$$

This would lead to

$$K_n \sim \frac{n}{\log n - \log \log n}.$$

These ideas were published in [175]. The study of the higher order approximations of the Lambert function was presented in the paper [176].

See also Section 4.3 for further results on K_n.

7.4 The asymptotics of the first kind Stirling numbers and Bell numbers

The estimation of the first kind Stirling numbers and the Bell numbers cannot be given as easily as we did for the second kind Stirling numbers.

Therefore, we restrict ourselves to merely present the results without giving proofs.

$$\frac{1}{(n-1)!}\begin{bmatrix} n \\ k \end{bmatrix} = \gamma_1 \frac{\log^{k-1} n}{(k-1)!} + \gamma_2 \frac{\log^{k-2} n}{(k-2)!} + \cdots + \gamma_k + O\left(\frac{\log^{k-2} n}{n}\right).$$

Here $n \to \infty$ and k is fixed, $O\left(\frac{\log^{k-2} n}{n}\right)$ encodes[10] an entity which is bounded divided by $\frac{\log^{k-2} n}{n}$, and the constants γ_n are the Taylor series coefficients of the *reciprocal Gamma function*:

$$\frac{1}{\Gamma(x)} = \sum_{i=n}^{\infty} \gamma_n x^n.$$

Thus, in particular, $\gamma_1 = 1$, $\gamma_2 = \gamma$ is the Euler-Mascheroni constant we have just met with as the special value of the Digamma function[11]. A proof of this result and some extensions to the polynomial

$$\frac{1}{n!} \sum_{k=0}^{n} \begin{bmatrix} n+1 \\ k+1 \end{bmatrix} x^k$$

can be found in the work [598].

For the Bell numbers, it is even more complicated to develop good approximations. By the so-called saddle point method, the following asymptotical result can be shown.

$$\frac{\log B_n}{n} = \log n - \log\log n - 1 + \frac{\log\log n}{\log n} + \frac{1}{\log n} + \frac{1}{2}\left(\frac{\log\log n}{\log n}\right)^2 + O\left(\frac{\log\log n}{\log^2 n}\right). \tag{7.10}$$

A detailed proof of this approximation can be found in the book of de Bruijn [190] (pp. 102-109).

It follows from deBruijn's result that for all $\varepsilon > 0$ there exists an n_0 (depending on the concrete value of ε) such that for all $n > n_0$

$$\left(\frac{n}{e\log n}\right)^n < B_n < \left(\frac{n}{e^{1-\varepsilon}\log n}\right)^n. \tag{7.11}$$

The drawback of these inequalities is that there is no simple way to determine the value of the index n_0 from which the approximation is valid. A 2010

---

[10]The first term in the approximation is the dominant, the other terms only slightly contribute to the main term. The last term $O\left(\frac{\log^{k-2} n}{n}\right)$ has the least contribution, as it tends to zero when n tends to infinity.

[11]Interestingly, there is a nice and recently discovered integral representation for the γ_n constants [228]:

$$\gamma_n = \frac{1}{\pi}\frac{(-1)^n}{n!}\int_0^\infty e^{-t}\Im[(\log t - i\pi)^n]dt.$$

result of Berend and Tassa [70] is the following upper bound which is valid for *all* $n > 0$:

$$B_n < \left(\frac{0.792n}{\log(n+1)}\right)^n.$$

This result is based on probabilistic arguments. To see the accuracy of this upper bound we present the following table:

n	1	5	10	20	50	100
B_n	1	52	115 975	$5.17 \cdot 10^{13}$	$1.86 \cdot 10^{47}$	$4.76 \cdot 10^{115}$
$\left(\frac{0.792n}{\log(n+1)}\right)^n$	1.14	52.73	154 508	$2.11 \cdot 10^{14}$	$1.43 \cdot 10^{50}$	$2.85 \cdot 10^{123}$

n	500	1000	5000
B_n	$1.06 \cdot 10^{843}$	$2.99 \cdot 10^{1927}$	$7.98 \cdot 10^{12\,543}$
$\left(\frac{0.792n}{\log(n+1)}\right)^n$	$1.18 \cdot 10^{902}$	$2.12 \cdot 10^{2059}$	$8.76 \cdot 10^{13\,336}$

An asymptotical formula for the B_n sequence using the Lambert W function is in the Problem book of L. Lovász [382, p. 17]:

$$B_n \sim \frac{1}{\sqrt{n}}\left(\frac{n}{W(n)}\right)^{n+1/2} \exp\left(\frac{n}{W(n)} - n - 1\right).$$

7.5   The asymptotics of the Fubini numbers

Luckily enough, the case of the Fubini numbers is easier to manage than that of the Bell numbers. We are going to prove the below inequalities:

$$\frac{n!}{2\ln^n(2)} \le F_n \le \frac{n!}{\ln^n(2)} \quad (n \ge 0). \tag{7.12}$$

Recall recursion (6.2):

$$F_n = \sum_{k=1}^{n}\binom{n}{k}F_{n-k}.$$

Write out the terms explicitly:

$$F_n = n!\left(\frac{F_{n-1}}{1!(n-1)!} + \frac{F_{n-2}}{2!(n-2)!} + \cdots + \frac{F_0}{n!0!}\right).$$

The terms of the form $F_k/k!$ suggest that it is beneficial to introduce the notation $G_k = F_k/k!$, so that $G_0 = 1$, and

$$G_n = \frac{G_{n-1}}{1!} + \frac{G_{n-2}}{2!} + \cdots + \frac{G_0}{n!} \quad (n > 0).$$

We use this equality to prove (7.12) by induction. For $n = 0$ the statement clearly holds. Supposing that the statement holds true for numbers smaller than n, we prove the statement for n.

$$G_n = \frac{G_{n-1}}{1!} + \frac{G_{n-2}}{2!} + \cdots + \frac{G_0}{n!} \leq \frac{1}{\ln^{n-1}(2)} + \frac{1}{2!\ln^{n-2}(2)} + \cdots + \frac{1}{n!\ln^0(2)} =$$

$$\frac{1}{\ln^n(2)}\left(\ln(2) + \frac{\ln^2(2)}{2!} + \cdots + \frac{\ln^n(2)}{n!}\right) \leq \frac{1}{\ln^n(2)}\left(\sum_{k=1}^{\infty} \frac{\ln^k(2)}{k!}\right) =$$

$$\frac{1}{\ln^n(2)}(e^{\ln(2)} - 1) = \frac{1}{\ln^n(2)},$$

which confirms the right-hand side of (7.12).

To confirm the left-hand side inequality, we use the same induction hypothesis (substituting $G_0 = 1$ in the last term):

$$G_n = \frac{G_{n-1}}{1!} + \frac{G_{n-2}}{2!} + \cdots + \frac{G_0}{n!} \geq$$

$$\frac{1}{2\ln^{n-1}(2)} + \frac{1}{2!2\ln^{n-2}(2)} + \cdots + \frac{1}{(n-1)!2\ln^1(2)} + \frac{1}{n!} =$$

$$\frac{1}{2\ln^n(2)}\left(\ln(2) + \frac{\ln^2(2)}{2!} + \cdots + \frac{\ln^{n-1}(2)}{(n-1)!} + \frac{2\ln^n(2)}{n!}\right).$$

One simple observation must be made: it is possible to find a $0 < c < \ln(2)$ such that

$$e^{\ln(2)} = 1 + \ln(2) + \frac{\ln^2(2)}{2!} + \cdots + \frac{\ln^{n-1}(2)}{(n-1)!} + \frac{e^c \ln^n(2)}{n!},$$

that is smaller than the same expression substituting the maximal $\ln(2)$ in place of c. So

$$e^{\ln(2)} \leq 1 + \ln(2) + \frac{\ln^2(2)}{2!} + \cdots + \frac{\ln^{n-1}(2)}{(n-1)!} + \frac{2\ln^n(2)}{n!}.$$

Rearranging:

$$\ln(2) + \frac{\ln^2(2)}{2!} + \cdots + \frac{\ln^{n-1}(2)}{(n-1)!} + \frac{2\ln^n(2)}{n!} \geq e^{\ln(2)} - 1 = 1.$$

Now let us put this into the expression for G_n. Taking into account that $F_n = G_n/n!$, we have proven inequality (7.12).

It can be shown by integral approximation methods that [571]

$$F_n \sim \frac{n!}{2\ln^{n+1}(2)}.$$

General approximation methods are available and can be used for many sequences known in set partition theory and permutation theory. Such methods are the saddle point method and Hayman's method. More can be read about these powerful tools in [190, 235, 597].

7.6 Inequalities

We have seen that the Bonferroni inequality results in (7.3), a two-sided inequality for the Stirling numbers $\left\{{n \atop k}\right\}$. Another inequality for the same numbers is (6.10). There are many ad-hoc methods to find combinatorial inequalities, but there also is a more-or-less systematic way, employing the theory of polynomials. Given some properties of a polynomial (like reality of zeros, zeros concentrated in a radius, etc.), many inequalities can be deduced with respect to their coefficients. Thanks to the fact that many counting sequences possess attached polynomials with real roots, this interrelationship proves to be useful in finding inequalities. In what follows, we present some theorems with respect to polynomials, which are interesting in themselves, and we show how they can be applied to deduce inequalities for our sequences. Other examples than the ones presented here can easily be worked out by the reader for other sequences.

7.6.1 Polynomials with roots inside the unit disk

Let[12]

$$p(x) = a_0 + a_1 x + a_2 x^2 + \cdots + a_n x^n$$

be a polynomial such that all of its zeros are less than one in absolute value. It is a result of D. J. Newman[13] that in this case the following inequality is valid:

$$\frac{\sum\limits_{k=0}^{n} k |a_k|^2}{\sum\limits_{k=0}^{n} |a_k|^2} > \frac{n}{2}.$$

Knowing that the Fubini polynomials have all their zeros in $]-1,0]$, it immediately comes that

$$\frac{\sum\limits_{k=0}^{n} k k!^2 \left\{{n \atop k}\right\}^2}{\sum\limits_{k=0}^{n} k!^2 \left\{{n \atop k}\right\}^2} > \frac{n}{2}.$$

Computer calculations show that the sequence

$$a_n = \frac{1}{n} \left(\frac{\sum\limits_{k=0}^{n} k k!^2 \left\{{n \atop k}\right\}^2}{\sum\limits_{k=0}^{n} k!^2 \left\{{n \atop k}\right\}^2} - \frac{n}{2} \right)$$

---

[12]We will always assume that the coefficient of the highest power of the indeterminate is not zero when we write out a given polynomial. This coefficient is called *leading coefficient*.

[13]The problem was proposed as Problem 5040 in the *American Mathematical Monthly* [452], and the solution was given by several authors in a later issue [97]. See also [440, 3.3.7].

possibly tends to a positive finite limit:

$$a_{10} = 0.249087\ldots$$
$$a_{100} = 0.224036\ldots$$
$$a_{1\,000} = 0.221616\ldots$$
$$a_{10\,000} = 0.221374\ldots$$

Therefore, we ask the question: does the sequence a_n converge to a finite limit? If so, can we find a closed expression for this limit?

7.6.2   Polynomials and interlacing zeros

Let two real[14] polynomials be defined such that

$$p(x) = a_0 + a_1 x + a_2 x^2 + \cdots + a_n x^n,$$

and

$$q(x) = b_0 + b_1 x + b_2 x^2 + \cdots + b_m x^m,$$

and their leading coefficients are positive. Let us suppose, moreover, that their zeros are real, and interlacing in the sense of Section 3.2. Then, if $m = n$, then

$$a_{k-1} b_k > a_k b_{k-1} \quad (k = 1, \ldots, n),$$

and, if $m = n - 1$, then

$$a_{k-1} b_k < a_k b_{k-1} \quad (k = 1, \ldots, n - 1).$$

These inequalities were presented by Constantinescu [161].

In the case of our sequences, the interlacing property is often present. Taking the Bell polynomials, for example, such that $p(x) = B_n(x)$, and $q(x) = B_{n-1}(x)$, we have $a_k = \left\{ {n \atop k} \right\}$ and $b_k = \left\{ {n-1 \atop k} \right\}$, with $m = n - 1$. The second inequality of Constantinescu results in the following:

$$\left\{ {n \atop k-1} \right\} \left\{ {n-1 \atop k} \right\} < \left\{ {n \atop k} \right\} \left\{ {n-1 \atop k-1} \right\}.$$

7.6.3   Estimations on the ratio of two consecutive sequence elements

There are some variants on the theme of the estimations for

$$\frac{|p'(x)|}{\max\limits_{a \le x \le b} |p(x)|}.$$

---

[14] Real polynomial that the coefficients of the polynomial are real.

The following result is due to P. Erdős [440, 3.3.31]: if the degree of $p(x)$ is n, and p has no zeros in $]a, b[$ but otherwise all the zeros are real, then

$$\frac{|p'(y)|}{\max_{a \le x \le b} |p(x)|} < e \frac{n}{b-a} \quad (a \le y \le b).$$

Many of our polynomials, like $B_n(x)$, $L_n(x)$, $F_n(x)$, have only real and negative roots, thus the most natural choice is $a = 0$ and $b = 1$. Also, let $y = 1$, because the derivatives are easily calculable at the unity (see exercises in Chapters 3, 4, and 6, respectively, for $B_n(x)$, $L_n(x)$, $F_n(x)$). Then, since all the polynomials in question are positive for positive arguments together with their derivatives, and are increasing as well,

$$\frac{p'(1)}{p(1)} < en. \tag{7.13}$$

For example, by the fact that $B_n'(1) = B_{n+1} - B_n$,

$$\frac{B_{n+1}}{B_n} < en + 1.$$

Such exact[15] estimations can be handy sometimes.

Similarly, since $L_n'(1) = L_{n+1} - (n+1)L_n$, and $F_n'(1) = \frac{F_{n+1} - F_n}{2}$, we get that

$$\frac{L_{n+1}}{L_n} < en + (n+1) = (e+1)n + 1,$$

$$\frac{F_{n+1}}{F_n} < 2en + 1.$$

We remark, that a result of Soble [525] permits us to delete the constant e under some conditions which our polynomials surely meet. Schoble's theorem says that if $p(x)$ has only real and non-negative coefficients, then

$$\frac{p'(x)}{p(x)} \le \frac{n}{x} \quad (x > 0).$$

Hence, it is also true that

$$\frac{B_{n+1}}{B_n} \le n + 1, \quad \frac{L_{n+1}}{L_n} \le 2n + 1, \quad \frac{F_{n+1}}{F_n} \le 2n + 1. \tag{7.14}$$

It would be interesting to find better estimations.

---

[15]We use the noun "exact" here as the opposite of asymptotic estimation.

7.6.4   Dixon's theorem

By log-concavity we know that in many triangles of counting sequences there is a peak or plateau in each row. We have also seen that it is not always easy to localize this maximum/maxima. Therefore, we might want to get estimations to these indices. Dixon's theorem [199] helps to do this[16]. It says that if

$$p(x) = a_0 + a_1 x + \cdots + a_n x^n$$

is a polynomial of degree n with only real zeros, then the following two estimations are valid:

$$a_0 + a_1 + \cdots + a_n \leq \alpha_n \max_{0 \leq k \leq n} a_k, \tag{7.15}$$

and

$$\min_{0 \leq k \leq n} a_k \leq \beta_n \max_{0 \leq k \leq n} a_k. \tag{7.16}$$

Here

$$\alpha_n = \frac{(n+1)^n}{\binom{n}{s}(n-s)^{n-s}(s+1)^s},$$

$$\beta_n = \frac{1}{\binom{n}{s}},$$

and $s = \left[\frac{n}{2}\right]$ is the integer part of $n/2$.

Lower estimations for $\left\{{n \atop K_n}\right\}$ and $\left[{n \atop K_n}\right]$

Applying the second part (7.16) of Dixon's theorem for the Bell polynomials

$$B_n(x)/x = \left\{{n \atop 1}\right\} + \left\{{n \atop 2}\right\}x + \cdots + \left\{{n \atop n}\right\}x^{n-1}$$

(which is famous for having only real zeros), we have that

$$\min_{1 \leq k \leq n} \left\{{n \atop k}\right\} = 1 \leq \beta_{n-1} \max_{1 \leq k \leq n} \left\{{n \atop k}\right\} = \left\{{n \atop K_n}\right\}.$$

That is,

$$\left\{{n \atop K_n}\right\} \geq \binom{n-1}{\left[\frac{n-1}{2}\right]}.$$

This means that the maximal Stirling number in the row n is at least as big as the central binomial coefficient in the $(n-1)$-th row.

---

[16]Dixon's theorem is a generalization of a result of L. Moser and J. R. Pounder. They gave similar result for a general second degree polynomial [446].

Better but a bit more complex estimation comes, however, from (7.15):

$$\max_{1 \le k \le n} \left\{ {n \atop k} \right\} = \left\{ {n \atop K_n} \right\} \ge \frac{1}{\alpha_{n-1}} \left(\left\{ {n \atop 1} \right\} + \left\{ {n \atop 2} \right\} + \cdots + \left\{ {n \atop n} \right\} \right).$$

(The shift in the index in α_{n-1} comes from the fact that we work with $B_n(x)/x$ instead of $B_n(x)$.) This, putting it in simple form, says that

$$\left\{ {n \atop K_n} \right\} \ge \frac{B_n}{\alpha_{n-1}}.$$

For large n this estimation is rather good as the following table shows.

n	10	100	500
$\left\{ {n \atop K_n} \right\}$	42 525	$7.77 \cdot 10^{114}$	$1.52 \cdot 10^{842}$
$\frac{B_n}{\alpha_{n-1}}$	28 540.7	$3.79 \cdot 10^{114}$	$5.73 \cdot 10^{841}$

Everything works, mutatis mutandis, with respect to the first kind Stirling numbers. Repeating the above argument, we have

$$\left[{n \atop K_n} \right] \ge \frac{n!}{\alpha_{n-1}}.$$

Here, naturally, K_n is the maximizing index in the first kind case. The estimation has a very similar performance as in the table above.

Note that, with respect to the binomial coefficients, Dixon's theorem gives the trivial result that the central binomial coefficient(s) is (are) greater than or equal to itself (themselves).

7.6.5 Colucci's theorem and the Samuelson–Laguerre theorem and estimations for the leftmost zeros of polynomials

Again we mention that many of our polynomials $B_n(x)$, $F_n(x)$, $E_n(x)$, etc. have only real and negative zeros. It is interesting to ask where the *leftmost* of these roots is located either by giving asymptotical results or by giving exact estimations.

To address these questions, a theorem of Colucci [157] will be useful. This theorem states that if

$$p(x) = a_0 + a_1 x + \cdots + a_n x^n$$

is a polynomial with real or complex coefficients, and the absolute values of the zeros are bounded by the positive real number M, then the absolute value of the kth derivative of p can be estimated as follows:

$$|p^{(k)}(x)| \leq k! \binom{n}{k} |a_n|(|x| + M)^{n-k}, \tag{7.17}$$

where k can be $0, 1, \ldots, n$. We use this inequality not to bound $p^{(k)}(x)$ but to find estimation for M.

Estimation for the leftmost zero of the Bell polynomials

Let us denote the leftmost zero of the nth Bell polynomial by z_n^*. Colucci's theorem immediately gives, with $k = 0$, and $x = 1$, that

$$B_n \leq 0! \binom{n}{0} \left\{ {n \atop n} \right\} (1 + |z_n^*|)^n;$$

in other words,

$$|z_n^*| \geq \sqrt[n]{B_n} - 1.$$

This is a pretty inequality, but even better can be given if we take $k = n - 1$ instead. Note that

$$B_n^{(n-1)}(x) = \left(\left\{ {n \atop 1} \right\} x + \cdots + \left\{ {n \atop n-1} \right\} x^{n-1} + \left\{ {n \atop n} \right\} x^n \right)^{(n-1)} =$$

$$\left(\left\{ {n \atop n-1} \right\} x^{n-1} + \left\{ {n \atop n} \right\} x^n \right)^{(n-1)} = (n-1)! \left\{ {n \atop n-1} \right\} + n! \left\{ {n \atop n} \right\} x.$$

Noting that $\left\{ {n \atop n} \right\} = 1$, we substitute this into (7.17),

$$(n-1)! \left\{ {n \atop n-1} \right\} + n! \left\{ {n \atop n} \right\} x \leq (n-1)! \binom{n}{n-1} \left\{ {n \atop n} \right\} (|x| + |z_n^*|)^1.$$

Rearranging, and recalling that

$$\left\{ {n \atop n-1} \right\} = \binom{n}{2} = \frac{n(n-1)}{2},$$

we get

$$\frac{n-1}{2} + x - |x| \leq |z_n^*|.$$

For the best result, we choose positive x, so we get the result that

$$\frac{n-1}{2} \leq |z_n^*|$$

for all n. It is worth it to remark that $\frac{n-1}{2}$ is much better estimation than $\sqrt[n]{B_n} - 1$.

It was proved in [175] that for large n

$$|z_n^*| < \frac{1}{2}\sqrt{\frac{5}{3}}n^{\frac{3}{2}}.$$

By this and the above result, we conclude[17] that there is a real number

$$\alpha \in \left[1, \frac{3}{2}\right],$$

and a positive real number c such that

$$\lim_{n \to \infty} \frac{|z_n^*|}{n^\alpha} = c.$$

Numerical calculations show [175] that $\alpha = 1$ is a probable value, and it is also possible that, for this α, $c = e$. A proof of these two statements would be welcome.

Estimation for the leftmost zero of the Eulerian polynomials

The question that which root estimation theorem works best for a given class of polynomials is mainly an issue of experimenting[18]. Colucci's theorem seems to be very sharp for the Bell polynomials. Another theorem, the Laguerre–Samuelson theorem, seems to be more appropriate for, for instance, the Eulerian polynomials.

The Laguerre–Samuelson theorem[19] states that for a polynomial

$$p(x) = a_0 + a_1 x + \cdots + a_n x^n$$

with only real zeros, the interval $[x_-, x_+]$ bounds the zeros of p, where

$$x_\pm = -\frac{a_{n-1}}{n a_n} \pm \frac{n-1}{n a_n}\sqrt{a_{n-1}^2 - \frac{2n}{n-1}a_{n-2}a_n}.$$

We are going to apply this interval estimation with respect to the Eulerian polynomials $E_{n+1}(x)$ to get an upper estimation for the leftmost root. (Note that $E_n(x)$ is an $(n-1)$-degree polynomial.) We do not need to deal with x_+, because we know that all the zeros are negative; we are only interested in x_-. Denoting the leftmost zero of $E_{n+1}(x)$ by x_{n+1}^*, we have (since $a_n = \left\langle {n+1 \atop n} \right\rangle = 1$) that

$$|x_{n+1}^*| \le |x_-| = \frac{a_{n-1}}{n} + \frac{n-1}{n}\sqrt{a_{n-1}^2 - \frac{2n}{n-1}a_{n-2}} =$$

---

[17]It is also necessary to know for the existence of α that $|z_n^*|$ monotone increases with n. But this fact follows from the interlacing property of the zeros of the Bell polynomials.

[18]Many root estimations exist, see, for example, [130, 237, 339, 434, 546] for the original works, and [471, Chapter 1] for a compilation.

[19]This theorem was first discovered by E. Laguerre [349] in 1880, and P. Samuelson rediscovered it in 1968 [506]. See also [302] for a detailed treatise on the theorem and its applications in statistics.

$$\frac{1}{n}\left\langle{n+1 \atop n-1}\right\rangle + \frac{n-1}{n}\sqrt{\left\langle{n+1 \atop n-1}\right\rangle^2 - \frac{2n}{n-1}\left\langle{n+1 \atop n-2}\right\rangle}. \tag{7.18}$$

This could be left as it is now, but a simpler form can be given if n is large. Exercise 6 of Chapter 6 gives that

$$\left\langle{n \atop n-2}\right\rangle = 2^n - n - 1,$$

$$\left\langle{n \atop n-3}\right\rangle = 3^n - (n+1)2^n + \frac{1}{2}n(n+1).$$

If n is large, these give that

$$\left\langle{n \atop n-2}\right\rangle \sim 2^n, \quad \left\langle{n \atop n-3}\right\rangle \sim 3^n,$$

whence

$$|x^*_{n+1}| < \frac{1}{n}O(2^{n+1}) + O(1)\sqrt{O(2^{2n+2}) - O(1)O(3^{n+1})} =$$

$$O\left(\frac{2^{n+1}}{n}\right) + \sqrt{O(2^{2n+2})} = O\left(\frac{2^{n+1}}{n}\right) + O(2^{n+1}) = O(2^{n+1}).$$

That is,

$$|x^*_n| < O(2^n).$$

If we use (7.18) in its exact form, we get surprisingly good estimations. For

$$x^*_5 = -23.2,$$

the Laguerre-Samuelson estimate gives -23.28. If $n = 10$, we have

$$x^*_{10} = -963.85,$$

while (7.18) predicts -964.48.

One can check that the Colucci estimation would give the lower bound

$$|x^*_n| > \frac{2^n}{n-1} - 1 - \frac{2}{n-1}.$$

These make us wondering about a constant $\beta \in [0,1]$ and a constant $d > 0$ such that

$$\lim_{n\to\infty}\frac{n^\beta|x^*_n|}{2^n} = d.$$

7.6.6 An estimation between two consecutive Fubini numbers via Lax's theorem

The property of the Fubini polynomials that their zeros concentrate on $[-1, 0]$ results in an interesting relation, thanks to a theorem of P. Lax [440, 3.3.31]. This theorem states the following. Let p be a polynomial of degree n having the bound $|p(x)| \leq 1$ on the closed unit disk, and having no zeros on the open unit disk, then

$$|p'(x)| \leq \frac{n}{2} \quad \text{(for } |x| \leq 1). \tag{7.19}$$

The Fubini polynomials have only real zeros, all belonging to $]-1, 0]$. Therefore, the polynomial $x^n F_n\left(\frac{1}{x}\right)$ has zeros only *outside* of the unit interval. One additional requirement we still have to meet, and this is the normalization: the polynomial must have absolute value at most one for $|x| < 1$. We can reach this aim by dividing $x^n F_n\left(\frac{1}{x}\right)$ with its maximal value on the unit disk, which is certainly F_n. Thus, the polynomial we will use is

$$p(x) = \frac{1}{F_n} x^n F_n\left(\frac{1}{x}\right).$$

The derivative of p is

$$p'(x) = \frac{1}{F_n}\left(n x^{n-1} F_n\left(\frac{1}{x}\right) + x^n F_n'\left(\frac{1}{x}\right) \cdot \left(-\frac{1}{x^2}\right)\right).$$

We employ (7.19) with $x = 1$:

$$\left|\frac{1}{F_n}(n F_n - F_n'(1))\right| \leq \frac{n}{2}.$$

Recall that $F_n'(1) = F_{n+1} - F_n$. After multiplying by $2F_n$ this gives that

$$|(2n+1)F_n - F_{n+1}| \leq n F_n.$$

From our former studies (see (7.14)) we know that $F_{n+1} < (2n+1)F_n$, thus the absolute value can be omitted. We finally get that

$$F_{n+1} \geq (n+1)F_n.$$

Putting this and (7.14) together, we get the nice result

$$(n+1)F_n \leq F_{n+1} \leq (2n+1)F_n$$

for all $n \geq 0$.

For these last inequalities, there is a combinatorial proof[20]. Recall that the

---

[20] This proof was told to the author by Daniel Ullman, problems editor of *The American Mathematical Monthly*.

objects that are counted by the Fubini numbers are the ordered partitions. The number $(n + 1)F_n$ counts the ordered partitions on $n + 1$ symbols that have a single element in the first block. This is clearly less than F_{n+1}; thus the left-hand side inequality follows[21]. Now Let $F_{n,k}$ be the number of ordered partitions on n symbols with k blocks. Then

$$F_{n+1} = \sum_{k=1}^{n} (2k + 1)F_{n,k},$$

since there are $2k+1$ places to insert the symbol $n+1$ into an ordered partition on n symbols with k blocks. This is clearly less than

$$\sum_{k=1}^{n} (2n + 1)F_{n,k} = (2n + 1)F_n,$$

which results in the upper bound above.

---

[21] Prof. Ullman noted that the lower bound can further be improved to $2(\sqrt{e} - 1)nF_n$; and numerical evidence suggests that even this can be improved to $\frac{1}{\log 2} nF_n$, which is compatible with the asymptotic estimation (7.12).

Exercises

1. Take a look at (7.2) and try to prove (7.4) directly. (This is the most rapidly deducible, but less precise, estimation from the three we have encountered so far.)

2. Prove that the H_n harmonic numbers behave asymptotically like the logarithm. That is, prove that the limit

$$\lim_{n \to \infty} (H_n - \log n)$$

exists. This limit is actually γ and is the very definition of the Euler-Mascheroni constant. (Hint: use the $1/x$ function, and estimate its integral on the $[1, n]$ intervals, and compare this to the harmonic numbers. Two proofs of this exercise plus a sketch of a third is contained in [90], where many other properties of the γ constant appear.)

3. Prove the validity of the following inequality:

$$n^m \left\{ {n \atop m} \right\} \geq m^n \binom{n}{m}.$$

(This is Problem 11957 in a 2017 issue of *The American Mathematical Monthly* [466].)

4. Show that

$$B_n \geq \frac{k^n}{k!} \quad (k = 0, 1, \dots).$$

This is Theorem 2 in [599], which seems to be surprising at first sight. Note that, however, the sequence on the right-hand side has a peak when k runs; therefore, it is always bounded from above. One bound is the nth Bell number. Thus, it might be more interesting to read the inequality "from right to left."

5. Prove (7.5) by using the integral definition of Γ.

6. Why can't the Gamma function be defined at zero and at the negative integers?

7. Show that

$$\frac{B_{n+1}}{B_n} \sim \frac{n+1}{\log(n+1)}.$$

(Hint: use (7.11).)

8. Prove the following asymptotics for the Eulerian numbers:

$$\left\langle {n \atop k} \right\rangle \sim (k+1)^n,$$

for fixed k. (Hint: use Exercise 5 of Chapter 6.) (This asymptotic result would suggest that the Eulerian numbers grow exponentially when n is fixed and k runs between 0 and n. But this is far from being true: we know that the Eulerian numbers are log-concave for fixed n, so they grow, reach a peak or plateau and then decrease. By symmetry the maximizing index is around the center. The argument breaks down because the above asymptotics you are asked to find in this exercise works only when k is fixed and n grows. More on the asymptotics of the Eulerian numbers can be found in [300].)

9. By using the recurrence (3.18) for the idempotent numbers, show by induction that

$$\sqrt{n} \le \frac{I_n}{I_{n-1}} \le \sqrt{n} + 1 \quad (n \ge 1).$$

This shows that $\frac{I_n}{I_{n-1}} \sim \sqrt{n}$. See [151] for the simple proof and for a better approximation, and also the Outlook of Chapter 10 in this book.

10. For short, let us introduce the notation

$$A_n(x) = \sum_{k=0}^{n} \left[{n \atop k} \right] x^k = x^{\overline{n}},$$

based on (2.44). Show that

$$A'_n(x) = A_n(x) \sum_{k=0}^{n-1} \frac{1}{x+k}.$$

In particular,

$$A'_n(1) = n! H_n \quad (n \ge 0).$$

11. Based on the previous exercise, show that the particular case (7.13) of Erdős' theorem, applied to $A_n(x)$, yields

$$H_n < en,$$

which is definitely not the best, since H_n grows logarithmically. Schoble's theorem gives the still very weak $H_n < n$.

Outlook

1. For more on the asymptotics of the sequences appearing in this book, see also the Outlook of Chapter 10.

2. Hsu [291] proved that, as n tends to infinity and k is fixed, then

$$\left\{ {n+k \atop n} \right\} = \frac{n^{2k}}{2^k k!} \left(1 + \frac{f_1(k)}{n} + \frac{f_2(k)}{n^2} + \cdots + \frac{f_t(k)}{n^t} + O\left(\frac{1}{n^{t+1}}\right) \right),$$

where $f_i(k)$ is a polynomial of degree $2i$. In particular,

$$f_1(k) = \frac{1}{3}(2k^2 + k), \quad f_2(k) = \frac{1}{18}(4k^4 - k^2 - 3k).$$

By a change of indices,

$$\left\{ {n \atop n-k} \right\} = \frac{n^{2k}}{2^k k!} \left(1 - \frac{k(4k-1)}{3n} + O\left(\frac{1}{n^2}\right) \right).$$

See also [365].

3. The asymptotics of the L_n horizontal sum of the Lah numbers was recently studied in a manuscript of Mező [430]:

$$L_n \sim \frac{1}{2} \sqrt{\frac{2}{e}} \frac{1}{n^{1/4}} \exp\left(n \log(n) - n + 2\sqrt{n} + \frac{1}{12\sqrt{n}} \right) \left(1 - \frac{3}{16\sqrt{n}} \right).$$

4. A profound analysis of the asymptotics of the Bell polynomials was carried out in [212, 213].

5. Inequalities can be found for the Fubini numbers in [616]. It is found there that

$$2^n < F_n < (n+1)^n,$$

the left-hand side is valid for $n \geq 3$, and the right-hand side is for all $n \geq 1$. These inequalities are not better than (7.12), but sometimes they might be more easy to use. In the same paper it was shown that $(\sqrt[n]{F_n})$ is a strictly increasing sequence.

Part II

Generalizations of our counting sequences

Chapter 8

Prohibiting elements from being together

After having presented the theory of Stirling numbers, we now turn to some of their generalizations. We shall see that there are many different ways one can choose to generalize the Stirling numbers by considering some additional constraints put on permutations and partitions[1].

Here we study three main generalizations when: (1) several elements are restricted to be in different cycles/blocks, and (2) the size of the cycles/blocks is limited from above or below.

8.1 Partitions with restrictions – second kind r-Stirling numbers

Going back to a former example, let us imagine that we need to put six prisoners in four cells. But there are two of them who are accomplices and cannot be put into the same cell. How many ways are there to perform this task? If we had no restriction, we could know the answer immediately: $\left\{6 \atop 4\right\} = 65$. But the restriction complicates things a bit. Before solving the problem, we continue with a definition.

Definition 8.1.1. *The number of those k-partitions of an n element set in which r elements are in different blocks is denoted by $\left\{n \atop k\right\}_r$, and we call this number the r-Stirling number of the second kind[2] with parameters n and k.*

---

[1] Even Quantum Field Theory inspired some Stirling-like numbers. See [394] where it is explained how Stirling numbers and more general numbers appear when one considers quantum mechanical creation and annihilation operators. We will restrict ourselves, however, to generalizations having a permutation theoretical or partition theoretical meaning.

[2] Sometimes *non-central Stirling number* is also used. These numbers were introduced by N. Nielsen in 1906 [454] whose name was already mentioned on p. 7. L. Carlitz studied the r-Stirling numbers, too [126, 127] and named them as *weighted Stirling numbers*. However, the combinatorial study on these numbers only appeared in 1984 [100]. The r-Stirling numbers were rediscovered by Merris [408] who called them p-Stirling numbers. More on the history of Stirling and r-Stirling numbers can be found in Charalambides' book [134, p. 319-321]. Another good source for Stirling and r-Stirling numbers is [135] where they are applied in

After having this definition, we can solve the above problem symbolically. The answer is $\left\{{6 \atop 4}\right\}_2$. Let us calculate this 2-Stirling number! We can apply the following argument. Take all the $\left\{{6 \atop 4}\right\} = 65$ cases and subtract the undesirable possibilities, i.e., those prison distributions where the two accomplices share the cell. The order of the cells does not matter, so we can suppose that they are, say, in the first cell. If so, there is at most one other prisoner that can be put into the first cell, as we can leave none of the other three cells empty. We therefore choose one prisoner from four to be put into the first cell: 4 possibilities. In this case, the three other cells contain $1 - 1 - 1$ prisoners, and this is only one case (not considering the order of the cells). If there is no other prisoner in the first cell, then the other three cells will host the other four prisoners. The number of such configurations is $\left\{{4 \atop 3}\right\} = 6$. Altogether there are $4 + 6 = 10$ cases we need to exclude. Hence, the solution is $65 - 10 = 55$.

Recursion for $\left\{{n \atop k}\right\}_r$

Another approach can be taken that will sound familiar. Let now our accomplices be the ones with indices 1 and 2. We suppose that we have hosted all our prisoners except the last, sixth, in the four cells, including 1 and 2. The sixth prisoner still must be hosted somewhere. It can happen that he goes into a cell and does not share it with others. Then the other three cells must be filled by the five prisoners in $\left\{{5 \atop 3}\right\}_2$ ways – taking into account the restriction on the accomplices. If prisoner number 6 does share a cell with one or more, then five prisoners must go into four cells not leaving an empty cell. This gives $\left\{{5 \atop 4}\right\}_2$ cases. Then prisoner number 6 can go to any of the four cells so we have $4\left\{{5 \atop 4}\right\}_2$ cases. Hence,

$$\left\{{6 \atop 4}\right\}_2 = \left\{{5 \atop 3}\right\}_2 + 4\left\{{5 \atop 4}\right\}_2.$$

This should look familiar. The above counting matches with the proof of recursion (1.4). Hence, the recursion of the second kind r-Stirling and Stirling numbers is the same! Concretely,

$$\left\{{n \atop k}\right\}_r = \left\{{n - 1 \atop k - 1}\right\}_r + k\left\{{n - 1 \atop k}\right\}_r. \tag{8.1}$$

How is it possible? Should they not give the same number sequence then? Obviously not, since we have an extra restriction here so the numbers cannot be the same (there must be less configurations when we pose extra restrictions). In particular, $\left\{{6 \atop 4}\right\}_2 < \left\{{6 \atop 4}\right\}$. We saw this, the difference is 10. What we have to take into account is the *initial value* of the recursions. By agreement, $\left\{{0 \atop 0}\right\} = 1$, and $\left\{{n \atop 0}\right\} = 0$. Now our recursion must be supplied with somewhat

---

probability theory. In [134] and [135] the r-Stirling numbers are called "non-central Stirling numbers."

different initial values:

$$\left\{ {n \atop k} \right\}_r = 0 \quad \text{if} \quad n < r,$$

$$\left\{ {n \atop k} \right\}_r = 1 \quad \text{if} \quad n = k = r,$$

$$\left\{ {n \atop k} \right\}_r = 0 \quad \text{if} \quad n = r \text{ and } k \neq r.$$

If $n > r$, then (8.1) applies.

Minimal elements in blocks

We now give another definition for the r-Stirling numbers of the second kind. If we partition a set into blocks, then every block will contain a minimal element that will be called, more than obviously, *minimal element* or, more elegantly, *block leader*. These lead to the following observation.

The number of those k-partitions of an n element set in which $1, 2, \ldots, r$ are all block leaders is the *r-Stirling number of the second kind with parameters n, k, and r.*

In what follows, we will use the *distinguished element* or *special element* expressions for the first r elements.

We use the above interpretation to prove the following statement.

$$\left\{ {n \atop k} \right\}_r = \left\{ {n \atop k} \right\}_{r-1} - (r-1) \left\{ {n-1 \atop k} \right\}_{r-1} \qquad (n \geq r \geq 1). \tag{8.2}$$

To prove this identity, first rearrange it:

$$(r-1) \left\{ {n-1 \atop k} \right\}_{r-1} = \left\{ {n \atop k} \right\}_{r-1} - \left\{ {n \atop k} \right\}_r.$$

On the right-hand side, there are the k-partitions of $1, \ldots, n$ where the first $r-1$ elements are minimal, but r is *not*. Such partitions can be formed such that we leave out r and we construct the k-block on $n-1$ elements in $\left\{ {n-1 \atop k} \right\}_{r-1}$ ways, and then we insert r appropriately. As r must not be minimal by our assumption, it can be put into blocks that already contain one of the $1, \ldots, r-1$ elements. All of these are in pairwise different $r-1$ blocks. So r can be inserted in one of these. This results in $(r-1)\left\{ {n-1 \atop k} \right\}_{r-1}$ possibilities, the quantity presented on the left-hand side.

A relation connecting second kind Stirling and r-Stirling numbers

We present a simple formula between the ordinary and r-Stirling numbers. Let us suppose that we want to distribute $n + r$ elements into $k + r$ blocks such that the first r elements are restricted to be in different blocks (that is, they are block leaders). By definition, there are $\left\{{n+r \atop k+r}\right\}_r$ many of these. Such partitions can also be formed as follows. First we construct k blocks containing no distinguished elements by filling them with j elements from that of n ($j \geq k$). Choose these j elements in $\binom{n}{j}$ ways. These j elements can be put into the k blocks in $\left\{{j \atop k}\right\}$ ways. Now we consider the remaining r blocks – these contain the distinguished elements – and we must put the remaining $n - j$ elements in these. The r blocks then must be chosen $n - j$ times independently and one block can be chosen repeatedly. There are r^{n-j} choices. At the end, we have to sum with respect to the possible values of j. The formula hence reads as

$$\left\{{n+r \atop k+r}\right\}_r = \sum_{j=k}^{n} \binom{n}{j} \left\{{j \atop k}\right\} r^{n-j}. \tag{8.3}$$

8.2 Generating functions of the r-Stirling numbers

Many of the formulas given for the Stirling numbers can be transferred to the r-Stirling case with tiny modifications. To avoid superfluous arguments, we try to prove the formulas of the subsequent sections in another way.

A formula splitting up the distinguished elements into two groups

In Chapter 2, we gave a large number of evidence proving that the generating functions are so useful. Now we present additional results proven by the generating function technique. We thus need the generating functions of the $\left\{{n \atop k}\right\}_r$ numbers. To have it, we first prove a nice identity:

$$\left\{{n \atop k}\right\}_r = \sum_{m=0}^{n-r} \binom{n-r}{m} \left\{{n-p-m \atop k-p}\right\}_{r-p} p^m \quad (r \geq p \geq 0). \tag{8.4}$$

We choose m elements greater than r to put them besides the first $p \leq r$ elements. We construct p-partitions such that the first p elements belong to separate blocks. We get $\binom{n-r}{m}\left\{{p+m \atop p}\right\}_p$ different cases. The remaining $n-p-m$ elements go to $k - p$ blocks such that $p + 1, \ldots, r$ distinguished elements go to separate blocks. This results in $\left\{{n-p-m \atop k-p}\right\}_{r-p}$ cases. We then multiply this number by the above because the allocations are independent of each other. Thus, for a fixed m there are

$$\binom{n-r}{m} \left\{{p+m \atop p}\right\}_p \left\{{n-p-m \atop k-p}\right\}_{r-p}$$

cases. As m is arbitrary, we must sum on it. Finally, taking into account the straightforward special value formula

$$\left\{{p+m \atop p}\right\}_p = p^m,$$

identity 8.4 is proven.

The exponential generating function of $\left\{{n+r \atop k+r}\right\}_r$

How do we get the generating function from (8.4)? Let us substitute $p = r$ and write $n + r, k + r$ in place of n, k:

$$\left\{{n+r \atop k+r}\right\}_r = \sum_{m=0}^{n} \binom{n}{m} \left\{{n+r-r-m \atop k+r-r}\right\}_{r-r} r^m =$$

$$\sum_{m=0}^{n} \binom{n}{m} \left\{{n-m \atop k}\right\} r^m = \sum_{m=0}^{n} \binom{n}{m} r^m \left\{{m \atop k}\right\}. \tag{8.5}$$

The binomial coefficient should ring the bell: the exponential generating function of $\left\{{n+r \atop k+r}\right\}_r$ can be written as the product of two exponential generating functions. One of them is the exponential generating function of r^m, the other is of $\left\{{n \atop k}\right\}$. We know both very well, so the result comes without effort:

$$\sum_{n=k}^{\infty} \left\{{n+r \atop k+r}\right\}_r \frac{x^n}{n!} = \frac{1}{k!} e^{rx} (e^x - 1)^k. \tag{8.6}$$

Note that (8.4) generalizes the formula (1.5) we proved earlier (see also Exercise 4 in Chapter 1). Just put $r = 1$ in (8.5).

A polynomial formula for $\left\{{n+r \atop k+r}\right\}_r$

Having the exponential generating function in our hand, we can determine the generalization of

$$y^n = \sum_{k=0}^{n} \left\{{n \atop k}\right\} y^{\underline{k}},$$

too. This task was done in no less than three ways for the Stirling numbers (see pages 17, 48 and 61)! We could choose the simplest one of these but, instead, we present another proof.

By (8.6) and by Exercise 24 of Chapter 2,

$$\sum_{n=0}^{\infty} \frac{[x(r+y)]^n}{n!} = e^{x(r+y)} = e^{rx}(1 + (e^x - 1))^y = e^{rx} \sum_{k=0}^{\infty} (e^x - 1)^k \frac{y^{\underline{k}}}{k!} =$$

$$\sum_{k=0}^{\infty} y^{\underline{k}} \sum_{n=0}^{\infty} \left\{ {n+r \atop k+r} \right\}_r \frac{x^n}{n!} = \sum_{n=0}^{\infty} \left(\sum_{k=0}^{n} \left\{ {n+r \atop k+r} \right\}_r y^{\underline{k}} \right) \frac{x^n}{n!}.$$

Compare the coefficients of x^n on the far left and far right sides:

$$(y+r)^n = \sum_{k=0}^{n} \left\{ {n+r \atop k+r} \right\}_r y^{\underline{k}}. \tag{8.7}$$

The ordinary generating function of $\left\{ {n+r \atop k+r} \right\}_r$

Combinatorial argument can be used to determine the ordinary generating function as well. To this end, we need another identity.

$$\left\{ {n+m \atop n} \right\}_r = \sum_{r \le i_1 \le \cdots \le i_m \le n} i_1 i_2 \cdots i_m. \tag{8.8}$$

Taking the set $\{1, 2, \ldots, n, n+1, \ldots, n+m\}$ we count the n-partitions when the usual restriction on special elements is considered. As there are n blocks, there are n minimal elements. The remaining $n+m-n = m$ elements are not minimal. Temporarily, we introduce the $x_1, \ldots, x_m$ notation for these elements. We suppose that by this indexing the elements are in ascending order, that is, $x_1 < x_2 < \cdots < x_m$. Let i_j denote the number of *minimal* elements smaller than x_j ($j = 1, \ldots, m$). As the x_js are ordered, the chain of relations $i_1 \le i_2 \le \cdots \le i_m$ also holds. Therefore, once the minimal elements are fixed, the non-minimal x_js can be put into blocks that already contain minimal elements smaller than they are. The number of these is $i_1, i_2, \ldots, i_m$, respectively. For example, x_1 can go to i_1 different blocks. The total number of cases is then $i_1 i_2 \cdots i_m$. Still, we have not taken into account that $1, \ldots, r$ are minimal elements. This restriction can be fulfilled by supposing simply that $i_1 \ge r$. (Clearly, x_j is not minimal while $1, \ldots, r$ is, so x_j must be bigger than r for all possible js. Having at least r minimal elements smaller than x_j, we know that $i_j \ge r$. But, as the i_js are increasing, it is enough to put this restriction only on i_1.) Summing all over the possible set of minimal elements, we are done.

Note one interesting thing. If we drop the restriction that $1, \ldots, r$ are minimal, we stay with the $i_1 \ge 1$ restriction and we get back the original Stirling numbers. This new relation was not presented in the first part:

$$\left\{ {n+m \atop n} \right\} = \sum_{1 \le i_1 \le \cdots \le i_m \le n} i_1 i_2 \cdots i_m. \tag{8.9}$$

Finally, we can deduce the ordinary generating function we were looking for:

$$\sum_{n=k}^{\infty} \left\{ {n \atop k} \right\}_r x^n = \frac{x^k}{(1-rx)(1-(r+1)x)\cdots(1-kx)}. \tag{8.10}$$

This can be gotten from (8.8) after dividing by x^k and re-parametrization:

$$\sum_{n=0}^{\infty} \left\{ {n+k \atop k} \right\}_r x^n = \frac{1}{(1-rx)(1-(r+1)x)\cdots(1-kx)}.$$

Take the $\frac{1}{1-rx}\frac{1}{1-(r+1)x}$ product first, then multiply it by $\frac{1}{1-(r+2)x}$, and so on, to see the validity of (8.10).

Some consequences of the previous results

Let us make use of the recently proven generating functions. The right-hand side of (8.10) can be split into two factors:

$$\frac{1}{(1-rx)(1-(r+1)x)\cdots(1-kx)} =$$

$$\frac{1}{(1-rx)\cdots(1-px)} \frac{1}{(1-(p+1)x)\cdots(1-kx)}.$$

These are two r-Stirling generating functions; in the first, the lower parameter is now p, in the second, $r = p+1$. Hence, the right-hand side is

$$\left(\sum_{n=0}^{\infty} \left\{ {n+p \atop p} \right\}_r x^n \right) \left(\sum_{n=0}^{\infty} \left\{ {n+k \atop k} \right\}_{p+1} x^n \right).$$

This can easily be elaborated by the multiplication rule (2.5). The formula we gain then reads as

$$\left\{ {n+k \atop k} \right\}_r = \sum_{m=0}^{n} \left\{ {m+p \atop p} \right\}_r \left\{ {n-m+k \atop k} \right\}_{p+1}. \tag{8.11}$$

The same trick of splitting can be performed in the exponential generating function case, too:

$$\frac{1}{k!}e^{rx}(e^x-1)^k \frac{1}{m!}e^{sx}(e^x-1)^m = \frac{(k+m)!}{(k+m)!} \frac{1}{k!m!} e^{(r+s)x}(e^x-1)^{k+m} =$$

$$\binom{k+m}{m} \frac{1}{(k+m)!} e^{(r+s)x}(e^x-1)^{k+m}.$$

The expansion of the initial product does not cause difficulty, and the last expression is the exponential generating function of $\left\{ {n+r+s \atop k+m+r+s} \right\}_{r+s}$ times a binomial coefficient. If we write all of these out, we should arrive at this identity:

$$\binom{k+m}{m} \left\{ {n+r+s \atop k+m+r+s} \right\}_{r+s} = \sum_{l=0}^{n} \binom{n}{l} \left\{ {l+r \atop k+r} \right\}_r \left\{ {n-l+s \atop m+s} \right\}_s. \tag{8.12}$$

This nice formula can be proven combinatorially as follows. First, we decipher the meaning of the left-hand side. Take the set $\{1, \dots, n+r+s\}$. We look for the decomposition of this set that contains $k + m + r + s$ blocks, and the first $r + s$ elements are in different blocks. Distinguish the blocks containing the first r elements and the blocks of the $r + 1, \dots, s$ elements by painting them, say, blue and red. Paint the remaining blocks also with one of these colors independently such that altogether there are $k + r$ blue and $m + s$ red blocks. To perform this latter painting, take the $k+m+r+s-r-s = k+m$ unpainted blocks. We need to choose the blocks to be painted blue. We have $\binom{k+m}{k} = \binom{k+m}{m}$ choices. So the left-hand side counts all of such blue-red-painted partitions. On the other hand, the number of such partitions can be gotten by determining which l elements go to the blue blocks. We need to choose these l elements from $n + r + s - r - s = n$: $\binom{n}{l}$ possibilities. As r elements are already in the blue blocks, and there are $k + r$ blue blocks in total, these constitute $\left\{ {l+r \atop k+r} \right\}_r$ cases. The remaining $n+r+s-(l+r) = n-l+s$ elements will go to the red blocks from which there are $m+s$, resulting $\left\{ {n-l+s \atop m+s} \right\}_s$ cases. We get the right-hand side by summing over l.

8.3   The r-Bell numbers and polynomials

The Bell numbers gave us the number of partitions on n elements. If we constrain ourselves to put r distinguished elements into different blocks, then we get a new notion: the notion of the r-Bell numbers.

Definition 8.3.1. *The nth r-Bell number $B_{n,r}$ gives how many partitions there are on $n + r$ elements such that r fixed elements belong to separate blocks[3].*

Once we know the r-Stirling numbers we just sum them to get the r-Bell numbers:

$$B_{n,r} = \sum_{k=0}^{n} \left\{ {n + r \atop k + r} \right\}_r.$$

The *r-Bell polynomials* are defined as usual:

$$B_{n,r}(x) = \sum_{k=0}^{n} \left\{ {n + r \atop k + r} \right\}_r x^k.$$

It can now be seen why the parametrization of the r-Bell numbers was given, so we can start the summation from $k = 0$ and the minimal value of n is also

---

[3]Note that $B_{n,r}$ counts partitions on $n + r$ and not on n elements.

zero. If we would not have given the definition as it is, the first $r - 1$ Bell numbers and polynomials were zero.

Let us take an example[4]:

$$B_{2,2} = \left\{ {4 \atop 2} \right\}_2 + \left\{ {4 \atop 3} \right\}_2 + \left\{ {4 \atop 4} \right\}_2.$$

$\left\{ {4 \atop 2} \right\}_2$ gives the ways 4 elements can be put into 2 blocks, while the first 2 (special) elements are in different blocks:

$$\{1, 3, 4\}, \{2\} \quad ; \quad \{1\}, \{2, 3, 4\} \quad ; \quad \{1, 3\}, \{2, 4\} \quad ; \quad \{1, 4\}, \{2, 3\}.$$

Similarly, for $\left\{ {4 \atop 3} \right\}_2$:

$$\{1\}, \{2\}, \{3, 4\} \quad ; \quad \{1, 3\}, \{2\}, \{4\} \quad ; \quad \{1, 4\}, \{2\}, \{3\} \quad ;$$

$$\{1\}, \{2, 3\}, \{4\} \quad ; \quad \{1\}, \{2, 4\}, \{3\}.$$

Finally $\left\{ {4 \atop 4} \right\}_2$ enumerates the single case contribution

$$\{1\}, \{2\}, \{3\}, \{4\}.$$

So

$$B_{2,2} = \left\{ {4 \atop 2} \right\}_2 + \left\{ {4 \atop 3} \right\}_2 + \left\{ {4 \atop 4} \right\}_2 = 4 + 5 + 1 = 10,$$

meaning that 4 elements can be partitioned in 10 ways if we want to avoid 1 and 2 to be together.

Basic properties of the r-Bell polynomials

We shortly summarize the most basic properties of the r-Bell polynomials. The results of this short summary are taken from [412] where the proofs are also presented. Since these proofs do not need new techniques or ideas, we omit them.

The exponential generating function can be determined from (8.6) (changing x to y to keep x for the polynomial variable, multiplying by $x^k/k!$ and sum over k). The result is

$$\sum_{n=0}^{\infty} B_{n,r}(x) \frac{y^n}{n!} = e^{x(e^y - 1) + ry}.$$

A corollary of this is the representation

$$B_{n,r}(x) = \sum_{k=0}^{n} \binom{n}{k} B_k(x) r^{n-k}. \tag{8.13}$$

---

[4]The r-Bell numbers appear in a paper by Whitehead who studied chromatic polynomials of graphs [596]. The first systematic study of the $B_{n,r}$ numbers and $B_{n,r}(x)$ polynomials was given in [412].

The proof is similar to that of (8.3), we factor the exponential into two:

$$e^{x(e^y-1)+ry} = e^{ry}e^{x(e^y-1)},$$

and consider the Cauchy product. One twist can result in another formula. Write $r = s + t$. Then

$$B_{n,r}(x) = \sum_{k=0}^{n} \binom{n}{k} B_{k,t}(x) s^{n-k} \quad (s+t=r). \tag{8.14}$$

Two recursions are of importance:

$$B_{n,r}(x) = x\left(\frac{d}{dx}B_{n-1,r}(x) + B_{n-1,r}(x)\right) + rB_{n-1,r}(x),$$

$$e^x x^r B_{n,r}(x) = x\frac{d}{dx}\left(e^x x^r B_{n-1,r}(x)\right)$$

from the first, for example, it can be seen that

$$B_{n,r}(0) = r^n.$$

The other can be used to show that all the zeros of the $B_{n,r}(x)$ polynomials are real (and, of course, non-positive as the coefficients of $B_{n,r}(x)$ are positive). This real zero property is an important one: remember Chapter 4 which implies that the second kind r-Stirling numbers are strictly log-concave. The location of the maximum can be estimated (see [414] for the details):

$$\frac{n}{\ln(n)} < K_{n+r,r} < \frac{n}{\ln(n) - \ln(\ln(n))} \quad (n > \max\{18, \ln(2)/\ln(1+1/r)\}).$$

Here $K_{n+r,r}$ is the *smallest* of those *one or two* parameters[5] for which

$$\left\{ \begin{matrix} n+r \\ K_{n+r,r} \end{matrix} \right\}_r \geq \left\{ \begin{matrix} n+r \\ k+r \end{matrix} \right\}_r \quad \text{for all } k = 0, 1, \ldots, n.$$

From the recursion (8.2) it follows that

$$B_{n,r}(x) = xB_{n-1,r+1}(x) + rB_{n-1,r}(x).$$

The Dobiński-formula can also be generalized:

$$B_{n,r}(x) = \frac{1}{e^x} \sum_{k=0}^{\infty} \frac{(k+r)^n}{k!} x^k.$$

In particular,

$$B_{n,r} = \frac{1}{e} \sum_{k=0}^{\infty} \frac{(k+r)^n}{k!}.$$

The Hankel determinants for the r-Bell polynomials are the same as the Bell polynomials, thanks to (8.13), and the remarks around (2.58).

---

[5]Remember that log-concavity assures that there are at most two maximizing indices.

8.4   The generating function of the r-Bell polynomials

The only proof we present in detail is the deduction of the generating function of the r-Bell polynomials. To this end, we need a new notion.

8.4.1   The hypergeometric function

The function

$$
{}_pF_q\left(\begin{array}{cccc} a_1, & a_2, & \ldots, & a_p \\ b_1, & b_2, & \ldots, & b_q \end{array}\middle|\ x\right) = \sum_{n=0}^{\infty} \frac{a_1^{\overline{n}} a_2^{\overline{n}} \cdots a_p^{\overline{n}}}{b_1^{\overline{n}} b_2^{\overline{n}} \cdots b_q^{\overline{n}}} \frac{x^n}{n!}
$$

is called *hypergeometric function*. We shall see that this function is not as abstract as it looks. In fact, most of the known functions are hypergeometric. Let us see some examples! The exponential function

$$
e^x = \sum_{n=0}^{\infty} \frac{x^n}{n!}
$$

is the simplest instance of a hypergeometric function: it contains no rising factorials in its expansion, so $p = q = 0$:

$$
e^x = {}_0F_0\left(\ \middle|\ x\right).
$$

If $a_1 = 1$ and there are no other parameters, then by $1^{\overline{n}} = n!$

$$
{}_1F_0\left(\ 1\ \middle|\ x\right) = \sum_{n=0}^{\infty} n! \frac{x^n}{n!} = \sum_{n=0}^{\infty} x^n = \frac{1}{1-x}.
$$

So the simple $\frac{1}{1-x}$ function is also hypergeometric.

The sine function is a hypergeometric function

Now let us see a more complex example! Is the $\sin(x)$ function hypergeometric? First, we write down its generating function. To this end, we use the knowledge given at the end of Section 2.2. What we get is

$$
\sin(x) = \sum_{n=0}^{\infty} \frac{(-1)^n}{(2n+1)!} x^{2n+1}.
$$

The $2n + 1$ exponent is disturbing, so we first divide by x then substitute $\sqrt{x}$ in place of x:

$$
\frac{\sin(\sqrt{x})}{\sqrt{x}} = \sum_{n=0}^{\infty} \frac{(-1)^n}{(2n+1)!} x^n.
$$

Now let us go back to the definition of the hypergeometric function. If we take the ratio of two consecutive members of their defining series, then we see that the initial scary fraction can be simplified considerably:

$$\frac{\frac{a_1^{\overline{n+1}}a_2^{\overline{n+1}}\cdots a_p^{\overline{n+1}}}{b_1^{\overline{n+1}}b_2^{\overline{n+1}}\cdots b_q^{\overline{n+1}}}\frac{x^{n+1}}{(n+1)!}}{\frac{a_1^{\overline{n}}a_2^{\overline{n}}\cdots a_p^{\overline{n}}}{b_1^{\overline{n}}b_2^{\overline{n}}\cdots b_q^{\overline{n}}}\frac{x^n}{n!}} = \frac{(n+a_1)(n+a_2)\cdots(n+a_p)}{(n+b_1)(n+b_2)\cdots(n+b_q)}\frac{x}{n+1}.$$

This tells us that the ratio of two consecutive terms in *any* hypergeometric function must have this form. And the reverse is also true: if in an expansion the ratio of two consecutive terms has the above form, then it is necessarily hypergeometric with parameters $a_1, \ldots, a_p, b_1, \ldots, b_q$.

Let us check what this argument says for the $\frac{\sin(\sqrt{x})}{\sqrt{x}}$ function.

$$\frac{\frac{(-1)^{n+1}}{(2(n+1)+1)!}x^{n+1}}{\frac{(-1)^n}{(2n+1)!}x^n} = \frac{-1}{(2n+2)(2n+3)}x.$$

This is almost a hypergeometric ratio! To gain full similarity we rewrite the right-hand side:

$$\frac{-1}{(2n+2)(2n+3)}x = \frac{1}{(n+3/2)}\frac{-x/4}{n+1}.$$

The final result can be read out directly: there is no upper parameter and there is only one lower parameter, $b_1 = 3/2$. Hence,

$$\frac{\sin(\sqrt{x})}{\sqrt{x}} = {}_0F_1\left(\begin{array}{c} \\ 3/2 \end{array} \middle| -\frac{x}{4} \right).$$

To express $\sin(x)$ we multiply by $\sqrt{x}$ and substitute x^2 in place of x.

$$\sin(x) = x \cdot {}_0F_1\left(\begin{array}{c} \\ 3/2 \end{array} \middle| -\frac{x^2}{4} \right).$$

The $\sin(x)$ function – more precisely $\sin(x)/x$ – is another instance of a hypergeometric function!

The generating function of the r-Bell polynomials

Now we are ready to prove that the generating function of the r-Bell polynomials can be expressed in hypergeometric function terms. Concretely,

$$\sum_{n=0}^{\infty} B_{n,r}(x)z^n = \frac{-1}{rz-1}\frac{1}{e^x}{}_1F_1\left(\begin{array}{c} \frac{rz-1}{z} \\ \frac{rz+z-1}{z} \end{array} \middle| x \right).$$

Our initial point is the (8.10) generating function of the second kind r-Stirling numbers. After a re-parametrization, it reads as

$$\sum_{n=0}^{\infty} \begin{Bmatrix} n+r \\ k+r \end{Bmatrix}_r z^n = \frac{z^k}{(1-rz)(1-(r+1)z)\cdots(1-(k+r)z)}.$$

The denominator can be written in terms of falling factorials:

$$(1-rz)(1-(r+1)z)\cdots(1-(k+r)z) =$$

$$\frac{(1-z)(1-2z)\cdots(1-(k+r)z)}{(1-z)(1-2z)\cdots(1-(r-1)z)} = \frac{z^{k+1}\left(\frac{1}{z}\right)^{\underline{k+r+1}}}{\left(\frac{1}{z}\right)^{\underline{r}}}.$$

Hence,

$$\sum_{n=k}^{\infty} \begin{Bmatrix} n+r \\ k+r \end{Bmatrix}_r z^n = \frac{1}{z}\left(\frac{1}{z}\right)^{\underline{r}} \frac{1}{\left(\frac{1}{z}\right)^{\underline{k+r+1}}}.$$

The

$$x^{\underline{n}} = (-1)^n (-x)^{\overline{n}}$$

reflection formula is applied to the falling factorials to turn them into rising factorials:

$$\left(\frac{1}{z}\right)^{\underline{k+r+1}} = (-1)^{k+r+1}\left(-\frac{1}{z}\right)^{\overline{k+r+1}}$$

$$= (-1)^{k+r+1}\left(-\frac{1}{z}\right)^{\overline{r+1}}\left(-\frac{1}{z}+r+1\right)^{\overline{k}}.$$

Consequently,

$$\sum_{n-k}^{\infty} \begin{Bmatrix} n+r \\ k+r \end{Bmatrix}_r z^n = \frac{1}{z}\frac{\left(\frac{1}{z}\right)^{\underline{r}}}{\left(-\frac{1}{z}\right)^{\overline{r+1}}}\frac{(-1)^{k+r+1}}{\left(\frac{rz+z-1}{z}\right)^{\overline{k}}}.$$

Since

$$\frac{\left(\frac{1}{z}\right)^{\underline{r}}}{\left(-\frac{1}{z}\right)^{\overline{r+1}}} = (-1)^r \frac{z}{rz-1},$$

we get that

$$\sum_{n=k}^{\infty} \begin{Bmatrix} n+r \\ k+r \end{Bmatrix}_r z^n = \frac{-1}{rz-1}\frac{(-1)^k}{\left(\frac{rz+z-1}{z}\right)^{\overline{k}}}.$$

Multiplying by x^k and summing over k,

$$\sum_{n=0}^{\infty} B_{n,r}(x)z^n = \frac{-1}{rz-1}\sum_{k=0}^{\infty}\frac{(-x)^k}{\left(\frac{rz+z-1}{z}\right)^{\overline{k}}} = \frac{-1}{rz-1}\,{}_1F_1\left(\begin{array}{c|c} 1 \\ \frac{rz+z-1}{z} \end{array} -x\right).$$

This is already what we wanted (expressing the generating function with the hypergeometric function) but we perform another transformation. Applying the

$$e^{-x}\,{}_1F_1\left(\begin{array}{c} a \\ b \end{array}\Big|\,x\right) = {}_1F_1\left(\begin{array}{c} b-a \\ b \end{array}\Big|\,-x\right)$$

Kummer[6]-formula with $b = \frac{rz+z-1}{z}$ and $a = \frac{rz-1}{z}$ we get the ultimate form of the generating function.

Putting $r = 0$ we get the generating function of the Bell polynomials:

$$\sum_{n=0}^{\infty} B_n(x)z^n = \frac{1}{e^x}\,{}_1F_1\left(\begin{array}{c} \frac{-1}{z} \\ \frac{z-1}{z} \end{array}\Big|\,x\right).$$

8.5   The r-Fubini numbers and r-Eulerian numbers

It is an interesting question what kind of counting sequence we get when we take the order of the blocks into account in partitions where the usual restriction on special elements is applied. This question leads to the definition of the r-Fubini numbers; generalizing the Fubini numbers of Chapter 6.

Definition 8.5.1. *The nth r-Fubini number $F_{n,r}$ gives the number of partitions of an $(n + r)$-element set such that r fixed elements go to separate partitions and the order of the blocks matters.*

By this definition, the basic formula for the $F_{n,r}$ numbers reads as

$$F_{n,r} = \sum_{k=0}^{n}(k + r)!\begin{Bmatrix} n+r \\ k+r \end{Bmatrix}_r,$$

while the *r-Fubini polynomials* are defined as

$$F_{n,r}(x) = \sum_{k=0}^{n}(k + r)!\begin{Bmatrix} n+r \\ k+r \end{Bmatrix}_r x^k.$$

Note that $F_{n,r}$ counts partitions on $n + r$ elements and not on n elements, similarly to the r-Bell numbers.

The exponential generating function of the r-Fubini polynomials

We begin the study of the r-Fubini numbers and polynomials by the determination of their exponential generating function. We are going to show

---

[6]Ernst Eduard Kummer (1810-1893), German mathematician.

that

$$\sum_{n=0}^{\infty} F_{n,r}(x)\frac{t^n}{n!} = \frac{r!e^{rt}}{(1-x(e^t-1))^{r+1}} = r!e^{rt}\left(\sum_{n=0}^{\infty} F_n(x)\frac{t^n}{n!}\right)^{r+1}.$$

Here $F_n(x) = F_{n,0}(x)$ is the classical Fubini polynomial (see Section 6.1).

To deduce this result, we apply a tricky idea[7]. As $x^{\underline{k}}$ is a polynomial of degree k (constant polynomials are of degree 0), any polynomial can be written as a linear combination of the polynomials $(x^{\underline{k}})_{k=0}^{\infty}$ (see Chapter 2). We say that the polynomials $(x^{\underline{k}})_{k=0}^{\infty}$ form a *base* in the vector space of polynomials (say, of real coefficients).

Let L be a linear operator acting on the vector space of real polynomials[8]. Here we prescribe the effect of the operator L on the falling factorial base:

$$L(x^{\underline{k}}) = L(x(x-1)\cdots(x-k+1)) := (k+r)!x^k.$$

Because of (8.7)

$$(x+r)^n = \sum_{k=0}^{n} \left\{\begin{matrix} n+r \\ k+r \end{matrix}\right\}_r x^{\underline{k}},$$

it holds true that

$$L((x+r)^n) = L\left(\sum_{k=0}^{n}\left\{\begin{matrix} n+r \\ k+r \end{matrix}\right\}_r x^{\underline{k}}\right) = \sum_{k=0}^{n}\left\{\begin{matrix} n+r \\ k+r \end{matrix}\right\}_r L(x^{\underline{k}}) =$$

$$\sum_{k=0}^{n}\left\{\begin{matrix} n+r \\ k+r \end{matrix}\right\}_r (k+r)!x^k = F_{n,r}(x).$$

Thus,

$$\sum_{n=0}^{\infty} F_{n,r}(x)\frac{t^n}{n!} = \sum_{n=0}^{\infty} L((x+r)^n)\frac{t^n}{n!} = L\left(\sum_{n=0}^{\infty}\frac{((x+r)t)^n}{n!}\right) = e^{rt}L(e^{xt}).$$

Write $e^t = 1+v$. Then $e^{xt} = (e^t)^x = (1+v)^x$, and apply the general binomial theorem

$$(1+v)^x = \sum_{n=0}^{\infty}\binom{x}{n}v^n = \sum_{n=0}^{\infty}\frac{x^{\underline{n}}}{n!}v^n.$$

By the definition of L,

$$L(e^{xt}) = L((1+v)^x) = \sum_{n=0}^{\infty}\frac{L(x^{\underline{n}})}{n!}v^n = \sum_{n=0}^{\infty}\frac{(n+r)!}{n!}(xv)^n.$$

---

[7]This proof was done by Stephen M. Tanny in [558] with respect to the Fubini polynomials, but the idea goes back to at least Gian-Carlo Rota [498].

[8]An operator is linear if it preserves the multiplication by constant and it also preserves the addition of polynomials. For example, f L is linear, then it must hold true that $L(5x) = 5L(x)$ and $L(x^2 + x) = L(x^2) + L(x)$. Therefore, it is enough to prescribe how L acts on the non-negative powers of x. Also, it is enough to prescribe how L acts on any other base.

The last sum can already be given in closed form:

$$\sum_{n=0}^{\infty} \frac{(n+r)!}{n!}(xv)^n = \frac{r!}{(1-xv)^{r+1}}.$$

Substituting $v = e^t - 1$, the statement is proven. From the (6.7) exponential generating function, we justify the rightmost equality, too.

If you would like to know more on polynomial bases and operators on them, the very well written and interesting book of Martin Aigner who is an Austrian mathematician [10] is a good source.

Results coming from the generating function

We deduce some important consequences of the above exponential generating function. Of course, the most straightforward of all these is the exponential generating function of the r-Fubini numbers:

$$\sum_{n=0}^{\infty} F_{n,r} \frac{t^n}{n!} = \frac{r!e^{rt}}{(2-e^t)^{r+1}}. \tag{8.15}$$

It can also be verified that

$$F_{n,r}(-1) = r!(-1)^n. \tag{8.16}$$

The Cauchy product gives another result:

$$F_{n,r+1}(x) = (r+1) \sum_{k=0}^{n} \binom{n}{k} \sum_{l=0}^{k} \binom{k}{l} F_l(x) F_{k-l,r}(x).$$

The proof of this goes as follows:

$$\sum_{n=0}^{\infty} F_{n,r+1}(x) \frac{t^n}{n!} = \frac{(r+1)e^t}{(1-x(e^t-1))} \frac{r!e^{rt}}{(1-x(e^t-1))^{r+1}} =$$

$$(r+1)\left(\sum_{n=0}^{\infty} \frac{t^n}{n!}\right)\left(\sum_{n=0}^{\infty} F_n(x)\frac{t^n}{n!}\right)\left(\sum_{n=0}^{\infty} F_{n,r}(x)\frac{t^n}{n!}\right) =$$

$$(r+1)\left(\sum_{n=0}^{\infty} \frac{t^n}{n!}\right)\sum_{n=0}^{\infty}\left(\sum_{l=0}^{n} \binom{n}{l} F_l(x) F_{n-l,r}(x)\right)\frac{t^n}{n!} =$$

$$(r+1)\sum_{n=0}^{\infty}\sum_{k=0}^{n} \binom{n}{k}\left(\sum_{l=0}^{k} \binom{k}{l} F_l(x) F_{k-l,r}(x)\right)\frac{t^n}{n!}.$$

Then we compare the coefficients to finalize the proof.

We prove the following recursion in details:

$$F_{n,r}(x) = x[(r+1)F_{n-1,r}(x) + (1+x)F'_{n-1,r}(x)] + rF_{n-1,r}(x). \qquad (8.17)$$

By using the recursion of the r-Stirling numbers,

$$F_{n,r}(x) = \sum_{k=0}^{n}(k+r)!(k+r)\left\{{n-1+r \atop k+r}\right\}_r x^k + \sum_{k=0}^{n}(k+r)!\left\{{n-1+r \atop k-1+r}\right\}_r x^k.$$

For the first sum

$$\sum_{k=0}^{n}(k+r)!(k+r)\left\{{n-1+r \atop k+r}\right\}_r x^k =$$

$$\frac{1}{x^{r-1}}\left(\sum_{k=0}^{n-1}(k+r)!\left\{{n-1+r \atop k+r}\right\}_r x^{k+r}\right)' = \frac{1}{x^{r-1}}(x^r F_{n-1,r}(x))' =$$

$$rF_{n-1,r}(x) + xF_{n-1,r}(x).$$

(The prime denotes derivation with respect to the variable x.) For the second sum

$$\sum_{k=0}^{n}(k+r)!\left\{{n-1+r \atop k-1+r}\right\}_r x^k =$$

$$\frac{1}{x^{r-1}}\left(\sum_{k=0}^{n-1}(k+r-1)!\left\{{n-1+r \atop k+r}\right\}_r x^{k+r}\right)' = \frac{1}{x^{r-1}}(x^{r+1}F_{n-1,r}(x))' =$$

$$(r+1)xF_{n-1,r}(x) + x^2 F'_{n-1,r}(x).$$

This results in the recursion. It also follows that

$$x^{1-r}\left[(x^{r+1} + x^r)F_{n-1,r}(x)\right]' = F_{n,r}(x). \qquad (8.18)$$

In particular,

$$F_n(x) = x((1+x)F_{n-1}(x))'.$$

This was proven earlier (see (6.8)).

The real root property of the r-Fubini polynomials

From the recursion (8.18) above, one can prove that all the zeros of $F_{n,r}(x)$ are real. What is more, these zeros all belong to the interval $]-1,0[$.

$F_{n,r}(x)$ cannot have positive real zeros, as its coefficients are positive so it takes positive values for positive x. Moreover, for positive integer r, x^{1-r} cannot be zero for finite x (but it is zero when $r = 0$ and $x = 0$), so $F_{n,r}(x)$ is 0 if and only if

$$[x^r(x+1)F_{n-1,r}(x)]' = 0.$$

From this point, we prove our statement by induction. The statement is true for the polynomial $F_{1,r}(x) = (r+1)!x + rr!$. Let us suppose that $F_{n-1,r}(x)$ has zeros that all satisfy the statement. If so, the polynomial $x^r(x+1)F_{n-1,r}(x)$ has a zero in $x = 0$ of multiplicity r, and $x = -1$ is another zero of multiplicity one. Moreover, it has $n - 1$ zeros in the interval $]-1, 0[$. It follows from the Rolle theorem that the derivative has n roots in $]-1, 0[$: there are $n - 2$ zeros among the zeros of $F_{n-1,r}(x)$; because of the factor $x + 1$, there is a zero on the left-hand side of the zeros of $F_{n-1,r}(x)$, but to the right of -1. In addition, because of the factor x^r, there must be another zero to the right of all the zeros of $F_{n-1,r}(x)$, but to the left of 0. These are n zeros in the interval $]-1, 0[$ as we stated.

"Ordered" Dobiński formula

At the end of this section, we present an interesting infinite sum representation for the r-Fubini numbers that can be considered as an "ordered version" of Dobiński's (2.17) formula:

$$F_{n,r} = \sum_{k=0}^{\infty} \frac{(k+r)^n}{2^{k+r+1}} \frac{(k+r)!}{k!}.$$

Upon setting $r = 0$, we get the already known formula (6.6)

$$F_n = \sum_{k=0}^{\infty} \frac{k^n}{2^{k+1}}.$$

To prove this, we cannot apply (6.6) because the exponential generating function is more complicated. Thus, we start with the Dobiński-formula given for the r-Bell numbers.

$$B_{n,r}(x) = \sum_{k=0}^{n} \left\{ {n+r \atop k+r} \right\}_r x^k = \sum_{k=0}^{\infty} \frac{(k+r)^n}{k!} \frac{x^k}{e^x}.$$

Multiply by $\frac{x^r}{e^x}$ and integrate from 0 to infinity:

$$\sum_{k=0}^{n} \left\{ {n+r \atop k+r} \right\}_r \int_0^{\infty} \frac{x^{k+r}}{e^x} = \sum_{k=0}^{\infty} \frac{(k+r)^n}{k!} \int_0^{\infty} \frac{x^{k+r}}{e^{2x}}. \tag{8.19}$$

These integrals are known (but you are asked to calculate these in the exercise at the end of this chapter):

$$\int_0^{\infty} \frac{x^{k+r}}{e^x} = (k+r)!, \qquad \int_0^{\infty} \frac{x^{k+r}}{e^{2x}} = (k+r)!2^{-(k+r+1)}. \tag{8.20}$$

Substituting these into (8.19) we are done.

8.6 The r-Eulerian numbers and polynomials

In Sections 6.3-6.5 we showed that the Fubini numbers are closely related to the Eulerian numbers. This can lead to the conjecture that, once we have the r-Fubini numbers, there should be some generalization of the Eulerian numbers, too, such that these are related in a similar fashion as the Fubini and Eulerian numbers are related. In this section we prove this conjecture[9].

The most important identities of the above-cited sections are the following:

$$F_n = \sum_{k=0}^{n} \left\langle \begin{matrix} n \\ k \end{matrix} \right\rangle 2^k, \tag{8.21}$$

and its polynomial version

$$F_n(x) = \sum_{k=0}^{n} \left\langle \begin{matrix} n \\ k \end{matrix} \right\rangle (x+1)^k x^{n-k}. \tag{8.22}$$

A relation between the Stirling and Eulerian numbers was also given:

$$k! \left\{ \begin{matrix} n \\ k \end{matrix} \right\} = \sum_{m=0}^{n} \left\langle \begin{matrix} n \\ m \end{matrix} \right\rangle \binom{m}{n-k}, \tag{8.23}$$

together with the Frobenius theorem:

$$\sum_{k=0}^{n} k! \left\{ \begin{matrix} n \\ k \end{matrix} \right\} (x-1)^{n-k} = \sum_{m=0}^{n} \left\langle \begin{matrix} n \\ m \end{matrix} \right\rangle x^m. \tag{8.24}$$

Symbolic generalization of the Eulerian numbers

We try to find the "r-version" of the above results. First, let us try the crudest method to find the generalization of the Eulerian numbers that are attached to the r-Fubini numbers. Write the r-Fubini polynomials to the left-hand side of (8.22) and denote the new coefficients on the right-hand side by $\left\langle \begin{matrix} n \\ k \end{matrix} \right\rangle_r$:

$$F_{n,r}(x) = \sum_{k=0}^{n} \left\langle \begin{matrix} n \\ k \end{matrix} \right\rangle_r (x+1)^k x^{n-k}. \tag{8.25}$$

By the definition of the $F_{n,r}(x)$ polynomials

$$\sum_{k=0}^{n} (k+r)! \left\{ \begin{matrix} n+r \\ k+r \end{matrix} \right\}_r x^k = \sum_{m=0}^{n} \left\langle \begin{matrix} n \\ m \end{matrix} \right\rangle_r (x+1)^m x^{n-m}.$$

---

[9]The r-Eulerian numbers were introduced in the thesis [419], see also [420].

Dividing by x^n and making the substitution $x \to 1/x$:

$$\sum_{k=0}^{n}(k+r)!\left\{{n+r \atop k+r}\right\}_r x^{n-k} = \sum_{m=0}^{n}\left\langle{n \atop m}\right\rangle_r (x+1)^m. \qquad (8.26)$$

To rewrite the right-hand side, we use the binomial theorem:

$$\sum_{m=0}^{n}\left\langle{n \atop m}\right\rangle_r (x+1)^m = \sum_{m=0}^{n}\left\langle{n \atop m}\right\rangle_r \sum_{k=0}^{n}\binom{m}{n-k}x^{n-k}.$$

Substituting this into the above,

$$\sum_{k=0}^{n}(k+r)!\left\{{n+r \atop k+r}\right\}_r x^{n-k} = \sum_{m=0}^{n}\left\langle{n \atop m}\right\rangle_r \sum_{k=0}^{n}\binom{m}{n-k}x^{n-k}.$$

Comparing the coefficients we have

$$(k+r)!\left\{{n+r \atop k+r}\right\}_r = \sum_{m=0}^{n}\left\langle{n \atop m}\right\rangle_r \binom{m}{n-k}.$$

This is the generalization of (8.23).

Frobenius theorem (8.24) also follows. In the intermediate equation (8.26) put $x-1$ in place of x:

$$\sum_{k=0}^{n}(k+r)!\left\{{n+r \atop k+r}\right\}_r (x-1)^{n-k} = \sum_{m=0}^{n}\left\langle{n \atop m}\right\rangle_r x^m.$$

After the above considerations, it is quite fair to call the $\left\langle{n \atop k}\right\rangle_r$ numbers as *r-Eulerian numbers*. Soon we find them a combinatorial meaning.

Our last identity can be transformed easily to find another relation. Just apply the binomial theorem to expand $(x-1)^{n-k}$ and compare the coefficients:

$$\left\langle{n \atop m}\right\rangle_r = \sum_{k=0}^{n}(k+r)!\left\{{n+r \atop k+r}\right\}_r \binom{n-k}{m}(-1)^{n-k-m}. \qquad (8.27)$$

From this, it is immediate to see that

$$\left\langle{n \atop n}\right\rangle_r = r!\left\{{n+r \atop r}\right\}_r = r!r^n,$$

$$\left\langle{n \atop 0}\right\rangle_r = r!.$$

To verify the first, just set $m=n$ and then note that the only surviving non-zero term is $k=0$, thanks to the binomial coefficient. For the second, we set $m=0$ and realize that $\binom{n-k}{0} = 1$, thus

$$\left\langle{n \atop 0}\right\rangle_r = (-1)^n \sum_{k=0}^{n}(k+r)!\left\{{n+r \atop k+r}\right\}_r (-1)^k = (-1)^n F_{n,r}(-1) = (-1)^n(-1)^n r!,$$

by (8.16).

The r-Eulerian polynomials

Definition 8.6.1. *We define the r-Eulerian polynomials*[10] *as*

$$E_{n,r}(x) := \sum_{m=0}^{n} \left\langle {n \atop m} \right\rangle_r x^m.$$

Equation (8.25) gives the following relation between the r-Eulerian polynomials and r-Fubini polynomials:

$$F_{n,r}(x) = x^n E_{n,r}\left(\frac{x+1}{x}\right),$$

$$E_{n,r}(x) = (x-1)^n F_{n,r}\left(\frac{1}{x-1}\right). \tag{8.28}$$

Recalling that the zeros of the r-Fubini polynomials lie in $]-1,0[$, a straight consequence of (8.28) is that all the zeros of $E_{n,r}(x)$ are real and belong to $]-\infty,0[$. The log-concavity of the $\left(\left\langle {n \atop m} \right\rangle_r\right)_{m=0}^{n}$ coefficients thus follows. That is,

$$\left\langle {n \atop m} \right\rangle_r^2 \ge \left\langle {n \atop m-1} \right\rangle_r \left\langle {n \atop m+1} \right\rangle_r.$$

The exponential generating function of the r-Eulerian polynomials is

$$\sum_{n=0}^{\infty} E_{n,r}(x)\frac{t^n}{n!} = r!e^{r(x-1)t}\left(\frac{x-1}{x-e^{(x-1)t}}\right)^{r+1} = r!e^{r(x-1)t}\left(\sum_{n=0}^{\infty} E_n(x)\frac{t^n}{n!}\right)^{r+1}.$$

Here $E_n(x) = E_{n,0}(x)$ is the usual Eulerian polynomial.

This generating function can be determined by (8.28). Namely,

$$\sum_{n=0}^{\infty} E_{n,r}(x)\frac{t^n}{n!} = \sum_{n=0}^{\infty} F_{n,r}\left(\frac{1}{x-1}\right)\frac{[(x-1)t]^n}{n!} = \frac{r!e^{r(x-1)t}}{\left[1 - \frac{1}{x-1}(e^{(x-1)t}-1)\right]^{r+1}}.$$

$$= r!e^{r(x-1)t}\left(\frac{x-1}{x-e^{(x-1)t}}\right)^{r+1}. \tag{8.29}$$

---

[10]We saw that the $E_n(x)$ classical Eulerian polynomials are of degree $n-1$. The $E_{n,r}(x)$ polynomials are of degree n when $r > 0$.

Generating function and recursions

Having the exponential generating function at our disposal, the basic recursion for the Eulerian numbers is readily found. This will be helpful when we want to find the combinatorial meaning of these numbers in the next section. Let

$$f(x,t) = \sum_{n=0}^{\infty} E_{n,r}(x)\frac{t^n}{n!}.$$

Then it can be seen that $f(x,t)$ satisfies a partial differential equation similar to the one found in Section 6.7:

$$(x - x^2)\frac{\partial f}{\partial x} + (tx - 1)\frac{\partial f}{\partial t} + (1 + rx)f = 0.$$

Recall Equation (6.26) given for the Eulerian polynomials:

$$(x - x^2)\frac{\partial f}{\partial x} + (tx - 1)\frac{\partial f}{\partial t} + f = 0.$$

The only difference is the additive rxf term. This term modifies the recursion in the following manner:

$$\left\langle {n \atop m} \right\rangle_r = (m+1)\left\langle {n-1 \atop m} \right\rangle_r + (n - m + r)\left\langle {n-1 \atop m-1} \right\rangle_r. \tag{8.30}$$

This result can be used to find a recursion for $E_{n,r}(x)$:

$$E_{n,r}(x) = (1 + (n + r - 1)x)E_{n-1,r}(x) + (x - x^2)E'_{n-1,r}(x).$$

Indeed,

$$E_{n,r}(x) =$$

$$\sum_{m=0}^{n} (m+1)\left\langle {n-1 \atop m} \right\rangle_r x^m + \sum_{m=0}^{n} (n - m)\left\langle {n-1 \atop m-1} \right\rangle_r x^m + r\sum_{m=0}^{n} \left\langle {n-1 \atop m-1} \right\rangle_r x^m.$$

The first two terms are the same as in the Eulerian polynomial case, the third term is $rxE_{n-1,r}(x)$. Using the recursion

$$E_n(x) = (1 + (n-1)x)E_{n-1}(x) + (x - x^2)E'_{n-1}(x)$$

given for the Eulerian polynomials, we are done.

We have derived a number of identities with respect to the r-Eulerian numbers and polynomials without knowing their combinatorial meaning. In the next section, we study this latter question.

8.7 The combinatorial interpretation of the r-Eulerian numbers

To find the combinatorial description of the r-Eulerian numbers, we go back to the combination lock of Section 6.4.

The generalized combinatorial lock problem

In the combination lock problem, there is a connection between simultaneous button presses and permutation runs, so it seems to be beneficial to go this direction again. It also seems obvious that – taking into consideration the restriction on our special elements – the simultaneously pressed buttons cannot contain more than one button indexed by a special element. Hence, it seems to be a good idea to modify the notion of a run as follows.

Definition 8.7.1. *A run is called r-run if in this run there is at most one distinguished element.*

For example, in the permutation

$$\begin{pmatrix} 1 & 2 & 3 & 4 & 5 & 6 & 7 & 8 \\ 1 & 2 & 3 & 6 & 4 & 5 & 7 & 8 \end{pmatrix}$$

there are four 4-runs: $1; 2; 3, 6; 4, 5, 7, 8$. (Hence, not any classical run is an r-run; classical runs must be broken where special elements appear.)

To keep the addition rules of the classical runs, ascents and descents valid (see (6.11) and the other two rules) we need to modify the latter two notions, too.

Definition 8.7.2. *An ascent is called r-ascent if not both of the members of the ascent are special elements. Moreover, every classical descent is an r-descent, and, in addition, the position between i_j, i_{j+1} is also an r-descent, if both i_j and i_{j+1} are special.*

If in a permutation on n elements there are k r-runs, then there are $k-1$ r-descents, because an r-run ends when the element is bigger than the next *or* both elements are special. As there are $n-1$ comparisons among n elements, there are $n-1-(k-1) = n-k$ r-ascents. Thus,

$$\begin{aligned} r\text{-ascents} + r\text{-descents} &= n-1, \\ r\text{-runs} &= r\text{-descents} + 1, \\ r\text{-runs} &= n - r\text{-ascents}. \end{aligned}$$

These relations show that the definitions are correct and these are proper generalizations of the classical notions of ascents, descents, and runs.

These new definitions make easier the generalization of the combination lock game. If we had classical runs in a lock combination, then additional separators must be put in the places where there are two distinguished elements. Such a button press sequence is the following when $r-3$:

$$1; 2; 3, 4, 5.$$

We must have a separator between 1 and 2, and between 2 and 3. This is automatized by the introduction of r-runs.

To see how these definitions work, let us take again the permutation

$$\begin{pmatrix} 1 & 2 & 3 & 4 & 5 & 6 & 7 & 8 \\ 1 & 2 & 3 & 6 & 4 & 5 & 7 & 8 \end{pmatrix},$$

and set $r = 4$. The 4-ascents are the following: $3, 6; 4, 5; 5, 7; 7, 8$, while the 4-descents: $1, 2; 2, 3; 6; 4$. (Unusual, but it follows from the definition that $1, 2$ is a descent whenever $r > 1$.) The 4-runs were already determined above.

We introduce the following definition.

Definition 8.7.3. *The number of permutations that act on $n + r$ elements and contain k r-ascents is given by the r-Eulerian number of parameters n and k. The notation for these numbers is $\left\langle {n \atop k} \right\rangle_r$ $(0 \le k \le n + r)$.*

If in the combination lock we do not permit distinguished buttons pressed simultaneously and in the resulting permutations runs are meant r-runs, then we get the generalization of the celebrated identity (6.20):

$$F_{n,r} = \sum_{k=0}^{n} \left\langle {n \atop k} \right\rangle_r 2^k.$$

Recursion for $\left\langle {n \atop k} \right\rangle_r$

After generalizing the notion of ascents, we can easily present the combinatorial proof of the (8.30) recursion

$$\left\langle {n \atop m} \right\rangle_r = (m+1)\left\langle {n-1 \atop m} \right\rangle_r + (n-m+r)\left\langle {n-1 \atop m-1} \right\rangle_r.$$

The proof given for the usual Eulerian number recursion can be repeated word for word (see (6.14)). The last element is put somewhere in the permutation such that we want or do not want a new ascent. As the last element is not distinguished if $n > 0$, the usual notion of ascent coincides with the r-ascent. The $n - m + r = n + r - m$ factor is different but only because we have $n + r$ elements in place of n. This is similar to the fact that the recursion for the Stirling numbers is unmodified when we take $r > 0$.

The initial values do change: $\left\langle {0 \atop 0} \right\rangle_r = 1$ and $\left\langle {n \atop 0} \right\rangle = r!$ if $n \ge 1$. The first is a convention and the second is clear: 0 r-ascent can be gotten only if we list our non-distinguished elements in a decreasing order in a permutation. The distinguished elements can never form r-ascents (they are always r-descents), so they can be put at the beginning of the permutation in any order. (But cannot be put to the end or somewhere else to keep zero ascents!) This results in $r!$ permutations with no r-ascents.

8.8 Permutations with restrictions – r-Stirling numbers of the first kind

Once we have second kind r-Stirling numbers, it is natural to consider the first kind Stirling numbers. It is obvious how these numbers should be defined. Later, it turns out that the first and second kind r-Stirling numbers are dual in the sense of Chapter 2.

Definition 8.8.1. *The number of k-permutations on n elements when r elements belong to separate cycles is given by the $\left[{n \atop k}\right]_r$ r-Stirling number of the first kind.*

Minimal elements and cycle leaders – The Riordan representation of a permutation

Another definition can be given by considering the notion of minimal elements. Cycles can always be written in a form that the smallest element of the cycle becomes the first (see p. 11.). Then this smallest element is called *cycle leader*. After putting the smallest elements in front of the cycles, we order the cycles according to the cycle leaders in *decreasing* order. In the resulting form, the permutation has k elements that are smaller than all the elements on their left-hand side. These elements are called *left-to-right minima*. Obviously, the leftmost element and 1 are always left-to-right minima. We will call such a rearrangement of a permutation a *Riordan representation* because it appeared in the classical book of Riordan [495].

Let us take an example. Below in the first step we do the cycle decomposition, then we shift the elements in the second cycle to have 2 in the cycle leader position, then we order the cycles according to the cycle leaders.

$$\begin{pmatrix} 1 & 2 & 3 & 4 & 5 & 6 & 7 & 8 & 9 \\ 3 & 7 & 5 & 2 & 1 & 4 & 6 & 8 & 9 \end{pmatrix} = (1 \quad 3 \quad 5)(4 \quad 2 \quad 7 \quad 6)(8)(9) \rightarrow$$

$$(1 \quad 3 \quad 5)(2 \quad 7 \quad 6 \quad 4)(8)(9) \rightarrow (9)(8)(2 \quad 7 \quad 6 \quad 4)(1 \quad 3 \quad 5).$$

Even the parentheses can be removed from $(9)(8)(2 \quad 7 \quad 6 \quad 4)(1 \quad 3 \quad 5)$. Indeed, writing simply 982764135, we can determine where were the parentheses. Write a left parenthesis in front of every element which is a left-to-right minima

$$982764135 \rightarrow (9(8(2764(135,$$

then close the parentheses appropriately:

$$(9(8(2764(135 \rightarrow (9)(8)(2764)(135).$$

We have already seen similar notations for permutations, the word representation on page 23.

We introduce the sub index R for the Riordan representation, and w for the word representation. So 375214689_w means that we have a word representation, while 982764135_R warns us that the given permutation is in the Riordan representation.

Let us go back now to the definition of the r-Stirling numbers. The distinguished elements $1, \ldots, r$ are certainly cycle leaders. But not only cycle leaders, they are left-to-right minima also. Hence, there are one-to-one correspondence between those permutations in which the last r left-to-right minima are $r, r-1, \ldots, 1$ and those in which the first r elements are in different cycles. Therefore, the next definition is equivalent to the former one.

Definition 8.8.2. *The number of those permutations of $1, \ldots, n$ in which there are k left-to-right minima and, in addition, the first r elements are left-to-right minima is counted by $\left[{n \atop k} \right]_r$.*

Two recursions for $\left[{n \atop k} \right]_r$

Applying the argument above, we can give a recursion for the r-Stirling numbers of the first kind.

$$\left[{n \atop k} \right]_r = \frac{1}{r-1} \left(\left[{n \atop k-1} \right]_{r-1} - \left[{n \atop k-1} \right]_r \right).$$

First rewrite this as

$$(r-1) \left[{n \atop k} \right]_r = \left[{n \atop k-1} \right]_{r-1} - \left[{n \atop k-1} \right]_r.$$

The right-hand side gives the number of those permutations which are constituted by $k-1$ cycles, the first $r-1$ elements are left-to-right minima but r is not. These permutations can be constructed from permutations with k cycles where the first r elements are left-to-right minima (there are $\left[{n \atop k} \right]_r$ many of these), but then we must put the cycle containing r *after* a cycle containing one of $1, \ldots, r-1$. This can be done in $r-1$ ways. (If it was put after another cycle, then it would be a minimum.)

The basic recurrence for $\left[{n \atop k} \right]_r$ is still missing. As for the second kind r-Stirling case, the recursion coincides with that of the classical Stirlings:

$$\left[{n \atop k} \right]_r = (n-1) \left[{n-1 \atop k} \right]_r + \left[{n-1 \atop k-1} \right]_r,$$

while the initial conditions are the following:

$$\begin{bmatrix} n \\ k \end{bmatrix}_r = 0, \quad \text{if} \quad n < r;$$

$$\begin{bmatrix} n \\ k \end{bmatrix}_r = 1, \quad \text{if} \quad n = r \text{ and } k = r;$$

$$\begin{bmatrix} n \\ k \end{bmatrix}_r = 0, \quad \text{if} \quad n = r \text{ and } k \neq r.$$

We give another proof, however, in which we use left-to-right minima. To count the permutations with k cycles on n elements, we can also count the permutations with k left-to-right minima. If we take a permutation on $n-1$ elements which already has k left-to-right minima, then nothing changes when we insert n after any element (in $n-1$ ways). On the other hand, we can take permutations in which there are $k-1$ left-to-right minima (there are $\begin{bmatrix} n-1 \\ k-1 \end{bmatrix}_r$ such permutations). To get another minimum, n needs to be a fixed point (otherwise, n is not a cycle leader), and this is only one possibility, giving the second term in the recursion.

The analogues of (8.8) **and** (8.3)

A representation similar to (8.8) is

$$\begin{bmatrix} n \\ n-m \end{bmatrix}_r = \sum_{r < i_1 < \cdots < i_m < n} i_1 i_2 \cdots i_m.$$

The lower parameter $n-m$ means that there are m non-left-to-right minima. Denote them in increasing order by $i_1 < i_2 < \cdots < i_m (\leq n)$. Moreover, write the left-to-right minima in decreasing order as we did at the beginning of the section. The number of left-to-right minima smaller than i_1 is $i_1 - 1$, the number of left-to-right minima smaller than i_2 is $i_2 - 2$ and, finally, the number of left-to-right minima smaller than i_m is $i_m - m$. Now put the $i_1, \ldots, i_m$ elements into the existing cycles. Begin with i_1. One can put these after the $i_1 - 1$ minima, so there are $i_1 - 1$ choices. Now insert i_2. There are $i_2 - 2$ minima and one can put i_2 after any of these ($i_2 - 2$ cases), and it can go after i_1 also (one case). Altogether $i_2 - 1$ cases. Do the same for $i_2, \ldots, i_{m-1}$. At the end, i_m can be inserted in $i_m - 1$ places. As $1, 2, \ldots, r$ are left-to-right minima we have that $i_1 > r$. We must sum all the possibilities when $r < i_1 < i_2 < \cdots < i_m \leq n$ and then the statement follows.

We show an example: let $n = 10$, and let the left-to-right minima be $1, 2, 4, 5, 6, 8, 9$. Then $(i_1, i_2, i_3) = (3, 7, 10)$. To the side of 9 and 8, only 10 can be inserted; by the elements $6, 5, 4$ we can insert $10, 7$, and by $2, 1$ we can put $3, 7, 10$:

$$(9 \quad .)(8 \quad .)(6 \quad . \quad .)(5 \quad . \quad .)(4 \quad . \quad .)(2 \quad . \quad .)(1 \quad . \quad . \quad .)$$

We insert 3 by the side of, say, 2:

$$(9 \quad .) \,(8 \quad .) \,(6 \quad . \quad .) \,(5 \quad . \quad .) \,(4 \quad . \quad .) \,(2 \quad 3 \quad .) \,(1 \quad . \quad)$$

Item 7 can go by the side of $1, 2, 4, 5, 6$ or after 3. We insert it in this example after 5:

$$(9 \quad .) \,(8 \quad .) \,(6 \quad .) \,(5 \quad 7) \,(4 \quad .) \,(2 \quad 3 \quad .) \,(1 \quad .)$$

10 can go after any cycle leader or after 3 or 7. Inserting it after 3, we have a full permutation on 10 elements:

$$(9) \,(8) \,(6) \,(5 \quad 7) \,(4) \,(2 \quad 3 \quad 10) \,(1) \,.$$

Three more identities

We prove the "first kind version" of (8.3):

$$\begin{bmatrix} n+r \\ k+r \end{bmatrix}_r = \sum_{j=k}^{n} \binom{n}{j} \begin{bmatrix} j \\ k \end{bmatrix} r^{\overline{n-j}}.$$

Until the point when we start to fill the distinguished blocks (with j elements) the proof is the same as the proof of (8.3). The remaining $n-j$ elements go to the distinguished blocks. The first element can go to r places, the second can go into $r+1$ positions, and so on. The $(n-j)$th element can go into $r+n-j-1$ positions. Multiplying these, we get the rising factorial $r^{\overline{n-j}}$, and the statement is proved.

We list some other identities without proof (as the proofs are similar to the former ones). The "first kind versions" of (8.4) and (8.12) are

$$\begin{bmatrix} n \\ k \end{bmatrix}_r = \sum_{m=0}^{n-r} \binom{n-r}{m} \begin{bmatrix} n-p-m \\ k-p \end{bmatrix}_{r-p} p^{\overline{m}} \quad (r \geq p \geq 0),$$

and

$$\binom{k+m}{m} \begin{bmatrix} n+r+s \\ k+m+r+s \end{bmatrix}_{r+s} = \sum_{l=0}^{n} \binom{n}{l} \begin{bmatrix} l+r \\ k+r \end{bmatrix}_r \begin{bmatrix} n-l+s \\ m+s \end{bmatrix}_s.$$

The horizontal generating function and the exponential generating function are

$$\sum_{k=r}^{n} \begin{bmatrix} n \\ k \end{bmatrix}_r x^k = x^r(x+r)(x+r+1)\cdots(x+n-1) = x^r(x+r)^{\overline{n-r}},$$

and

$$\sum_{n=0}^{\infty} \begin{bmatrix} n+r \\ k+r \end{bmatrix}_r \frac{x^k}{k!} = \frac{1}{k!} \left(\frac{1}{1-x} \right)^r \ln^m \left(\frac{1}{1-x} \right).$$

The first and second kind r-Stirling numbers are orthogonal to each other, similarly to the classical Stirling numbers:

$$\sum_{n=k}^{m} \overline{\begin{bmatrix} m \\ n \end{bmatrix}}_r \begin{Bmatrix} n \\ k \end{Bmatrix}_r = \begin{cases} 1 & \text{if } m = k \\ 0 & \text{otherwise} \end{cases} .$$

Here the overline denotes the signed Stirling numbers of the first kind:

$$\overline{\begin{bmatrix} n \\ k \end{bmatrix}}_r = (-1)^{n-k} \begin{bmatrix} n \\ k \end{bmatrix}_r .$$

8.9 The hyperharmonic numbers

At the very beginning of the book, we showed that the Stirling numbers of the first kind are related to the harmonic numbers, see p. 14.

This relation was easy to establish by the basic recursion of $\begin{bmatrix} n \\ k \end{bmatrix}$. To generalize that result for the $r > 0$ case via combinatorial arguments, we need some new ideas. This section is based on the work of A. T. Benjamin, D. Gaebler, and D. Gaebler [61].

The hyperharmonic numbers

Similarly to the fact that the r-Stirling numbers generalize the Stirling numbers, we need a generalization of the harmonic numbers to express the small lower parameter special values of the r-Stirling numbers. This generalization is introduced now. The numbers

$$H_n^0 := \frac{1}{n} \quad (H_0^0 = 0)$$

are called *harmonic numbers of order zero*, the numbers

$$H_n^1 := \sum_{k=1}^{n} \frac{1}{k} = \sum_{k=1}^{n} H_k^0$$

are the *harmonic numbers of order 1* or simply harmonic numbers, while the numbers

$$H_n^2 := \sum_{k=1}^{n} H_k^1$$

are called – as you probably expect – *second order harmonic numbers*, or *hyperharmonic numbers*. In general,

$$H_n^r := \sum_{k=1}^{n} H_k^{r-1} \quad (H_0^r = 0)$$

is the nth *hyperharmonic number of order r*. The hyperharmonic numbers are often denoted by $H_n^{(r)}$ to distinguish the order r from exponents[11]. For the case of harmonic number, we simply leave the order 1, and write H_n.

A relation between hyperharmonic numbers and first kind Stirling numbers

Our main goal is to show that

$$H_n^r = \frac{1}{n!}\begin{bmatrix} n+r \\ r+1 \end{bmatrix}_r. \tag{8.31}$$

First factor out a factorial in the denominator of H_n^r; and write it as

$$H_n^r = \frac{a_{n,r}}{n!}$$

As $H_0^r = 0$, we have $a_{0,r} = 0$. Moreover – since $H_n^0 = \frac{1}{n}$ – it also comes that $a_{n,0} = (n-1)!$. If, instead, both n and r are greater than 0, then, by the recursion

$$H_n^r = H_1^{r-1} + H_2^{r-1} + \cdots + H_{n-1}^{r-1} + H_n^{r-1} = H_{n-1}^r + H_n^{r-1}$$

we infer

$$\frac{a_{n,r}}{n!} = \frac{a_{n-1,r}}{(n-1)!} + \frac{a_{n,r-1}}{n!}.$$

This results in the recursion

$$a_{n,r} = n a_{n-1,r} + a_{n,r-1}. \tag{8.32}$$

This recursion will be important later on.

To proceed, we turn to the r-Stirling numbers of the first kind and prove a new recursion:

$$\begin{bmatrix} n+r \\ k \end{bmatrix}_r = n\begin{bmatrix} n+r-1 \\ k \end{bmatrix}_r + \begin{bmatrix} n+r-1 \\ k-1 \end{bmatrix}_{r-1}.$$

The proof of this recursion is based on a simple argument. Taking the first element, 1, we have two alternatives: (1) it is a fixed point, or (2) there are other elements in its cycle. If 1 is a fixed point, then we put the remaining $n+r-1$ elements into $k-1$ cycles such that $r-1$ elements are in different cycles. This results in the second term. If 1 shares its cycle with some other elements, then we can do the following. Let us determine which element will be the immediate successor of 1 in the cycle (by right shifts we can move 1 to

---

[11] A funny fact that the hyperharmonic numbers were firstly defined in *The Book of Numbers* by J. H. Conway and R. K. Guy – independently of the r-Stirling numbers, but incidentally the authors were using the same letter r for the order [162].

the first position, and then the immediate successor of 1 goes to the second position in the cycle) and denote it by s. There are n choices for this successor s. We construct a k-permutation on $n + r - 1$ elements such that the first r elements are in different cycles in $\left[{n+r-1 \atop r}\right]_r$ ways. In the final step, we insert s into its position.

Now we are ready to prove (8.31). For brevity, we introduce the abbreviation

$$A_{n,r} = \left[{n + r \atop r + 1}\right]_r.$$

Hence, $A_{0,r} = \left[{r \atop r+1}\right]_r = 0 = a_{0,r}$, while for $n \geq 1$ we have $A_{n,0} = \left[{n \atop 1}\right]_0 = (n-1)! = a_{n,0}$. By the recursion we have just proved

$$A_{n,r} = nA_{n-1,r} + A_{n,r-1},$$

and this agrees with 8.32. Since the initial values of $a_{n,r}$ and $A_{n,r}$ also agree, we get that the two sequences are identical. Hence, (8.31) is valid.

An expression for the hyperharmonic numbers

We prove another identity[12]:

$$H_n^r = \binom{n + r - 1}{r - 1}(H_{n+r-1} - H_{r-1}). \tag{8.33}$$

It is interesting that this is a relation connecting two non-integer sequences[13] but the proof still can be done combinatorially. We introduce the following notation. Let $T_{n,k,r}$ be the set in which there are k-permutations of an n-set such that r elements are in different cycles. By definition, there are $\left[{n \atop k}\right]_r$ elements in $T_{n,k,r}$. Cycles containing distinguished elements are also called distinguished.

Another fact is necessary to see. If $r + 1 \leq t \leq n + r$, then the number of permutations belonging to $T_{n+r,r+1,r}$ such that t is in the (unique) non-distinguished cycle *and* t is minimal equals to

$$\frac{(n + r - 1)!}{(r - 1)!(t - 1)}.$$

To see why this statement is true, note that, as t is minimal in its cycle, the

---

[12] This relation firstly appeared in the book [162] in 1996 without proof. An analytic proof of this identity can be found in [426] where infinite sums of hyperharmonic numbers are studied.

[13] It is known since 1915 that the H_n harmonic numbers are never integer, but whether there exists integer hyperharmonic numbers is still an open question. There is ongoing research on this topic, see the literature [20, 21, 22, 257, 415]. The latest result on the non-integerness of the hyperharmonic numbers is [16], in which it is shown that if there are integer hyperharmonic numbers, they must be "rare."

$t - 1$ non-distinguished elements go to the distinguished blocks. There are $\begin{bmatrix} t-1 \\ r \end{bmatrix}_r = \frac{(t-2)!}{(r-1)!}$ such allocations. Still we need to insert $t + 1, t + 2, \ldots, n + r$ somewhere. Rewriting the permutation such that the smallest elements in the cycles become the first elements in their cycles, the $t + 1, t + 2, \ldots, n + r$ elements can go to the *right-hand side* of the already inserted items. (All of them are greater than r so they cannot be cycle leaders.) Hence, $t + 1$ can go into t positions, $t + 2$ has $t + 1$ places to go, and finally the last element $n + r$ can go into $n + r - 1$ positions. These result in the product

$$\frac{(t - 2)!}{(r - 1)!} t(t + 1) \cdots (n + r - 1) = \frac{(t - 2)!}{(r - 1)!} \frac{(n + r - 1)!}{(t - 1)!} = \frac{(n + r - 1)!}{(r - 1)!(t - 1)}.$$

This is just the statement we gave.

We are one step away from (8.33). Recalling that

$$|T_{n+r,r+1,r}| = \begin{bmatrix} n + r \\ r + 1 \end{bmatrix}_r,$$

and noting that in the unique non-distinguished cycle any element t $(r + 1 \leq t \leq n + r)$ can be minimal we need to sum on these values of t. By (8.31)

$$H_n^r = \frac{1}{n!} \begin{bmatrix} n + r \\ r + 1 \end{bmatrix}_r = \frac{1}{n!} \sum_{t=r+1}^{n} \frac{(n + r - 1)!}{(r - 1)!(t - 1)} = \frac{(n + r - 1)!}{n!(r - 1)!} \sum_{t=r+1}^{n} \frac{1}{t - 1} =$$

$$\binom{n + r - 1}{r - 1} (H_{n+r-1} - H_{r-1}).$$

Another proof of (8.33)

Identity (8.33) can be proven in another, equally interesting way. Considering (8.31) and (8.33) we may write

$$\begin{bmatrix} n + r \\ r + 1 \end{bmatrix}_r = n! H_n^r = \frac{(n + r - 1)!}{(r - 1)!} (H_{n+r-1} - H_{r-1}),$$

from which

$$(r - 1)! \begin{bmatrix} n + r \\ r + 1 \end{bmatrix}_r = (n + r - 1)!(H_{n+r-1} - H_{r-1})$$

follows. Applying again (8.31) we have

$$(r - 1)! \begin{bmatrix} n + r \\ r + 1 \end{bmatrix}_r = \begin{bmatrix} n + r \\ 2 \end{bmatrix}_1 - \frac{(n + r - 1)!}{(r - 1)} \begin{bmatrix} r \\ 2 \end{bmatrix}_1. \tag{8.34}$$

This is a statement equivalent to (8.33). If we can prove (8.34) we have another proof for (8.33). In fact, much more is true than (8.34). Namely, for all $0 \leq r \leq m \leq n$ we have

$$\begin{bmatrix} n \\ k \end{bmatrix}_r = \sum_{t=r}^{k} \begin{bmatrix} m \\ t \end{bmatrix}_r \begin{bmatrix} n \\ k+m-t \end{bmatrix}_m.$$

This reduces to (8.34) when $r = 1$ and $k = 2$. Let us prove this latter statement. The left-hand side counts the elements of $T_{n,k,r}$. Elements of $T_{n,k,r}$ can be constructed such that we put the n elements not into k cycles but into $k + m - t$ cycles such that the first $m \geq r$ elements are in different cycles. There are $\begin{bmatrix} n \\ k+m-t \end{bmatrix}_r$ ways to do this. Since $m \geq r$ we need to decompose further the m cycles into t cycles by treating the m cycles as elements for a moment. Still we need to put the r elements into separate cycles so there are $\begin{bmatrix} m \\ t \end{bmatrix}_r$ possibilities. We sum over the possible values of t.

We note that the already mentioned paper [61] has many other identities with respect to the first kind r-Stirling numbers.

Exercises

1. Calculate $\left\{{6\atop4}\right\}_2$ by using (8.1).

2. Prove that $\left\{{n+r\atop r}\right\}_r = r^n$, and that $\left[{n+r\atop r}\right]_r = r^{\overline{n}}$.

3. Determine $\left\{{6\atop4}\right\}_2$ by (8.8).

4. Decompose the generating function of the r-Stirling numbers of the second kind into three factors and give the corresponding analogue of (8.11).

5. Verify the r-Stirling version of (2.20):

$$\left\{{n+r\atop k+r}\right\}_r = \frac{1}{k!}\sum_{l=0}^{k}\binom{k}{l}(l+r)^n(-1)^{k-l}.$$

6. How many surjective functions are there from an n-set to a k-set ($n \geq k$) such that the first r elements in the n-set are mapped into pairwise different values? That is, we ask for the cardinality of the set

$$\{f : \{1,\dots,n\} \to \{1,\dots,k\} \mid f \text{ is surj. and}$$

$$f(i) \neq f(j) \text{ if } i,j \in \{1,\dots,r\}, i \neq j\}.$$

What is the number of not necessarily surjective functions?

7. Prove (8.11) by combinatorial arguments.

8. Let ten balls be given. How many ways are there to put these into four boxes such that three boxes have capacity three and one has capacity four? How many possibilities are there if we suppose that from the ten balls seven is blue, three is red, and the red balls cannot share a box?

9. Express the polynomial $x^2 + x + 1$ in the base $(x^{\underline{k}})$.

10. Expand the polynomial $3x^{\underline{2}} - 2x^{\underline{1}} + 5x^{\underline{0}}$ in the usual $1, x, x^2, \dots$ base.

11. How many 4-ascents, 4-descents and 4-runs are there in the permutation

$$\begin{pmatrix} 1 & 2 & 3 & 4 & 5 & 6 & 7 & 8 \\ 7 & 8 & 1 & 3 & 2 & 4 & 6 & 5 \end{pmatrix}?$$

12. Show that the r-Eulerian numbers have the following special values

$$\left\langle {n \atop 0} \right\rangle_r = r!,$$

$$\left\langle {n \atop n} \right\rangle_r = r! r^n,$$

$$\left\langle {n \atop n-1} \right\rangle_r = r! \left[(r+1)((r+1)^n - r^n) - nr^n \right].$$

13. Prove that the generating function of the hyperharmonic numbers is

$$\sum_{n=0}^{\infty} H_n^r x^n = -\frac{\ln(1-x)}{(1-x)^r}.$$

(The exponential generating function is more complicated, see [181, 185] for the details with respect to the ordinary harmonic numbers and [425, 421] for the exponential generating function of the hyperharmonic numbers.)

14. Based on Section 7.2, prove the inequality

$$\frac{(k+r)^n}{k!} - \frac{(k-1+r)^n}{(k-1)!} < \left\{ {n+r \atop k+r} \right\}_r < \frac{(k+r)^n}{k!}$$

for all $n \geq m > 0$.

15. Show the validity of the identity

$$\left[{n+r \atop k} \right] = \sum_{l=0}^{k} \left[{r \atop l} \right] \left[{n+r \atop k-l+r} \right]_r.$$

16. Prove the r-Stirling version of the (11.27) Spivey formula and its dual (11.40):

$$B_{n+m+r,r} = \sum_{k=0}^{n} \sum_{j=0}^{m} (j+r)^{n-k} \left\{ {m+r \atop j+r} \right\}_r \binom{n}{k} B_k,$$

$$(r+1)^{\overline{n+m}} = \sum_{k=0}^{n} \sum_{j=0}^{m} (m+r)^{\overline{n-k}} \left[{m+r \atop j+r} \right]_r \binom{n}{k} (r+1)^{\overline{k}}.$$

(See [413].)

17. Show that

$$\left(e^{e^{iz}+riz} \right)^{(n)} = i^n e^{e^{iz}+riz} B_{n,r}(e^{iz}).$$

(Take the proof of (3.8) as a hint.)

18. Prove the generalization of Touchard's congruence (the Touchard congruence is studied in Section 11.7.2) with respect to the r-Bell numbers:

$$B_{n+p,r} \equiv rB_{n,r} + B_{n,r+1} + B_{n,p+r} \pmod{p}$$

for all $n, r \geq 0$. For the details and for a polynomial generalization see [427].

19. Show that
$$B_{n+p,r} \equiv (r+1)B_{n,r} + B_{n,r+1} \pmod{p}.$$

See [427].

20. Prove (8.13) and (8.14) for the r-Bell numbers (that is, when $x = 1$). Do not use analysis, only combinatorics.

21. Prove that

$$B_{n,r} = rB_{n-1,r} + \sum_{k=0}^{n-1} \binom{n-1}{k} B_{k,r} = rB_{n-1,r} + \sum_{k=0}^{n-1} \binom{n-1}{k} B_{n-1-k,r}.$$

22. Show that the r-Bell number sequence is log-convex. (Hint: use (8.13) and the binomial convolution's log-convexity preserving property mentioned in Point 6 in the Outlook of Chapter 4.)

23. Find the combinatorial proof of

$$\begin{Bmatrix} n \\ k \end{Bmatrix}_r = \sum_{i=0}^{r} \binom{r}{i} (k-i)^{r-i} \begin{Bmatrix} n-r \\ k-i \end{Bmatrix},$$

$$\begin{Bmatrix} n+r \\ k+r \end{Bmatrix}_r = \sum_{i=0}^{r} k^i \begin{Bmatrix} n+r-i \\ k+r \end{Bmatrix}_{r-i}.$$

What is the difficulty if we try to prove the corresponding identities for the first kind case?

24. Prove the following identity, where $B_n(x)$ now refers to the Bernoulli polynomials:

$$B_n(r) = \sum_{k=0}^{n} (-1)^k \frac{k!}{k+1} \begin{Bmatrix} n+r \\ k+r \end{Bmatrix}_r.$$

(The paper [269] contains a generating function proof.)

25. Calculate the (8.20) integrals by partial integration.

26. Write 4732561_w into Riordan and standard representations.

Outlook

1. We noted already in the Outlook of Chapter 11, that it is possible to generalize the Touchard congruence (originally given for the Bell numbers) to polynomials. In Exercise 18 of this chapter, we have seen that this congruence can be extended to the r-Bell numbers. Polynomial extension is also possible with respect to the r-Bell polynomials, see [427] for the details, and for other congruences of the r-Bell numbers and derangement numbers. In this paper, one can also find results with respect to the remainders of $B_{n,r}$ modulo primes as well as results for modulo 2-powers.

2. It is known for more than 100 years [561] that the H_n harmonic numbers are not integers except, obviously, $H_1 = 1$. (The reader can prove it full of hope; this is a standard textbook problem. See also Chapter 11 for a detailed discussion.) That the hyperharmonic numbers (except $H_1^r = 1$) can be integers or not – this is a much harder question. This question was first studied by the author in 2007 [415]. There it was proven that H_n^2 and H_n^3 are never integers (see, again, Chapter 11). In the subsequent years, a large number of results came out and a large class of hyperharmonic numbers is known to be non-integers [20, 21, 22, 195, 257].

3. For which $n_1 \neq n_2$ and $r_1 \neq r_2$ does the equality $H_{n_1}^{r_1} = H_{n_2}^{r_2}$ holds? This is another question from [415].

4. The r-*Lah numbers* generalize the Lah numbers in the spirit of the r-Stirlings. That is, the r-Lah number $\left\lfloor {n \atop k} \right\rfloor_r$ counts the number of partitions of a set with $n + r$ elements into $k + r$ non-empty subsets such that r distinguished elements have to be in distinct *ordered* blocks. These numbers were studied in [54, 142] and were thoroughly studied by G. Nyul and G. Rácz [456]. In [55] one can find the extension of the r-Lah numbers to the associated case. M. Shattuck [516] introduced a two-parameter generalization of the r-Lah numbers by set partition statistics.

5. The zigzag permutations can be subject to a restriction that in their cycles the first element is in different cycles. These are the r-zigzag permutations and were studied in [485].

6. The asymptotic behavior of the r-Stirling and r-Bell numbers was extensively studied in [163, 164, 174].

7. Cesàro's integral formula (p. 30) can be extended easily to the r-Bell numbers, see [412].

8. As we saw, the r-Bell polynomials have only real zeros, and even the corresponding result of (3.8) can be found for the r-Bell polynomials

(see the exercises). Note that Edrei's paper [210] determines the final set (in the sense of Section 3.2.4) not only for e^{-e^z} but also for any function of the form $h(z)e^{-e^z}$, where $h(z)$ is an admissible entire function taking real values for real z.

As we see from Exercise 17, we need the final set of $f(z) = e^{e^{iz}+riz}$. This function can be transformed into

$$f(-iz + \pi) = e^{e^{i(-iz+\pi)}+ri(-iz+\pi)} = e^{rz}(e^{i\pi})^r e^{i\pi} e^z = (-1)^r e^{rz} e^{-e^z}.$$

Hence, in this case $h(z) = (-1)^r e^{rz}$ is an admissible[14] function, so we get that the zeros of the r-Bell polynomials for any fixed non-negative integer r form a dense set on the negative real line.

9. An application of the r-Stirling numbers of the first kind appears in the study of permutation patterns. See [187] for the details. In this paper, some new formulas are also given for the first kind r-Stirling numbers. The reader needs extra caution during reading, because, strangely enough, in this paper the first kind r-Stirling numbers are denoted by $\left\{ {n \atop k} \right\}_r$.

10. There exists the notion of generalized Stirling numbers, due to Hsu and Shiue [292]. These generalized Stirling numbers unify the binomial coefficients, Stirling numbers, r-Stirlings and many other numbers. The asymptotic behavior of the generalized Stirling numbers of the second kind is described in [165].

11. The hypergeometric function of Section 8.4 can be used to represent the first kind Stirling numbers:

$$\left[{z \atop k} \right] = \frac{(-1)^{k+1}}{z^k \Gamma(1-z)} {}_{k+1}F_k \left(\begin{matrix} z, & z, & \cdots, & z \\ z+1, & z+1, & \cdots, & z+1 \end{matrix} \Big| 1 \right),$$

or, alternatively,

$$\left[{z \atop k} \right] = \frac{(-1)^{k+1} \sin(\pi z)}{\pi} \sum_{i=0}^{\infty} \frac{\Gamma(i+z)}{i!(i+z)^k}.$$

Here z can be any real or even complex number such that its real part is less than k. Γ is the Gamma function (Subsection 7.3.1).

An integral representation can also be deduced, where even the lower parameter can be complex:

$$\left[{z \atop w} \right] = \frac{1}{\Gamma(1-z)\Gamma(w)} \int_0^1 \frac{t^{z-1}}{(1-t)^z} \log^{w-1}(t)dt.$$

---

[14] Admissibility here means that the function does not vanish in some strip. Here this trivially satisfies, because the exponential is never zero.

Defined this way, we can calculate, for example, that

$$\begin{bmatrix} \frac{1}{2} \\ 2 \end{bmatrix} = \sqrt{\pi} \log\left(\frac{1}{4}\right).$$

These interesting representations were all given by V. Adamchik [3].

12. Lattices [10, 265] are special algebraic structures with many applications in combinatorics, number theory, geometry, linear algebra and even physics. Their structures are partially studied via the attached Whitney numbers. The Stirling numbers of the second kind are just the Whitney numbers of the set partition lattice (see p. 69-70. in [10]). Dowling lattices [203] are a particular family of lattices and their Whitney numbers are very similar to the Stirling numbers. Although their original motivation is not set partition theoretical, but it is possible to interpret them this way [417, 490]. The Whitney numbers of Dowling lattices were also studied in [64, 65, 66].

13. An interesting problem is the enumeration problem of not necessarily maximal chains in the partition lattice. This problem was first considered by T. Lengyel [367]. See [352] for a recent review and references.

14. The size of the largest antichain in the partition lattice is known, at least asymptotically. See [117, 118].

15. Modifications of the partition lattices are studied in [107, 215, 537, 503, 576] where the sizes of the blocks are congruent to a given fixed positive integer i modulo another integer k. Such partition lattices with $(i, k) = (0, 2)$ (that is, when the block sizes are even) were found to be useful in the theory of Ising ferromagnets [554].

16. The r-Whitney numbers $W_{r,m}(n, k)$ are generalizations of the Stirling and r-Stirling numbers and the above-mentioned Whitney numbers of the Dowling lattices. The quickest way to define them is via the connecting coefficients approach:

$$(mx + r)^n = \sum_{k=0}^{n} m^k W_{m,r}(n, k) x^{\underline{k}}$$

$$m^n x^{\underline{n}} = \sum_{k=0}^{n} w_{m,r}(n, k)(mx + r)^k.$$

The coefficients $W_{m,r}(n, k)$ and $w_{m,r}(n, k)$ give back the Stirling,

r-Stirling, and Whitney numbers, respectively:

$$W_{1,0}(n,k) = \left\{ {n \atop k} \right\} \equiv S^2(n,k),$$

$$W_{1,r}(n,k) = \left\{ {n+r \atop k+r} \right\}_r,$$

$$W_{m,0}(n,k) = W_m(n,k).$$

Note that $(mx+r)$ is a generalization of the first degree monomial x to a general first degree polynomial.

The r-Whitney numbers appeared under this name in [418] where they were used to study the Bernoulli polynomials and, independently, in [167] under the name (r,β)-Stirling number, and also in [500]. It is worth to note that the r-Whitney numbers belong to the Hsu-Shiue generalized Stirling number class. The asymptotics of the $W_{r,m}(n,k)$ and $w_{r,m}(n,k)$ numbers was studied in [166, 168, 172, 173], lattice theoretical connections can be found in [142], polynomial identities and real zero studies of r-Whitney-Fubini polynomials were done in [170], r-Whitney-Lah numbers were studied in [142, 273, 428]. In [273] a new combinatorial interpretation is also given for the r-Whitney and r-Whitney-Lah numbers. Generalization via partition statistics can be found in [393, 486, 487]. Further generalizations and identities for these numbers are contained in [56, 204, 388, 389, 390, 391, 404, 429, 436, 480, 607].

17. There exists a generalization of the r-Stirling numbers when r_1 numbers are restricted to be in different blocks, r_2 other elements are restricted to be in separate blocks (but can be mixed with the first r_1 elements), and so on. This generalization was studied in [385].

18. Recall that the Eulerian numbers help to evaluate some infinite sums: (6.25) says that

$$\sum_{i=0}^{\infty} i^n x^i = \frac{x^n}{(1-x)^{n+1}} E_n\left(\frac{1}{x}\right).$$

This can be generalized to the r-Eulerian number case [419, 420]:

$$\sum_{i=0}^{\infty} i^n i^{\overline{r}} x^i = \frac{x^{n+r}}{(1-x)^{n+r+1}} E_{n,r}\left(\frac{1}{x}\right).$$

Here $E_{n,r}$ is the *r-Eulerian polynomial*

$$E_{n,r}(x) = \sum_{k=0}^{n} \left\langle {n \atop k} \right\rangle_r x^k.$$

Even more general polynomials, the *r-Whitney-Eulerian polynomials* were studied in [488].

19. The Hankel transform of the r-Fubini polynomials cannot be as simply related to the Hankel transform of the classical Fubini polynomials as we did for the r-Bell and Bell polynomials. Indeed, for the latter the Hankel determinants coincide (see the end of Section 8.3); but it is not known how to express the Hankel transform of the r-Fubini numbers and polynomials in closed form at least when $r > 2$. By simple inspection, for $r = 2$ the $h_{n,2}(x)$ determinants for $F_{n,2}(x)$ and the $h_n(x)$ determinants of $F_n(x)$ are seemingly easily related:

$$h_{n,2} = n!(n+1)!h_n$$

(for the form of $h_n(x)$ see Exercise 15 of Chapter 6).

For $r = 3$ we have that

$$h_{1,3}(x) = 6$$
$$h_{2,3}(x) = 144x(1+x)$$
$$h_{3,3}(x) = 34\,560x^3(1+x)^3$$
$$h_{4,3}(x) = 149\,299\,200x^6(1+x)^6$$
$$h_{5,3}(x) = 18\,059\,231\,232\,000x^{10}(1+x)^{10}$$
$$h_{5,3}(x) = 87\,377\,784\,392\,908\,800\,000x^{15}(1+x)^{15}$$

For $r = 4$

$$h_{1,4}(x) = 24$$
$$h_{2,4}(x) = 2\,880x(1+x)$$
$$h_{3,4}(x) = 4\,147\,200x^3(1+x)^3$$
$$h_{4,4}(x) = 125\,411\,328\,000x^6(1+x)^6$$
$$h_{5,4}(x) = 121\,358\,033\,879\,040\,000x^{10}(1+x)^{10}$$
$$h_{6,4}(x) = 5\,284\,608\,400\,083\,124\,224\,000\,000x^{15}(1+x)^{15}$$

Thus, $h_{n,r}(x)$ supposedly has the form

$$h_{n,r}(x) = i_{n,r}x^{\binom{n}{2}}(1+x)^{\binom{n}{2}}.$$

J. L. Ramírez suggests that

$$i_{n,r} = r!\prod_{k=1}^{n-1}k!\prod_{k=1}^{n-1}(k+r)!.$$

Chapter 9

Avoidance of big substructures

In this and the next chapter, we start to study those partitions and permutations where the blocks/cycles are limited in size.

9.1 The Bessel numbers

We go back to the prison problem. Let us suppose again that a guard must put six prisoners into four cells. But now the *size* of the cells is limited: one cell can host a maximum of two prisoners. Under this restriction, there is only one partition pattern: two cells host two prisoners, and the other two host one and one. This gives

$$\binom{6}{2}\binom{4}{2} = 90$$

possibilities for the guard, but, it must also be taken into account that the order of the cells does not count. The binomial coefficients already guaranteed that the order of the prisoners in the cells does not matter, so the order of the one-sized cells is already considered. We divide only by the order of the two two-sized cells which is $2! = 2$. Hence, the final result is 45.

If the guard knows about the Bessel numbers, he could say: the solution to my problem is the Bessel number of parameters 6 and 2, i.e., 45. We do not know the Bessel numbers yet, so the time has come to introduce them.

Definition 9.1.1. *If n elements are put into k blocks such that the size of the blocks is at most two, then the total number of allocations is given by the* $\left\{ {n \atop k} \right\}_{\leq 2}$ *Bessel[1] number[2] of parameters n and k.*

As far as we know, Bessel had nothing to do with the restricted partition problem above. The Bessel numbers are related to the Bessel polynomials and these are what Bessel was interested in. See Section 9.5 for this polynomial

---

[1] Friedrich Wilhelm Bessel (1784-1846), German astronomer and mathematician. He was the first who managed to measure the distance between the Earth and a star (the 61 Cygni).

[2] Usually the notation, $B(n,k)$ is used for the Bessel numbers but we think that the above notation is more expressive. And it is apt for further generalization.

relation. These numbers were relatively recently studied by a number of authors [146, 147, 148, 149, 215], see also [225] for an application and [416] for divisibility questions.

Closed formula for the Bessel numbers

If an n element set is partitioned into k blocks with the above restriction, then, obviously, there are $2k - n$ singletons and $n - k$ blocks of size two,

$$2k - n + 2(n - k) = 2k - n + 2n - 2k = n.$$

Knowing the partition pattern formula (2.31) we can give a simple expression for the Bessel numbers:

$$\left\{ {n \atop k} \right\}_{\leq 2} = \frac{n!}{(2k - n)! 2^{n-k}(n - k)!}. \tag{9.1}$$

Clearly, k cannot be smaller than $\lceil n/2 \rceil$ (this is the upper integer part of $n/2$). It can also be seen that the Bessel numbers satisfy the recursion

$$\left\{ {n \atop k} \right\}_{\leq 2} = \left\{ {n - 1 \atop k - 1} \right\}_{\leq 2} + (n - 1)\left\{ {n - 2 \atop k - 1} \right\}_{\leq 2}. \tag{9.2}$$

This is not proven here, as we will present a more general treatment from which this will immediately follow.

9.2   The generating functions of the Bessel numbers

The horizontal generating function of $\left\{ {n \atop k} \right\}_{\leq 2}$

Recall the (2.46) identity for the Stirling numbers of the second kind:

$$\sum_{k=0}^{n} \left\{ {n \atop k} \right\} x^{\underline{k}} = x^n.$$

We complicate the right-hand side a bit. e^{tx} is simply the exponential generating function of x^n. Because of this fact, based on (2.9), we know that the nth derivative of e^{tx} with respect to t at $t = 0$ equals x^n. Therefore

$$\left. (e^{tx})^{(n)} \right|_{t=0} = \sum_{k=0}^{n} \left\{ {n \atop k} \right\} x^{\underline{k}}.$$

The exponential is

$$e^{tx} = (e^t)^x = \left(1 + t + \frac{t^2}{2!} + \frac{t^3}{3!} + \cdots\right)^x.$$

If we truncate this series drastically, then we get the horizontal generating function of the Bessel numbers:

$$f(t) = \left(1 + t + \frac{t^2}{2!}\right)^x.$$

Our statement, to be proven below, is that

$$f^{(n)}(0) = \sum_{k=1}^{n} \left\{n \atop k\right\}_{\leq 2} x^{\underline{k}}. \tag{9.3}$$

Note one interesting thing. We permit blocks of size at most two and, at the same time, in the horizontal generating function the maximal exponent of t is two! This is an appearance of a more general phenomenon to be fully developed in Sections 10.3-10.4.

We prove (9.3). By Exercise 24 of Chapter 2

$$f(t) = \left[1 + \left(t + \frac{t^2}{2!}\right)\right]^x = \sum_{k=0}^{\infty} \frac{x^{\underline{k}}}{k!}\left(t + \frac{t^2}{2!}\right)^k =$$

$$\sum_{k=0}^{\infty} \frac{x^{\underline{k}}}{k!}\left(\sum_{l=0}^{k} \binom{k}{l} t^{k-l}\left(\frac{t^2}{2!}\right)^l\right) =$$

$$\sum_{k=0}^{\infty}\sum_{l=0}^{k} \frac{x^{\underline{k}} t^{k+l}}{2^l l!(k-l)!} = \sum_{n=0}^{\infty}\sum_{k=\lceil n/2\rceil}^{n} \frac{x^{\underline{k}}}{2^{n-k}(n-k)!(2k-n)!} t^n.$$

In the last step, we re-indexed such that $k + l = n$. Since

$$f(t) = \sum_{k=0}^{\infty} f^{(n)}(0)\frac{t^n}{n!},$$

Our statement (9.3) indeed holds by (9.1).

The exponential generating function of $\left\{n \atop k\right\}_{\leq 2}$

Now we determine the exponential generating function of $\left\{n \atop k\right\}_{\leq 2}$. It is already apparent from (9.1) that the sum

$$\sum_{n=0}^{\infty} \left\{n \atop k\right\}_{\leq 2} \frac{x^n}{n!}$$

has non-zero terms only for values when $n \geq k$ and $n \leq 2k$. Therefore, the function we are looking for is actually a *finite* sum:

$$\sum_{n=k}^{2k} \left\{ {n \atop k} \right\}_{\leq 2} \frac{x^n}{n!}. \tag{9.4}$$

For the Stirling and r-Stirling numbers, it was true that the exponential generating function is of the form $f^k(x)/k!$, so we insist that for some $f(x)$

$$\frac{f^k(x)}{k!} = \sum_{n=k}^{2k} \left\{ {n \atop k} \right\}_{\leq 2} \frac{x^n}{n!}.$$

Let us determine $f(x)$. Rewrite the recursion (9.2):

$$\left\{ {n+1 \atop k+1} \right\}_{\leq 2} = \left\{ {n \atop k} \right\}_{\leq 2} + n \left\{ {n-1 \atop k} \right\}_{\leq 2}.$$

This shifting is useful to avoid second order derivatives. By the observations made on p. 38, it can be seen that this modified recursion holds true only if the exponential generating function of $\left\{ {n \atop k} \right\}_{\leq 2}$ satisfies the equation

$$\frac{f^k(x)}{k!} f'(x) = \frac{f^k(x)}{k!} + x \frac{f^k(x)}{k!}.$$

We can simplify to end up with

$$f'(x) = 1 + x.$$

Thus,

$$f(x) = x + \frac{x^2}{2} + c$$

As $f(x)$ does not contain constant term (as it is seen from (9.4)) we have that $c = 0$. Therefore,

$$\sum_{n=k}^{2k} \left\{ {n \atop k} \right\}_{\leq 2} \frac{x^n}{n!} = \frac{1}{k!} \left(x + \frac{x^2}{2} \right)^k. \tag{9.5}$$

This is the first case when we encounter a polynomial exponential generating function of a counting sequence. Observe that on both sides we have a $2k$-degree polynomial of x.

9.3 The number of partitions with blocks of size at most two

The Bessel numbers are very similar to the second kind Stirling numbers, except the restriction on the size of the blocks. Summing over the number of the blocks we got the Bell numbers before. Summing now the Bessel numbers we get another interesting counting sequence.

Definition 9.3.1. *A number of partitions of an n element set where the blocks are restricted to be singletons or size two are given by the* 2-restricted Bell *numbers. These numbers are denoted*[3] *by $B_{n,\leq 2}$.*

Exponential generating function of the 2-restricted Bell numbers and polynomials

Clearly, the 2-restricted Bell numbers can be calculated via the Bessel numbers:

$$B_{n,\leq 2} = \sum_{k=\lceil n/2 \rceil}^{n} \left\{ {n \atop k} \right\}_{\leq 2}.$$

This relation helps us find the exponential generating function of $B_{n,\leq 2}$. But it is better to be a bit more general and find the exponential generating function of the polynomial

$$B_{n,\leq 2}(y) = \sum_{k=\lceil n/2 \rceil}^{n} \left\{ {n \atop k} \right\}_{\leq 2} y^k.$$

Multiply both sides of (9.5) by y^k, sum over k and interchange the order of summation:

$$\sum_{n=0}^{\infty} B_{n\leq 2}(y) \frac{x^n}{n!} = e^{y\left(x + \frac{x^2}{2}\right)}.$$

In particular, for the 2-restricted Bell numbers:

$$\sum_{n=0}^{\infty} B_{n,\leq 2} \frac{x^n}{n!} = e^{x + \frac{x^2}{2}}.$$

Compare this to (3.19), the exponential generating function of involutions. The two functions are the same. This results in the – otherwise obvious – fact that the number of involutions is equal to the number of 2-partitions (on the same set). In short,

$$B_{n,\leq 2} = I_n.$$

The involutions are constituted by fixed points and transpositions. These correspond to singletons and 2-blocks, respectively.

These numbers appear in many contexts, see the definitions given in the Sloane On-Line Encyclopedia of Integer Sequences http://oeis.org/A000085, and also [600] for connection with matrix theory. We now show another interesting appearance of the $B_{n,\leq 2}$ (and I_n) numbers.

---

[3]There is no standard notation for these numbers, although they already appeared in the literature no later than in 1800! See some historical details in [438].

9.4   Young diagrams and Young tableaux

The *Young*[4] diagram often pops up in different contexts, not only in combinatorics but also in physics. Young diagrams are very easy constructions but lead to highly non-trivial results.

We put some squares in a row and then we repeat this process in new rows such that the squares touch each other in the rows and columns and, in addition, the number of the squares decreases row-by-row. A permitted configuration is

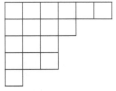

The following is also a correct configuration:

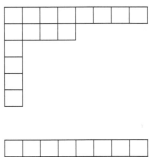

However,

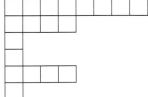

is not permitted. These diagrams are often used to describe partitions of positive integers into sums [25].

Young diagrams become *Young tableaux*[5] when we fill in the squares with positive integers such that the rows from left to right and the columns from the top down contain the numbers increasingly. Take the very first Young diagram

---

[4] Alfred Young (1873-1940), British mathematician.

[5] The word "tableaux" comes from the French word meaning "board", and it is the plural form of "tableau".

above which contains 17 squares, and fill it to make it a Young tableau[6]:

1	3	5	6	8	9
2	4	7	10		
11	13	15			
12	14	16			
17					

There are many solutions, another one is the following:

1	4	6	7	8	10
2	5	9	11		
3	12	14			
13	15	17			
16					

This enjoyable simple game is not as innocent as it looks; it is applied in Quantum Theory, for example[7] Here we are interested in the number of Young tableaux pairs to be introduced soon.

We study now the so-called *bumping algorithm* through an example. Let the number of squares in the Young tableau be fixed: $n = 10$ in our example. Let us give an arbitrary permutation on these 10 numbers.

$$\begin{pmatrix} 1 & 2 & 3 & 4 & 5 & 6 & 7 & 8 & 9 & 10 \\ 6 & 8 & 5 & 2 & 3 & 9 & 7 & 4 & 1 & 10 \end{pmatrix}.$$

The bumping algorithm assigns a Young tableau to this permutation in the following manner. We start the tableau with the first element of the permutation (the first in the second line):

6

Then we successively fill the row of 6 with the subsequent elements of the permutation but *only if* it results in a permitted tableau (i.e., if the new element is greater than the former ones in the row). If this is not the case, then the first greater element in the row (counted from the left) is *bumped out* from its position and it goes to the next row. Before inserting this bumped element, we must check whether there is an element in this row to be bumped.

In our example 6 is put, so the next element, 8 comes. As $8 > 6$ it is just put to the right of 6:

6	8

---

[6]Normally it is permitted to choose a number repeatedly. If no repetition is allowed, the tableau is called *standard*. We do not consider non-standard tableaux, so we do not use the attribute "standard."

[7]In Quantum Theory, a central question is the description of symmetries of molecules. Young diagrams help to describe the decomposition of such symmetries into simpler, "irreducible" ones. More on this topic can be found in [238, 354].

Now 5 comes. As 5 is smaller than 6, 6 is bumped out and goes to the next row, and 5 goes into the position of 6:

5	8
6	

The following permutation element is 2 which bumps 5 out. Hence, 5 would go to the next line, but $6 > 5$ is already there so 6 is bumped out:

2	8
5	
6	

The rest of the steps is here:

2	3
5	8
6	

2	3	9
5	8	
6		

2	3	7
5	8	9
6		

2	3	4
5	7	9
6	8	

1	3	4
2	7	9
5	8	
6		

The algorithm terminates after inserting 10 (simply in the first line as there are no greater elements):

1	3	4	10
2	7	9	
5	8		
6			

Note that the output of the algorithm is a Young tableau with increasing rows and columns.

In the next step, we extend this algorithm by constructing a second Young tableau parallel to the first one. In the kth step, we put down the kth element of the permutation as we did before and, in addition, we attach squares to the second Young tableau such that it follows the shape of the first one. The new square in the second tableau is to be filled with k. This table is called *insertion table*.

The first step in this extended algorithm results in the pair

| 6 | | 1 |

In the second step, the first table extends to the right so we extend the second table similarly:

| 6 | 8 | | 1 | 2 |

In the third step, a second line is inserted so we modify the second table accordingly:

5	8		1	2
---	---		---	---
6			3	

and so on. The final result is the Young tableau pair

1	3	4	10
2	7	9	
5	8		
6			

1	2	6	10
3	5	7	
4	8		
9			

It is crucial that the initial permutation can be traced back from the pair. This fact is called *Robinson*[8] *– Schensted*[9]*-correspondence*.

Now we explain how these Young tableaux are related to the restricted Bell numbers. Because of the Robinson–Schensted-correspondence, there is a bijective relation between permutations and Young tableaux pairs. Moreover, if the permutation π results in a Young tableau pair (Y_1, Y_2), then the inverse of π results in (Y_2, Y_1). In particular, if $\pi^{-1} = \pi$, i.e., $\pi^2 = \pi$, then this permutation results in a standard Young tableau pair of identical tableaux, (Y, Y). By the bijection of the Robinson–Schensted-correspondence, we get that there are as many Y Young tableaux as many involutions.

9.5 The differential equation of the Bessel polynomials

In physical problems, the

$$x^2 y_n''(x) + (2x + 2)y_n'(x) = n(n+1)y_n(x) \qquad (9.6)$$

Bessel differential equation often appears [77, 268]. We show that the solutions of this equation are polynomials and the coefficients of these are the Bessel numbers. This is where the name of the $B(n,k) = \left\{ {n \atop k} \right\}_{\leq 2}$ numbers comes[10].

The solutions of the Bessel differential equation are the

$$y_n(x) = \sum_{k=0}^{n} \frac{(n+k)!}{2^k k!(n-k)!} x^k$$

Bessel polynomials. It is easy to show that they indeed satisfy Equation (9.6)[11]. The terms of the equation, taking the explicit form of $y_n(x)$ into

---

[8] Gilbert de Beauregard Robinson (1906-1992), Canadian mathematician.

[9] Craige Eugene Schensted (1927), physicist. He described the above insertion algorithm 23 years after Robinson, but it was Schensted who made the correspondence widely known. Interestingly, Schensted changed his name to Ea Ea in 1999. Ea comes from the name of the Sumerian god Enki.

[10] And the Bessel functions are named so by H. L. Krall and O. Frink [344].

[11] If we would not have the explicit coefficients of the polynomials provided, still we could find them. We should write down a function with unknown coefficients and substitute this function into the differential equation. Performing the derivatives, we would get a recursion for the coefficients and then we could show that these coefficients are the ones given above.

account, is

$$x^2 y_n''(x) = \sum_{k=0}^{n} \frac{(n+k)!}{2^k (k-2)!(n-k)!} x^k,$$

$$2xy_n'(x) = \sum_{k=0}^{n} \frac{(n+k)!}{2^{k-1}(k-1)!(n-k)!} x^k,$$

$$2y_n'(x) = \sum_{k=1}^{n} \frac{(n+k)!}{2^{k-1}(k-1)!(n-k)!} x^{k-1} = \sum_{k=0}^{n} \frac{(n+k+1)!}{2^k k!(n-k-1)!} x^k,$$

$$n(n+1)y_n(x) = \sum_{k=0}^{n} n(n+1) \frac{(n+k)!}{2^k k!(n-k)!} x^k.$$

Summing the first three of these, we need to show that this sum agrees with the right-hand side. As on both sides we have polynomials we simply compare the coefficients.

$$\frac{(n+k)!}{2^k (k-2)!(n-k)!} + \frac{(n+k)!}{2^{k-1}(k-1)!(n-k)!} + \frac{(n+k+1)!}{2^k k!(n-k-1)!} =$$

$$n(n+1) \frac{(n+k)!}{2^k k!(n-k)!}$$

Multiplying by $2^k k!(n-k)!$ and dividing by $(n+k)!$ the left-hand side simplifies to

$$k(k-1) + 2k + (n+k+1)(n-k) = n(n+1),$$

so the Bessel polynomials are indeed the solutions of (9.6), as we stated.

To relate these $y_n(x)$ polynomials to the Bessel numbers, we make the following transformation[12].

$$x^n y_n \left(\frac{1}{x}\right) = \sum_{k=0}^{n} \frac{(n+k)!}{2^k k!(n-k)!} x^{n-k} = \sum_{k=0}^{n} \frac{(n+n-k)!}{2^{n-k}(n-k)!(n-(n-k))!} x^k =$$

$$\sum_{k=0}^{n} \frac{(2n-k)!}{2^{n-k}(n-k)!k!} x^k.$$

If we introduce the temporary notation

$$B_{n,k} = \frac{(2n-k)!}{2^{n-k}(n-k)!k!},$$

then we get the connection we were looking for:

$$\left\{ n \atop k \right\}_{\leq 2} = B(n,k) = B_{k,2k-n}.$$

This yields the identification between the Bessel numbers, and Bessel polynomial coefficients. More on the Bessel polynomials can be found in [268].

---

[12]The transformed polynomial $x^n y_n \left(\frac{1}{x}\right)$ is also often used, it is usually denoted by $\theta_n(x)$, and called *reverse Bessel polynomial* [268].

9.6 Blocks of maximal size m

There is an obvious way for the generalization of the Bessel numbers.

Definition 9.6.1. *The number of those k-partitions of n elements which contain blocks of size at most m is given by the* m*-restricted Stirling number of the second kind*[13] $\left\{{n \atop k}\right\}_{\leq m}$.

Note that the number k cannot be smaller than $\lceil n/m \rceil$.

For the 2-restricted second kind Stirling numbers (i.e., the Bessel numbers) we could easily give a closed form (see (9.1)) but when the blocks can be longer than two, things get complicated. Therefore, we cannot expect a simple closed form for $\left\{{n \atop k}\right\}_{\leq m}$ for a general m.

The notation $\left\{{n \atop k}\right\}_{\leq m}$ is still not very widely known and accepted, but we insist to be the advocates of this notation as we feel it very expressive.

We postponed the proof of (9.2)) saying that it will be proven in a more general form later. The time has now come. The m-restricted second kind Stirling numbers satisfy the recursion

$$\left\{{n \atop k}\right\}_{\leq m} = \left\{{n-1 \atop k-1}\right\}_{\leq m} + k\left\{{n-1 \atop k}\right\}_{\leq m} - \binom{n-1}{m}\left\{{n-m-1 \atop k-1}\right\}_{\leq m}. \quad (9.7)$$

It is worth it to compare this to the recursion (1.4) of the classical Stirling numbers. Here we have a third term. A set of n elements can recursively be partitioned into k blocks of maximal size m such that (1) either we construct a $(k-1)$-partition on $n-1$ elements, and we put the last element n into a new block ($\left\{{n-1 \atop k-1}\right\}_{\leq m}$ cases); or (2) the element n goes to an already existing block, which results $k\left\{{n-1 \atop k}\right\}_{\leq m}$ cases. But in the second case we must be careful as putting down n we can exceed the size limit of the block where n goes. This can happen only when there are $k-1$ blocks on $n-1-m$ elements of size not exceeding m *and* another block of size m. This latter block can be formed in $\binom{n-1}{m}$ ways, and the rest of the partition can be constructed in $\left\{{n-1-m \atop k-1}\right\}_{\leq m}$ ways. The number of these unwanted cases must be subtracted from the above possibilities.

The initial conditions are, for $n \geq 1$,

$$\left\{{0 \atop 0}\right\}_{\leq m} = 1 \quad \text{and} \quad \left\{{n \atop 0}\right\}_{\leq m} = \left\{{0 \atop n}\right\}_{\leq m} = 0.$$

---

[13] In some works, the 2-restricted Stirlings are also called bi-restricted, the 3-restricted Stirlings are called tri-restricted and, in general, named as multi-restricted Stirling numbers.

The analogue of (1.5) and the exponential generating function

We are now going to prove the analogue of (1.5) which we present here for the convenience of the reader

$$\left\{ {n+1 \atop k+1} \right\} = \sum_{i=0}^{n} \binom{n}{i} \left\{ {i \atop k} \right\}.$$

The m-restricted version is

$$\left\{ {n+1 \atop k+1} \right\}_{\leq m} = \sum_{i=0}^{m-1} \binom{n}{i} \left\{ {n-i \atop k} \right\}_{\leq m}$$

Consider a set of $n+1$ elements and pick an element, say, the first one. Then, by our restriction, this element can share its block with $i = 0, 1, \ldots, m-1$ other elements. Thus, the block of 1 can be filled in $\binom{n}{i}$ ways, while the remaining $n + 1 - 1 - i = n - i$ elements must go into k blocks to have $k + 1$ blocks in total. Summing on i the statement follows.

Later we prove in a general framework that the (9.5) exponential generating function generalizes to

$$\sum_{n=k}^{mk} \left\{ {n \atop k} \right\}_{\leq m} \frac{x^n}{n!} = \frac{1}{k!} \left(x + \frac{x^2}{2!} + \frac{x^3}{3!} + \cdots + \frac{x^m}{m!} \right)^k. \qquad (9.8)$$

Note the following: if m tends to infinity, we get $e^x - 1$ in the parenthesis and we arrive at the exponential generating function of the classical second kind Stirling numbers.

Some special values of $\left\{ {n \atop k} \right\}_{\leq m}$

It is obvious that

$$\left\{ {n \atop k} \right\}_{\leq m} = \left\{ {n \atop k} \right\}$$

for $k \leq n \leq mk$, while k is in the interval $k \leq n \leq m$. It is equally easy to see that

$$\left\{ {n \atop n-1} \right\}_{\leq m} = \binom{n}{2}$$

when $n, m \geq 2$. It is a bit more work to show that

$$\left\{ {n \atop n-2} \right\}_{\leq m} = \begin{cases} \frac{3n-5}{4} \binom{n}{3} & (n \geq 4,\ m \geq 3); \\ 3\binom{n}{4} & (n \geq 4,\ m = 2), \end{cases}.$$

To show this latter statement, first let $m = 2$, and the number of blocks be $k = n - 2$. Then, for the block structure we have the only one possibility

$$\underbrace{.\,|\,.\,|\cdots|\,.\,|\;.\,.\,|\,.\,.}_{n-4}$$

That is, there are $n - 4$ singletons and two blocks of length 2. There are $\frac{1}{2}\binom{4}{2}\binom{n}{4} = 3\binom{n}{4}$ such partitions: we have to choose those four elements going to the non-singleton blocks in $\binom{n}{4}$ ways. Then we put two of four elements into the first block and the other two go to the other block: $\binom{4}{2} = 6$ cases. Finally, we have to divide by two because the order of the blocks does not matter.

If $m = 3$ then we have one more possible distribution of blocks sizes apart from the above:

$$\underbrace{.|.|\cdots|.|}_{n-5} \cdots$$

Into the last block we have $\binom{n}{3}$ possible option to put 3 elements. So if $m = 3$ and $k = n - 2$, then we have $\binom{n}{3} + 3\binom{n}{4} = \frac{3n-5}{4}\binom{n}{3}$ cases in total.

If $m \geq 4$ there is no additional configuration, so the statement holds.

More special values of the restricted Stirling numbers can be found in [340].

9.7   The restricted Bell numbers

The next step after studying the $\left\{{n \atop k}\right\}_{\leq m}$ numbers is to define the horizontal sum of $\left\{{n \atop k}\right\}_{\leq m}$:

$$B_{n,\leq m} = \sum_{k=\lceil n/m \rceil}^{n} \left\{{n \atop k}\right\}_{\leq m}.$$

This number is the nth *m-restricted Bell number* or simply *restricted Bell number*. It counts the total number of partitions of an n element set where the blocks cannot contain more than m elements.

Two recursions for $B_{n,\leq m}$

Taking an n element set, the last element – if it is not in a singleton – can go to a block of size $k \in \{2, 3, \ldots, m - 1\}$. Choose these k elements in $\binom{n-1}{k}$ ways and put the rest of the elements (there are $n - 1 - k$) into a partition taking into account the restriction of the block size. We have $\binom{n-1}{k}B_{n-1-k,\leq m}$ possibilities. If our element is in a singleton, then the rest of the elements can be partitioned into $B_{n-1,\leq m}$ different configurations. Summing over k we get that

$$B_{n,\leq m} = B_{n-1,\leq m} + \binom{n-1}{1}B_{n-2,\leq m} + \cdots + \binom{n-1}{m-1}B_{n-m,\leq m} \quad (9.9)$$

Another way of counting such partitions leads to another formula. Let us suppose that s blocks are full, that is, they contain m elements ($0 \leq s \leq$

$\lfloor n/m \rfloor$). To fill these blocks, we have

$$\binom{n}{m}\binom{n-m}{m}\binom{n-2m}{m}\cdots\binom{n-(s-1)m}{m}$$

different possibilities. We need to divide by $s!$, as the order of these blocks does not matter:

$$\frac{1}{s!}\binom{n}{m}\binom{n-m}{m}\binom{n-2m}{m}\cdots\binom{n-(s-1)m}{m} = \frac{n!}{s!m!^s(n-(s-1)m)!}.$$

The remaining (partially filled) blocks must contain $n - ms$ elements and all the blocks are of maximal size $m - 1$. There are $B_{n-ms,\leq m-1}$ ways to reach this state. Finally, we sum on s:

$$B_{n,\leq m} = \sum_{s=0}^{\lfloor n/m \rfloor} \frac{n!}{s!m!^s(n-(s-1)m)!} B_{n-ms,\leq m-1}. \tag{9.10}$$

The exponential generating function of the m-restricted Bell numbers comes immediately by (9.8):

$$\sum_{n=0}^{\infty} B_{n,\leq m}\frac{x^n}{n!} = \exp\left(x + \frac{x^2}{2!} + \frac{x^3}{3!} + \cdots + \frac{x^m}{m!}\right).$$

Or, for the polynomials

$$B_{n,\leq m}(y) = \sum_{k=\lceil n/m \rceil}^{n} \begin{Bmatrix} n \\ k \end{Bmatrix}_{\leq m} y^k$$

it reads as

$$\sum_{n=0}^{\infty} B_{n,\leq m}(y)\frac{x^n}{n!} = \exp\left[y\left(x + \frac{x^2}{2!} + \frac{x^3}{3!} + \cdots + \frac{x^m}{m!}\right)\right].$$

9.8   The gift exchange problem

The m-restricted Stirling numbers of the second kind appear in an interesting context – in the so-called gift exchange problem. This problem (and the description in the below paragraph verbatim) is taken from [33] (see also the newer version [30]).

We have an evening party with n guests. Each guest brings a wrapped gift, the gifts are placed on a table (this is the "pool" of gifts), and slips of paper containing the numbers 1 to n are distributed randomly among the guests.

A number m is fixed by consensus. The host calls out the numbers 1 through n in order. When the number you have been given is called, you can either choose one of the wrapped (and so unknown) gifts remaining in the pool, or you can take (or "steal") a gift that some earlier person has unwrapped, subject to the restriction that no gift can be "stolen" more than a total of m times. If you choose a gift from the pool, you unwrap it and show it to everyone. If a person's gift is stolen from them, they immediately get another turn, and can either take a gift from the pool, or can steal someone else's gift, subject always to the limit of m thefts per gift. The game ends when someone takes the last (nth) gift. The problem is to determine the number of possible ways the game can be played out, for given values of m and n.

Let us take the example from the original text! We have three players, A, B, C, and fix m to be 1, that is, an already taken gift can be stolen at most once. Then a possible output is the following: guest A chooses a gift, B chooses another one and guest C chooses a third gift. If there is someone who steals, then, for instance, A can choose gift 1, B chooses gift 2, but C steals gift 1. Then, as 1 was previously chosen by A, he or she must choose again. As a gift cannot be stolen more than once, gift 1 cannot be chosen anymore. If A chooses, say, 3, then the game terminates as all the gifts are chosen. If A chooses 2, then B follows as 2 is his/her gift. Now B can choose neither 1 nor 2 as these were stolen already. B ends up with the only one choice, the third gift. The game terminates. These processes can shortly be described by numbers and letters:

A1	B2	C3		
A1	B2	C1	A3	
A1	B2	C1	A2	B3

It is not hard to see that there are four other possible outputs:

A1	B2	C2	B3	
A1	B2	C2	B1	A3
A1	B1	A2	C3	
A1	B1	A2	C2	A3

Altogether there are seven different game outputs. To simplify life a bit, we suppose that the guest who is about to choose either steals or chooses the gift which is one greater than the greatest already chosen gift. (If the gifts already chosen by the previous guests are $3, 7, 5$, then he or she chooses 8.) This is not a restriction as we can label the gifts in any order we want.

Another simplification can be done. If the players do not change their seats during the game and always A begins we can simply leave out the letters. The list

1	2	3		
1	2	1	3	
1	2	1	2	3
1	2	2	3	
1	2	2	1	3
1	1	2	3	
1	1	2	2	3

still describes the game flow uniquely.

The game would be uninteresting if $m = 0$ as there was only one output (taking into account the simplifications we have made so far). Hence, we suppose that $m > 0$. The following considerations can be given. The output lists always begin with 1 and always terminate with n. The length of a list is always at least n and at most $(n-1)(m+1)+1$. Each element between 1 and $n-1$ appears at least once, at most m times; n always appears only once (at the end of the list). Because of the claim on the order a gift i can appear in a list only if $i-1$ has already appeared. We get the following simplified list after leaving out the last element.

1	2		
1	2	1	
1	2	1	2
1	2	2	
1	2	2	1
1	1	2	
1	1	2	2

For concreteness, we fix the length of the list to be k (so that $n - 1 \leq k \leq (n-1)(m+1)$). From this list one can construct an $(n-1)$-partition on k elements in the following way: the first block contains the positions of 1, the second block contains the positions of 2, and so on. Finally, the last block contains the positions of the penultimate gift $n-1$.

In the above example, we get the below $3 - 1 = 2$-partitions:

1	2
1,3	2
1,3	2,4
1	2,3
1,4	2,3
1,2	3
1,2	3,4

This is nothing else but the possible decompositions of a set of two, three, or four elements into two blocks where all the blocks are of size at most $m = 2$. This statement is true in its full generality: a gift exchange output encodes a partition on sets of at least $n - 1$ and at most $(n - 1)(m + 1)$ elements, where there are $n-1$ blocks, and each block contains at most $m+1$ elements. Summing over the possible k length of lists we get that there are

$$\sum_{k=n-1}^{(n-1)(m+1)} \left\{ {k \atop n-1} \right\}_{\leq m+1}$$

possible gift exchange game outputs.

9.9 The restricted Stirling numbers of the first kind

Problems with duality

Since the duality of the *signed* first kind and second kind Stirling numbers so nicely works and even generalizes smoothly to the r-Stirling numbers, we would think that some duality should exist for the restricted Stirling numbers as well. This is not so. The table of the first 3-restricted second kind Stirling numbers begins

$\left(\left\{ {n \atop k} \right\}_{\leq 3}\right)_{6\times 6}$	$k=1$	$k=2$	$k=3$	$k=4$	$k=5$	$k=6$
$n=1$	1					
$n=2$	1	1				
$n=3$	1	3	1			
$n=4$	0	7	6	1		
$n=5$	0	10	25	10	1	
$n=6$	0	10	75	65	15	1

while the corresponding inverse matrix is

$\left(\left\{ {n \atop k} \right\}_{\leq 3}\right)_{6\times 6}^{-1}$	$k=1$	$k=2$	$k=3$	$k=4$	$k=5$	$k=6$
$n=1$	1					
$n=2$	-1	1				
$n=3$	2	-3	1			
$n=4$	-5	11	-6	1		
$n=5$	10	-45	35	-10	1	
$n=6$	35	175	-210	85	-15	1

In the last line of the inverse, we see that the sign does *not* alternate. Ji Young Choi et al. called this disturbing phenomenon an anomalous sign behavior [149]. This behavior is an obstacle for *direct* combinatorial interpretations of these sequences of the rows in the inverses. Although it is obvious how to define the $\left[{n \atop k}\right]_{\leq m}$ numbers or, better, the signed version $(-1)^{n-k}\left[{n \atop k}\right]_{\leq m}$ but these will not coincide with the inverse matrix elements[14].

The problem of anomalous sign behavior was solved recently. It turned out that the items of the inverse matrix count *differences* in sizes of two *labelled Schröder forests*. It also came out that for even m the sign behavior is regular, and the numbers $(-1)^{n-k}\left[{n \atop k}\right]_{\leq m}$ are always positive and count specific Schröder forests. These findings were published in [215] in 2016.

Definition and exponential generating function of $\left[{n \atop k}\right]_{\leq m}$

We do not insist that the first kind restricted Stirling numbers be dual with $\left\{{n \atop k}\right\}_{\leq m}$, instead, we do require them to have the expected combinatorial meaning.

Definition 9.9.1. *The number of permutations of an n element set where all the k cycles are restricted to be of length at most m is given by the m-restricted (or simply restricted) Stirling number of the first kind. We denote these numbers by $\left[{n \atop k}\right]_{\leq m}$ $(k \geq \lceil n/m \rceil)$.*

After reaching this point in the text, you might have sufficient routine to determine the exponential generating function of $\left[{n \atop k}\right]_{\leq m}$:

$$\sum_{n=k}^{mk} \left[{n \atop k}\right]_{\leq m} \frac{x^n}{n!} = \frac{1}{k!}\left(x + \frac{x^2}{2} + \frac{x^3}{3} + \cdots + \frac{x^m}{m}\right)^k. \qquad (9.11)$$

If m tends to infinity, the sum in the parenthesis tends to $\ln\left(\frac{1}{1-x}\right)$ and we get back the exponential generating function of the first kind Stirling numbers. This and (9.8) show very nicely that the only difference between the generating functions is in the denominators. In the first kind case, there are simply integers and in the second kind case, we have factorials. This is not independent of the fact that in a cycle of length l there are l possible identical configurations (by the l right shifts) and in a block of size l there are $l!$ equivalent order of the elements. We already saw similar arguments in Section 2.5. In the next chapter, we study this revelation in its full generality. More on symbolic combinatorial computations of this kind can be found in [235].

If $m = 2$, then (9.8) and (9.11) coincide and

$$\left[{n \atop k}\right]_{\leq 2} = \left\{{n \atop k}\right\}_{\leq 2}.$$

---

[14]In the particular case when $m = 2$ the inverse matrix is still regular in sign alternation and the elements are the Bessel numbers.

Two recursions for $\left[\begin{smallmatrix} n \\ k \end{smallmatrix}\right]_{n,\leq m}$

The basic recursion for the first kind restricted Stirling numbers reads as follows [341].

$$\begin{bmatrix} n+1 \\ k \end{bmatrix}_{\leq m} = n \begin{bmatrix} n \\ k \end{bmatrix}_{\leq m} + \begin{bmatrix} n \\ k-1 \end{bmatrix}_{\leq m} - \frac{n!}{(n-m)!} \begin{bmatrix} n-m \\ k-1 \end{bmatrix}_{\leq m}.$$

We have two disjoint possibilities during constructing all the permutations that the left-hand side counts:

(1) The last element, $n+1$, goes into an already constructed cycle. First we construct a permutation on n elements into k cycles in $\left[\begin{smallmatrix} n \\ k \end{smallmatrix}\right]_{\leq m}$ ways and then insert the last element among the other elements. There are n different places, so this results in $n\left[\begin{smallmatrix} n \\ k \end{smallmatrix}\right]_{\leq m}$ cases. We must be careful not to exceed the maximal cycle size m. Therefore, we have to subtract the number of the cases when the last element goes into a cycle of m elements. There are $m!\binom{n}{m}\left[\begin{smallmatrix} n-m \\ k-1 \end{smallmatrix}\right]_{\leq m}$ permutations with one m-sized cycle, and this number must be subtracted (the binomial coefficient does not count the order of the m elements but we must consider this $(m-1)!$ order; moreover, the last element can go into m places and this explains the $(m-1)!m = m!$ factor).

(2) This case is simple: the last element goes into a singleton, and the rest of the elements form a permutation with $k-1$ cycles: $\left[\begin{smallmatrix} n \\ k-1 \end{smallmatrix}\right]_{\leq m}$.

Another identity is the following, for $k > 0$:

$$\begin{bmatrix} n+1 \\ k \end{bmatrix}_{\leq m} = \sum_{i=0}^{m-1} \frac{n!}{(n-i)!} \begin{bmatrix} n-i \\ k-1 \end{bmatrix}_{\leq m}. \tag{9.12}$$

The left-hand side counts the restricted permutations of $n+1$ elements with k cycles such that each cycle contains at most m elements. Such permutations can be constructed such that we first form a restricted permutation on $n-i$ elements with $k-1$ cycles, and then we construct an additional cycle which contains the element $n+1$ in order to have k cycles in total. The first part of the construction can be realized in $\left[\begin{smallmatrix} n-i \\ k-1 \end{smallmatrix}\right]_{\leq m}$ ways. Into the kth cycle of the element $n+1$ we choose i elements in $\binom{n}{i}$ ways. Since the order of the elements in the cycles matters, we have $i!\binom{n}{i} = \frac{n!}{(n-i)!}$ possibilities. Altogether, for a fixed i, we have $\frac{n!}{(n-i)!}\left[\begin{smallmatrix} n-i \\ k-1 \end{smallmatrix}\right]_{\leq m}$ different permutations. Here i can run from 0 to $m-1$. Summing these disjoint possibilities, we are done.

Since

$$\sum_{i=1}^{n-k+1} \frac{n!}{(n-i)!} \begin{bmatrix} n-i \\ k-1 \end{bmatrix} = n \begin{bmatrix} n \\ k \end{bmatrix}$$

holds true (exercise), relation (9.12) reduces to the basic recursion of $\left[\begin{smallmatrix} n \\ k \end{smallmatrix}\right]$ if $m \geq n - k + 2$.

Some special values of $\left[\begin{smallmatrix} n \\ m \end{smallmatrix}\right]_{\le m}$

For $0 \le n \le k - 1$ or $n \ge km + 1$, we have

$$\begin{bmatrix} n \\ k \end{bmatrix}_{\le m} = 0.$$

If $k \le n \le km$, the restricted and classical Stirling numbers of the first kind coincide:

$$\begin{bmatrix} n \\ k \end{bmatrix}_{\le m} = \begin{bmatrix} n \\ k \end{bmatrix} \quad (k \le n \le m).$$

Moreover, $\left[\begin{smallmatrix} n \\ n \end{smallmatrix}\right]_{\le m} = 1$ and, also, $\left[\begin{smallmatrix} n \\ n-1 \end{smallmatrix}\right]_{\le m} = \binom{n}{2}$. Similarly,

$$\begin{bmatrix} n \\ 1 \end{bmatrix}_{\le m} = (n - 1)! \quad (1 \le n \le m).$$

Things get a bit more complicated for other special values:

$$\begin{bmatrix} n \\ n-2 \end{bmatrix}_{\le m} = \begin{cases} \frac{3n-1}{4}\binom{n}{3} & (m \ge 3); \\ 3\binom{n}{4} & (m = 2), \end{cases} \tag{9.13}$$

We prescribe the fixed points of a permutation and see how the lengthier cycles look like. First we prove the $m = 2$ case. The number of cycles is $k = n - 2$. Then for the cycle structure we have the only one possibility

$$\underbrace{(.)(.) \cdots (.)}_{n-4}(..)(..)$$

That is, there are $n - 4$ fixed points and two cycles of length 2. There are $\frac{1}{2}\binom{4}{2}\binom{n}{4} = 3\binom{n}{4}$ such permutations, as it is easy to see. Therefore, the $m = 2$ case of (9.13) follows.

If $m \ge 3$, then we have two possible distributions of cycle lengths:

$$\underbrace{(.)(.) \cdots (.)}_{n-4}(..)(..)$$

as before, or

$$\underbrace{(.)(.) \cdots (.)}_{n-3}(...)$$

There are $2\binom{n}{3}$ such permutations. Since $2\binom{n}{3} + 3\binom{n}{4} = \frac{3n-1}{4}\binom{n}{3}$, the first part of (9.13) also follows.

We remember the fact from p. 14 that the harmonic numbers and $\left[\begin{smallmatrix} n \\ 2 \end{smallmatrix}\right]$ numbers are related. Do we have some relation between $\left[\begin{smallmatrix} n \\ 2 \end{smallmatrix}\right]_{\le m}$ and harmonic (or harmonic-like) numbers? The answer is yes:

$$\begin{bmatrix} n \\ 2 \end{bmatrix}_{\le m} = \begin{cases} (n-1)!H_{n-1} & (2 \le n \le m+1); \\ (n-1)!(H_m - H_{n-m-1}) & (m+2 \le n \le 2m). \end{cases}$$

The simplest way to present a proof is via generating functions. If $k = 2$ and $2 \leq n \leq m + 1$, (9.11) simplifies to

$$\sum_{n=2}^{2m} \begin{bmatrix} n \\ 2 \end{bmatrix}_{\leq m} \frac{x^n}{n!} = \frac{1}{2!} \left(x + \frac{x^2}{2} + \frac{x^3}{3} + \cdots + \frac{x^m}{m} \right)^2.$$

We need to find the coefficient of x^n on the right-hand side. During squaring we multiply every member in the first factor by every member of the second factor. But now we need only those factors which result in the nth power of x. If a term in the first factor is x^k/k, then the second term must be $\frac{x^{n-k}}{n-k}$. Thus,

$$\begin{bmatrix} n \\ 2 \end{bmatrix}_{\leq m} \frac{1}{n!} = \frac{1}{2} \sum_{k=1}^{n-1} \frac{1}{k(n-k)}.$$

Hence, by the decomposition $\frac{1}{k(n-k)} = \frac{1}{n} \left(\frac{1}{k} + \frac{1}{n-k} \right)$ we infer that

$$\begin{bmatrix} n \\ 2 \end{bmatrix}_{\leq m} = \frac{n!}{2} \frac{1}{n} \sum_{k=1}^{n-1} \left(\frac{1}{k} + \frac{1}{n-k} \right) = (n-1)! H_{n-1}$$

holds.

Similarly, if $k = 2$ and $m + 2 \leq n \leq 2m$, by analyzing again the generating function, we have the intermediate result

$$\begin{bmatrix} n \\ 2 \end{bmatrix}_{\leq m} \frac{1}{n!} = \frac{1}{2} \sum_{k=0}^{2m-n} \frac{1}{(n-m+k)(m-k)},$$

from where

$$\begin{bmatrix} n \\ 2 \end{bmatrix}_{\leq m} = \frac{(n-1)!}{2} \sum_{k=1}^{n-1} \left(\frac{1}{n-m+k} + \frac{1}{m-k} \right)$$
$$= (n-1)! (H_m - H_{n-m-1}).$$

The restricted factorials $A_{n, \leq m}$

Summing all the $\begin{bmatrix} n \\ k \end{bmatrix}$ Stirling numbers of the first kind we simply get $n!$, the total number of permutations on n elements. If we sum the $\begin{bmatrix} n \\ k \end{bmatrix}_{\leq m}$ restricted Stirling numbers, the result is far from being so simple. The number

$$A_{n, \leq m} = \sum_{k=\lceil n/m \rceil}^{n} \begin{bmatrix} n \\ k \end{bmatrix}_{\leq m}$$

gives the total number of permutations on n elements such that in the cycle decomposition of the individual permutations the lengths are not longer than m.

Their recursion can similarly be proven as (9.9) for the m-restricted Bell numbers. The only difference that the binomial coefficients must be multiplied by the corresponding factorial as the order of the elements in the cycle counts. We get

$$A_{n,\leq m} = A_{n-1,\leq m} + (n-1)A_{n-2,\leq m} + (n-1)(n-2)A_{n-3,\leq m} + \cdots +$$

$$+(n-1)(n-2)\cdots(n-(m-1))A_{n-m,\leq m}. \tag{9.14}$$

As the $B_{n,\leq m}$ numbers are "restricted versions" of the Bell numbers, the $A_{n,\leq m}$ numbers are the "restricted versions" of the factorials. Indeed,

$$A_{n,\leq n} = n!.$$

Therefore, we call these numbers as *m-restricted factorials* or simply *restricted factorials*[15].

Note that
$$A_{n,\leq 2} = B_{n,\leq 2} = I_n.$$

The exponential generating function of these numbers is

$$\sum_{n=0}^{\infty} A_{n,\leq m}\frac{x^n}{n!} = \exp\left(x + \frac{x^2}{2} + \frac{x^3}{3} + \cdots + \frac{x^m}{m}\right).$$

See the Outlook of the next chapter to get information about the newest results with respect to the restricted Stirling and related numbers.

---

[15]There is no commonly used name for these sequences.

Exercises

1. In how many ways can we stick 10 photos into an album if each page has space for at most two pictures? How many possibilities are there if we want to use exactly 6 pages?

2. Construct the Young tableau pair from the permutation

$$\begin{pmatrix} 1 & 2 & 3 & 4 & 5 & 6 & 7 & 8 & 9 & 10 \\ 1 & 2 & 5 & 7 & 8 & 3 & 9 & 10 & 4 & 6 \end{pmatrix}.$$

3. Take the Young tableau pair

1	4	6	7
2	3	5	8
9			

1	2	3	4
5	7	8	9
6			

and decode the original permutation by the Robinson-Schensted correspondence. (Hint: observe from the second table that the last element we put down is 8, and 8 was bumped out necessarily by 7. As 7 is in the first row, we infer that 7 is the last element in the permutation. The other steps are similar.)

4. By using (4.1) show that the Bessel numbers form log-concave sequences $(n \geq 1)$. Hint: let

$$a_m = \frac{B(n, m)}{B(n, m - 1)} \quad (m = \lceil n/2 \rceil + 1, \dots, n),$$

and show that $u_m/u_{m+1} > 1$. Moreover, verify that $a_{\lceil n/2 \rceil + 1} > 1$ and $a_n < 1$ for $n > 7$. (This unimodality problem was studied in [147], and was proved combinatorially in [276].)

5. Use (9.5) to prove the (9.1) exact formula for the Bessel numbers. (Hint: use the binomial theorem.)

6. Prove combinatorially that

$$\left\{ \begin{matrix} n \\ k \end{matrix} \right\}_{\leq m} = \sum_{j=0}^{\lfloor n/m \rfloor} \frac{1}{j!} \underbrace{\binom{n}{m, m, \dots, m}}_{j} \left\{ \begin{matrix} n - mj \\ k - j \end{matrix} \right\}_{\leq m-1}.$$

For the definition of the multinomial coefficient, see the appendix. (The proof is simple but if you need help, the article [341] contains a proof in combinatorial flavor.)

7. Show the validity of the first kind version of the above identity:

$$\begin{bmatrix} n \\ k \end{bmatrix}_{\leq m} = \sum_{j=0}^{\lfloor n/m \rfloor} \frac{1}{j!} n^{\underline{m}} (n-m)^{\underline{m}} \cdots (n-(j-1)m)^{\underline{m}} \begin{bmatrix} n - mj \\ k - j \end{bmatrix}_{\leq m-1}.$$

8. Prove that

$$\sum_{i=1}^{n-k+1} \frac{n!}{(n-i)!} \begin{bmatrix} n-i \\ k-1 \end{bmatrix} = n \begin{bmatrix} n \\ k \end{bmatrix}$$

holds true.

9. Give a combinatorial reasoning why the following special values hold:

$$\begin{Bmatrix} n \\ n-3 \end{Bmatrix}_{\leq m} = \begin{cases} \binom{n}{4}\binom{n-2}{2} & (n \geq 4, m \geq 4); \\ 15\binom{n}{6} + 10\binom{n}{5} & (n \geq 4, m = 3); \\ 15\binom{n}{6} & (n \geq 4, m = 2). \end{cases}$$

10. The same task as above for the special values

$$\begin{bmatrix} n \\ n-3 \end{bmatrix}_{\leq m} = \begin{cases} \binom{n}{4}\binom{n}{2} & (m \geq 4); \\ \frac{5(n+3)}{2}\binom{n}{5} & (m = 3); \\ 15\binom{n}{6} & (m = 2). \end{cases}$$

11. Define the m-restricted Fubini numbers as it should be:

$$F_{n,\leq m} = \sum_{k=0}^{n} k! \begin{Bmatrix} n \\ k \end{Bmatrix}_{\leq m}.$$

Using (9.8), determine the exponential generating function of $F_{n,\leq m}$, and the convergence radius for $m = 2, 3$. (What is the proper lower limit for k in the summation?) One can find some determinantal identities for $F_{n,\leq m}$ in [342].

12. Prove that the above-defined restricted Fubini numbers are always even when $n > m \geq 1$.

13. Prove the recursion

$$F_{n,\leq m} = \binom{n}{1} F_{n-1,\leq m} + \binom{n}{2} F_{n-2,\leq m} + \cdots + \binom{n}{m} F_{n-m,\leq m} \quad (n > m).$$

(This and the previous problem come from the paper [416].)

14. Prove by induction and also by combinatorial methods that

$$A_{m+1,\leq m} = m \cdot m!.$$

15. Prove combinatorially that

$$B_{m+1,\leq m} = B_{m+1} - 1 \quad (m > 1),$$
$$B_{m+2,\leq m} = B_{m+2} - 1 - (m + 2) \quad (m > 1).$$

16. By combinatorial reasoning, show that for any $m > \left[\frac{n}{2}\right]$

$$B_{n,\leq m} = B_{n,\leq m-1} + \binom{n}{m} B_{n-m,\leq m}.$$

17. Generalize the formulas of the previous exercise:

$$B_{n,\leq m} = B_n - \sum_{k=1}^{n-m} \binom{n}{m+k} B_{n-m-k} \quad (m < n \leq 2m).$$

(See [416] for the details.)

18. By using (9.9), justify that

$$B_{n+p,\leq m} \equiv B_{n,\leq m} \quad (\text{mod } p).$$

Prove also that the congruences

$$B_{p,\leq m} \equiv 1 \quad (\text{mod } p)$$

are valid for any prime p for $m < p$. (The last two results can be found in [438].)

19. Based on (1.11) in Chapter 1, we can look for a sequence $N^m(n, k)$ such that

$$N^m(n, k) = (-1)^{n+k} M_2^m(-k, -n),$$

to insist duality. Prove that in this case, by using the recursion (9.7) for $\left\{\begin{matrix} n \\ k \end{matrix}\right\}_{\leq m}$, we have

$$N^m(n-1, k-1) - $$
$$N^m(n, k) + (n-1)N^m(n-1, k) - \binom{k+m-1}{m} N^m(n, k+m).$$

(See (1.10) in [149].)

20. Prove that

$$A_{n,\leq 2} = \sum_{j=0}^{\lfloor n/2 \rfloor} \binom{n}{2j} \frac{(2j)!}{j!2^j},$$

and

$$A_{n,\leq 3} = \sum_{j=0}^{\lfloor n/3 \rfloor} \binom{n}{3j} \frac{(3j)!}{j!3^j} A_{n-3j,\leq 2}.$$

(See the online short note of Dennis Walsh [580].)

21. Research problem: Is it possible to modify the combination lock problem of Section 6.4 to find m-restricted and m-associated Eulerian numbers $\left\langle {n \atop k} \right\rangle_{\leq m}$ and $\left\langle {n \atop k} \right\rangle_{\geq m}$ such that

$$F_{n,\leq m} = \sum_{k=0}^{n} \left\langle {n \atop k} \right\rangle_{\leq m},$$

and

$$F_{n,\geq m} = \sum_{k=0}^{n} \left\langle {n \atop k} \right\rangle_{\geq m}?$$

(For the associated partitions, see the next chapter.)

Outlook

1. The restricted Stirling numbers were used by B. Bényi [67] to enumerate special restricted *lonesum matrices*. A lonesum matrix is a matrix containing only values 0 and 1 such that it is uniquely reconstructible from its row and column sum vectors.

2. The restricted Bell numbers are connected to certain sequences of random-to-top shuffles, and sequences of box removals and additions on the Young diagrams of integer partitions [99].

3. The question whether the Bessel polynomials $y_n(x)$ are irreducible was raised in [268] (see Problem 9 on p. 163.). If the irreducibility cannot be proved in full generality, give a class of values of n for which the irreducibility holds.

4. The r-version (see the previous chapter) of the Bessel and restricted Bell numbers was studied recently in [307].

Chapter 10

Avoidance of small substructures

10.1   Associated Stirling numbers of the second kind

A binomial representation of $\left\{{n \atop n-k}\right\}$

To introduce the associated Stirling numbers of the second kind, we do not follow the usual way of the book (enumeration problem, then definition, then recursion). That has been done sufficiently many times. Instead, we first show an interesting representation of the classical Stirling numbers and then we prove that the occurring sequence enumerates what we expect – partitions of a given number of blocks where the size of the blocks is limited *from below*.

Gazing back to the recursion (1.4) of the second kind Stirling numbers, we can easily have a formula for the parameter pair $n, n - 1$:

$$
\begin{aligned}
\left\{{n \atop n-1}\right\} &= \left\{{n-1 \atop n-2}\right\} + (n-1)\left\{{n-1 \atop n-1}\right\} = \left\{{n-1 \atop n-2}\right\} + (n-1) \\
&= \left\{{n-2 \atop n-3}\right\} + (n-2)\left\{{n-2 \atop n-2}\right\} + (n-1) \\
&= \left\{{n-2 \atop n-3}\right\} + (n-2) + (n-1) = \cdots \\
&= 1 + 2 + \cdots + (n-2) + (n-1) = \binom{n}{2} \quad (n \geq 2).
\end{aligned}
$$

For the pair $n, n - 2$:

$$
\left\{{n \atop n-2}\right\} = \left\{{n-1 \atop n-3}\right\} + (n-2)\left\{{n-1 \atop n-2}\right\} = \left\{{n-1 \atop n-3}\right\} + (n-2)\binom{n-1}{2},
$$

or, rolling back to the lowest parameter values,

$$
\left\{{n \atop n-2}\right\} = (n-2)\binom{n-1}{2} + (n-3)\binom{n-2}{2} + \cdots + 1\binom{2}{2} =
$$

$$
\sum_{k=1}^{n-2} k\binom{k+1}{2}.
$$

Can we simplify this sum? It is very useful for us to note that

$$k\binom{k+1}{2} = \binom{k+1}{2} + 3\binom{k+1}{3}.$$

This observation results in

$$\sum_{k=1}^{n-2} k\binom{k+1}{2} = \sum_{k=2}^{n-1}\binom{k}{2} + 3\sum_{k=3}^{n-1}\binom{k}{3}.$$

Both sums are of the same form:

$$\sum_{k=c}^{n}\binom{k}{c}$$

It is worth it to remember this often occurring sum. It can be proven (exercise!)
that

$$\sum_{k=c}^{n}\binom{k}{c} = \binom{n+1}{c+1}. \tag{10.1}$$

By making use of this formula, we get the ultimate form

$$\left\{ \begin{matrix} n \\ n-2 \end{matrix} \right\} = \binom{n}{3} + 3\binom{n}{4} \quad (n \geq 4).$$

We will see immediately that it is always true that the Stirling numbers of the
second kind with parameters $n, n - k$ can be expressed as a sum of binomial
coefficients. The coefficients of the binomials will be studied soon. For the
moment, let us write this representation in the following form:

$$\left\{ \begin{matrix} n \\ n-k \end{matrix} \right\} = \sum_{j=0}^{k} \left\{ \begin{matrix} k+j \\ j \end{matrix} \right\}_{\geq 2} \binom{n}{k+j} \quad (n \geq k+1). \tag{10.2}$$

This notation is not too riddling, you can easily find out what is the reason
behind choosing this notation. By the above results we know that

$$\left\{ \begin{matrix} 1 \\ 0 \end{matrix} \right\}_{\geq 2} = \left\{ \begin{matrix} 2 \\ 0 \end{matrix} \right\}_{\geq 2} = 0, \quad \left\{ \begin{matrix} 2 \\ 1 \end{matrix} \right\}_{\geq 2} = \left\{ \begin{matrix} 3 \\ 1 \end{matrix} \right\}_{\geq 2} = 1, \quad \left\{ \begin{matrix} 4 \\ 2 \end{matrix} \right\}_{\geq 2} = 3.$$

The exponential generating function of $\left\{ \begin{smallmatrix} n \\ k \end{smallmatrix} \right\}_{\geq 2}$

With small effort we can find the exponential generating function of $\left\{ \begin{smallmatrix} n \\ k \end{smallmatrix} \right\}_{\geq 2}$,
and it will make obvious what is the meaning of these numbers. First let us
re-index (10.2)[1]:

$$\left\{ \begin{matrix} n \\ k \end{matrix} \right\} = \sum_{j=0}^{k} \left\{ \begin{matrix} n-j \\ k-j \end{matrix} \right\}_{\geq 2} \binom{n}{j}. \tag{10.3}$$

---

[1] From the re-indexed identity, it is actually very easy to find the combinatorial meaning
of $\left\{ \begin{smallmatrix} n \\ k \end{smallmatrix} \right\}_{\geq 2}$.

Multiplying both sides by $t^k x^n/n!$ and sum over $k = 0, 1, \ldots, n$ and over $n = 0, 1, \ldots$. The left-hand side is simply the exponential generating function of the Bell polynomials (see (3.3)), and so

$$e^{t(e^x-1)} = \sum_{n=0}^{\infty} \sum_{k=0}^{n} \sum_{j=0}^{k} \left\{ n-j \atop k-j \right\}_{\geq 2} \binom{n}{j} t^k \frac{x^n}{n!}.$$

Change the order of the summation:

$$\sum_{n=0}^{\infty} \frac{x^n}{n!} \sum_{j=0}^{n} \binom{n}{j} t^j \sum_{k=j}^{n} \left\{ n-j \atop k-j \right\}_{\geq 2} t^{k-j} =$$

$$\sum_{j=0}^{\infty} \frac{(xt)^j}{j!} \sum_{n=j}^{\infty} \frac{x^{n-j}}{(n-j)!} \sum_{k=j}^{n} \left\{ n-j \atop k-j \right\}_{\geq 2} t^{k-j}.$$

By the (2.7) multiplication rule of generating functions (Cauchy product), we see that the right-hand side is a product of two functions:

$$e^{xt} \cdot \left(\sum_{n=0}^{\infty} \frac{x^n}{n!} \sum_{k=0}^{n} \left\{ n \atop k \right\}_{\geq 2} t^k \right). \tag{10.4}$$

Recalling that the left-hand side is $e^{t(e^x-1)}$, we divide both sides by e^{xt} to get

$$\sum_{n=0}^{\infty} \frac{x^n}{n!} \sum_{k=0}^{n} \left\{ n \atop k \right\}_{\geq 2} t^k = e^{t(e^x-1-x)}. \tag{10.5}$$

This is the exponential generating function of the polynomials made up by the $\left\{ n \atop k \right\}_{\geq 2}$ numbers. From this the exponential generating function of $\left\{ n \atop k \right\}_{\geq 2}$ can also be recovered. The function on the right is represented as

$$e^{t(e^x-1-x)} = \sum_{k=0}^{\infty} \frac{[t(e^x-1-x)]^k}{k!},$$

and we extract the coefficients of t:

$$\sum_{n=k}^{\infty} \left\{ n \atop k \right\}_{\geq 2} \frac{x^n}{n!} = \frac{1}{k!} (e^x - 1 - x)^k. \tag{10.6}$$

It is very interesting to compare this to (9.5). We have

$$\frac{1}{k!} \left(x + \frac{x^2}{2!} \right)^k \quad \text{and} \quad \frac{1}{k!} \left(e^x - 1 - \frac{x^1}{1!} \right)^k,$$

so that these are "complement" of each other. Indeed, continuing the sum on the left-hand side parenthesis, or truncating the sum on the right parenthesis we get the same function $e^x - 1$.

Note also that the series in $\frac{1}{k!}\left(e^x - 1 - \frac{x^1}{1!}\right)^k$ starts with $\frac{x^2}{2!}$ and, after raising to the kth power the smallest power of x will be $2k$. Thus, $\left\{{n \atop k}\right\}_{\geq 2} = 0$ if $n < 2k$ (except when $k = 0$). Therefore, in (10.6) the summation could run from $n = 2k$.

We introduce the similar functions

$$\frac{1}{k!}\left(e^x - 1 - \frac{x^1}{1!} - \frac{x^2}{2!} - \frac{x^3}{3!} - \cdots - \frac{x^{m-1}}{(m-1)!}\right)^k.$$

Recall (9.8):

$$\frac{1}{k!}\left(x + \frac{x^2}{2!} + \frac{x^3}{3!} + \cdots + \frac{x^m}{m!}\right)^k.$$

It is clear that these functions are related. The coefficients of the former exponential generating functions are the *m-associated Stirling numbers of the second kind*:

$$\sum_{n=mk}^{\infty} \left\{{n \atop k}\right\}_{\geq m} \frac{x^n}{n!} = \frac{1}{k!}\left(e^x - 1 - \frac{x^1}{1!} - \frac{x^2}{2!} - \frac{x^3}{3!} - \cdots - \frac{x^{m-1}}{(m-1)!}\right)^k. \quad (10.7)$$

The basic recursion for $\left\{{n \atop k}\right\}_{\geq m}$

Now we derive the basic recursion for these numbers. Take the derivative on both sides of (10.7). The left-hand side becomes the exponential generating function of $\left\{{n+1 \atop k}\right\}_{\geq m}$, while the right-hand side transforms as follows:

$$\frac{1}{(k-1)!}\left(e^x - 1 - \frac{x^1}{1!} - \cdots - \frac{x^{m-1}}{(m-1)!}\right)^{k-1}.$$

$$\left(e^x - 1 - \frac{x^1}{1!} - \cdots - \frac{x^{m-2}}{(m-2)!}\right) =$$

$$\frac{1}{(k-1)!}\left(e^x - 1 - \frac{x^1}{1!} - \cdots - \frac{x^{m-1}}{(m-1)!}\right)^{k-1}.$$

$$\left(e^x - 1 - \frac{x^1}{1!} - \cdots - \frac{x^{m-2}}{(m-2)!} - \frac{x^{m-1}}{(m-1)!} + \frac{x^{m-1}}{(m-1)!}\right) =$$

$$k\frac{1}{k!}\left(e^x - 1 - \frac{x^1}{1!} - \cdots - \frac{x^{m-1}}{(m-1)!}\right)^k +$$

$$\frac{1}{(k-1)!}\frac{x^{m-1}}{(m-1)!}\left(e^x - 1 - \frac{x^1}{1!} - \cdots - \frac{x^{m-1}}{(m-1)!}\right)^{k-1}.$$

On the right-hand side, we have two generating functions and all of these result in the equality

$$\sum_{n=mk}^{\infty} \left\{ {n+1 \atop k} \right\}_{\geq m} \frac{x^n}{n!} =$$

$$k \sum_{n=mk}^{\infty} \left\{ {n \atop k} \right\}_{\geq m} \frac{x^n}{n!} + \frac{x^{m-1}}{(m-1)!} \frac{1}{(k-1)!} \sum_{n=m(k-1)}^{\infty} \left\{ {n \atop k-1} \right\}_{\geq m} \frac{x^n}{n!}.$$

In the last term, we carry the factor $\frac{x^{m-1}}{(m-1)!}$ behind the summation and re-index. The final result comes by comparing the coefficients:

$$\left\{ {n \atop k} \right\}_{\geq m} = k \left\{ {n-1 \atop k} \right\}_{\geq m} + \binom{n-1}{m-1} \left\{ {n-m \atop k-1} \right\}_{\geq m}. \tag{10.8}$$

The functional equalities also show us that this relation is valid for $n \geq mk$, for n smaller than this $\left\{ {n \atop k} \right\}_{\geq m} = 0$. Except that $\left\{ {0 \atop 0} \right\}_{\geq m} = 1$.

We close this section by the combinatorial proof of this recursion. If the last element, n, goes into an already existing block where there are at least m elements, then there are $n-1$ elements going to k blocks and we insert n into some of the k blocks. This explains the first term. If, in turn, the block where n goes is still not having m elements, then it must have $m-1$ elements, not less. To form such a block, we must choose $m-1$ elements from $n-1$ such that the order of the elements does not count ($\binom{n-1}{m-1}$ cases), and we put n into this block. The $n - (m-1) - 1 = n - m$ other elements must form $k-1$ blocks with the restriction that the blocks contain at least m elements: $\left\{ {n-m \atop k-1} \right\}_{\geq m}$ possibilities. This explains the second term.

Other formulas as well as special values can be found among the exercises.

10.1.1   The associated Bell numbers and polynomials

Definition 10.1.1. *The $B_{n,\geq m}$ number gives that how many partitions are there on n elements such that the partitions do not contain blocks smaller than of size m. They are called* m-associated Bell numbers *or simply* associated Bell numbers[2].

The associated Bell numbers are related to the $\left\{ {n \atop k} \right\}_{\geq m}$ numbers via

$$B_{n,\geq m} = \sum_{k=0}^{\lfloor n/m \rfloor} \left\{ {n \atop k} \right\}_{\geq m}.$$

---

[2]For these numbers, there is no widely used notation or name, so we feel free to use the "associated Bell number" term. This expresses that we are talking about numbers somehow related to the Bell numbers. The notation that we introduce here seems to be new.

And similarly we define the *associated Bell polynomials*:

$$B_{n,\geq m}(x) = \sum_{k=0}^{\lfloor n/m \rfloor} \left\{ {n \atop k} \right\}_{\geq m} x^k.$$

Basic recursion

For $m > 0$

$$B_{n+1,\geq m} = 1 + \sum_{k=m-1}^{n-m} \binom{n}{k} B_{n-k,\geq m}. \tag{10.9}$$

To see why this recursion is true, we consider the last element, $n + 1$, and choose some, say k, other elements into its block. We cannot choose less than $m - 1$ elements because then the block will contain less than $m - 1 + 1 = m$ elements, and this is not permitted. Similarly, we cannot choose more than $n - m$ elements because in that case the other blocks will contain less than $n - (n - m) = m$ elements. The $n - m$ elements also must form a partition with this restriction, and so they have $B_{n-k,\geq m}$ possibilities to be partitioned properly. Summing over k, we get the sum. And what possibility is that plus one responsible for? Note that we have not counted the case when all the partition consist of only one block.

Two relations between the associated and classical Bell numbers

In the particular case when $m = 2$ we have a simple way to relate $B_{n,\geq 2}$ to the classical Bell numbers:

$$B_n = B_{n,\geq 2} + B_{n+1,\geq 2}.$$

Any partition on n or $n+1$ elements without singletons can be identified with a partition on n elements. The partitions counted by $B_{n,\geq 2}$ are identified with themselves, and from the partitions counted by $B_{n+1,\geq 2}$ we remove $n+1$ and the elements of the block of $n + 1$ are separated into singletons. This process can be reversed, so the equality follows. This idea comes from M. Bóna [87, Proposition 2.].

It is possible to relate the Bell and 2-associated Bell numbers in another way:

$$B_n = \sum_{k=0}^{n} \binom{n}{k} B_{k,\geq 2}.$$

We decide first how many singletons we want in our partition, then the rest of the elements go into partitions with at least two elements in each block. This results in

$$B_n = \sum_{k=0}^{n} \binom{n}{k} B_{n-k, \geq 2},$$

where the previous identity comes from by re-indexing.

The associated Bell numbers are far less known than the classical Bell numbers. See an appearance of these numbers in [217].

10.2 The associated Stirling numbers of the first kind

The process of how we get the associated Stirling numbers of the first kind can be performed as we did at the beginning of the previous section, so we can have shortcuts. In the first kind case, it is similarly true that

$$\left[\begin{matrix} n \\ n-1 \end{matrix} \right] = \binom{n}{2}.$$

Applying (10.1) we conclude that

$$\left[\begin{matrix} n \\ n-2 \end{matrix} \right] = 2\binom{n}{3} + 3\binom{n}{4}.$$

In general,

$$\left[\begin{matrix} n \\ n-k \end{matrix} \right] = \sum_{j=0}^{k} \left[\begin{matrix} k+j \\ j \end{matrix} \right]_{\geq 2} \binom{n}{k+j} \quad (n \geq k+1).$$

Re-indexing this,

$$\left[\begin{matrix} n \\ k \end{matrix} \right] = \sum_{j=0}^{k} \left[\begin{matrix} n-j \\ k-j \end{matrix} \right]_{\geq 2} \binom{n}{j}. \tag{10.10}$$

Multiplying by t^k and summing over $k = 0, 1, \ldots, n$ we get the rising factorials on the left-hand side, see (2.41). Then multiplying by $x^n/n!$ and summing over n, we get the $(1-x)^{-t}$ exponential generating function of the rising factorial (use Exercise 23 in Chapter 2). The right-hand side is structurally the same as in the second kind case (10.4). Hence, the counterpart of (10.5) is

$$\sum_{n=0}^{\infty} \frac{x^n}{n!} \sum_{k=0}^{[n/2]} \left[\begin{matrix} n \\ k \end{matrix} \right]_{\geq 2} t^k = (1-x)^{-t} e^{-xt}.$$

The right-hand side can be rewritten by applying a simple trick:

$$(1 - x)^{-t} e^{-xt} = e^{t[-\ln(1-x)-x]}.$$

Expanding the exponential, we get the counterpart of (10.6):

$$\sum_{n=2k} \begin{bmatrix} n \\ k \end{bmatrix}_{\geq 2} \frac{x^n}{n!} = \frac{1}{k!} \left(-\ln(1-x) - x \right)^k .$$

From Exercise 5 of Chapter 2 we know that $-\ln(1-x)$ is the generating function of $1/n$:

$$-\ln(1-x) = \sum_{n=1}^{\infty} \frac{1}{n} x^n .$$

How can we put this into a generalized form? Subtract more terms from the series expansion of $-\ln(1-x)$. These lead to the *associated Stirling numbers of the first kind*:

$$\sum_{n=mk} \begin{bmatrix} n \\ k \end{bmatrix}_{\geq m} \frac{x^n}{n!} = \frac{1}{k!} \left(-\ln(1-x) - x - \frac{x}{2} - \cdots - \frac{x^{m-1}}{m-1} \right)^k . \qquad (10.11)$$

The recursion comes by the same method as in the second kind case.

$$\begin{bmatrix} n \\ k \end{bmatrix}_{\geq m} = (n-1) \begin{bmatrix} n-1 \\ k \end{bmatrix}_{\geq m} + (n-1)^{\underline{m-1}} \begin{bmatrix} n-m \\ k-1 \end{bmatrix}_{\geq m} , \qquad (10.12)$$

if $n \geq mk$.

Historical remarks on the associated Stirling numbers can be found in [287]. See also [158, p. 256.] for some additional citations. From this source one can learn, for example, that the recursion (10.12) was known (at least for $m = 2$) by Appell [32].

10.2.1   The associated factorials $A_{n,\geq m}$

The associated factorials are defined as follows.

Definition 10.2.1. *The $A_{n,\geq m}$ number gives that how many permutations there are on n elements such that the permutations do not contain cycles shorter than m. The $A_{n,\geq m}$ numbers are called* associated factorials[3].

Clearly,

$$A_{n,\geq m} = \sum_{k=0}^{\lfloor n/m \rfloor} \begin{bmatrix} n \\ k \end{bmatrix}_{\geq m} .$$

And similarly we define the *associated factorial polynomials*:

$$A_{n,\geq m}(x) = \sum_{k=0}^{\lfloor n/m \rfloor} \begin{bmatrix} n \\ k \end{bmatrix}_{\geq m} x^k . \qquad (10.13)$$

---

[3]Sometimes generalized derangements are also used.

The recursion corresponding to (10.9) reads as

$$A_{n+1,\geq m} = n! + \sum_{k=m-1}^{n-m} n^{\underline{k}} A_{n-k,\geq m}.$$

From (10.12) it is immediate that

$$A_{n,\geq m}(x) = (n-1)A_{n-1,\geq m}(x) + (n-1)^{\underline{m-1}} \cdot x \cdot A_{n-m,\geq m}(x),$$

while the consequence of (10.11) is the exponential generating function of the associated factorial polynomials:

$$\sum_{n=0}^{\infty} A_{n,\geq m}(x) \frac{y^n}{n!} = \exp\left[x\left(-\ln(1-y) - y - \frac{y}{2} - \cdots - \frac{y^{m-1}}{m-1} \right) \right]. \quad (10.14)$$

When $m = 2$ we get a particular case of the associated factorials which is, for some reason, much more known in the literature than the general case. The $A_{n,\geq 2}$ numbers are called "derangement numbers" or simply "derangements", and commonly denoted by D_n. In what follows, we also apply this notation to shorten the formulas and not carry "≥ 2" in the lower index.

10.2.2 The derangement numbers

It comes from (10.11) that the exponential generating function of the sequence $A_{n,\geq 2} = D_n$ is

$$\sum_{n=0}^{\infty} D_n \frac{x^n}{n!} = \exp\left(-\ln(1-x) - x \right) = \frac{e^{-x}}{1-x}.$$

This offers a simple way of calculating D_n. Multiplying e^{-x} by $1/(1-x)$ it is easy to see by Cauchy's product and by the comparison of the coefficients that

$$D_n = n! \sum_{i=0}^{n} (-1)^i \frac{1}{i!}. \quad (10.15)$$

A consequence of this fact is the nice and simple asymptotic relation:

$$\frac{D_n}{n!} \sim \frac{1}{e}. \quad (10.16)$$

This can be seen by noting that $\sum_{i=0}^{n}(-1)^i \frac{1}{i!}x^i$ is the partial sum of e^{-x}. Substituting $x = 1$ and tending to infinity with n we get (10.16).

The proof of (10.15)

Now we prove (10.15). The *inclusion-exclusion principle* of Section 7.1 will help us. Let Ω be the set of all the permutations on n elements, and let A_i be the set of permutations such that i is a fixed point. (Note that the elements of A_i may have other fixed points!) As

$$\bigcap_{i=1}^{n} A_i$$

is not empty (there is a permutation such that $i = 1, 2, \dots, n$ are all fixed points), we have that $m = n$, where m is introduced in (7.1) on p. 168. We determine

$$\left| \bigcap_{i \in T} A_i \right|,$$

where T is the index set of the prescribed fixed points. What positions we have fixed points at is irrelevant, only the number of the fixed points, $|T|$, matters. If $|T| = j$ we have that

$$\left| \bigcap_{i \in T} A_i \right| = (n - j)!.$$

This is so because we do not move those prescribed j points, indexed by T, and the rest of the elements, $n - j$, goes to a permutation in $(n - j)!$ ways.

By the inclusion-exclusion principle,

$$\left| \bigcup_{i=1}^{n} A_i \right| = \sum_{j=1}^{l} (-1)^{j-1} \sum_{\substack{T \subset \{1,2,\dots,n\} \\ |T|=j}} \left| \bigcap_{i \in T} A_i \right| = \sum_{j=1}^{l} (-1)^{j-1} \sum_{\substack{T \subset \{1,2,\dots,n\} \\ |T|=j}} (n-j)!.$$

Since there are $\binom{n}{j}$ different T sets, we have that

$$\left| \bigcup_{i=1}^{n} A_i \right| = \sum_{j=1}^{l} (-1)^{j-1} \binom{n}{j} (n-j)!.$$

On the left-hand side, we have those permutations which have some fixed points somewhere. This is exactly the number we have to subtract from $n!$ to get the fixed point free permutations:

$$D_n = n! - \sum_{j=1}^{l} (-1)^{j-1} \binom{n}{j} (n-j)!.$$

And this is the same as (10.15)

The recurrence $D_n = nD_{n-1} + (-1)^n$ and its combinatorial proof

Let us scrutinize (10.15) a bit further. Applying it to nD_{n-1}, we have

$$nD_{n-1} = n(n-1)! \sum_{i=0}^{n-1} \frac{(-1)^i}{i!} = n! \sum_{i=0}^{n} \frac{(-1)^i}{i!} - (-1)^n.$$

The sum on the right-hand side is just D_n, so we get the remarkably simple recursion

$$D_n = nD_{n-1} + (-1)^n.$$

This is somewhat different in flavor than we have gotten used to. The alternating term on the right-hand side suggests that some non-routine argument will be necessary to prove this recurrence combinatorially. The following proof was found by A. T. Benjamin and J. Ornstein [62][4].

Let us try to interpret the first term, nD_{n-1}, on the right-hand side hoping that it dictates what we should do. If we take one element from that of n and make it a fixed point, and put the rest into a fixed point free permutation, then we see that the number of such permutations is exactly nD_{n-1}. Let F be the set of permutations of n elements with exactly one fixed point and let D be the set of the derangements on these n elements. Then, as we have just said, $|F| = nD_{n-1}$, and $|D| = D_n$.

If we try to find a one-to-one correspondence between the elements of F and D, we see that there will either be one unmapped element of F or one unhit element of D, depending on the parity of n.

We write the permutations in cycle decomposition such that the cycle leaders (see p. 221) are the smallest elements, and the cycles are listed in increasing order of their smallest element. Note that now every cycle decomposition begins with 1.

First we describe the bijection between those elements of F and D which do not contain the cycle $(1 \quad 2)$. To this end we first fix a notation: let π be a permutation in F and let α be the (unique) fixed point of π. The bijection is based on the number of elements in the first cycle of π, and we consider four cases:

Case I. When the first cycle of π has three or more elements, so it looks like

$$\pi = \begin{pmatrix} 1 & a_1 & a_2 & \cdots a_j \end{pmatrix} \cdots (\alpha) \cdots$$

(here $j \geq 2$), then π is mapped to the derangement where α is inserted after 1. For example,

$$\pi = \begin{pmatrix} 1 & a_1 & a_2 & \cdots a_j \end{pmatrix} \cdots (\alpha) \cdots \to \begin{pmatrix} 1 & \alpha & a_1 & a_2 & \cdots a_j \end{pmatrix} \cdots \in D.$$

This mapping hits every derangement where the first cycle has four or more elements.

Case II. When the first cycle has two elements (it is of the form $\begin{pmatrix} 1 & a_1 \end{pmatrix}$, where $a_1 \neq 2$, then we proceed as before: α go after 1. This way we hit every

---

[4]J. B. Remmel gave a proof earlier via the so-called weighted inversions [489].

derangement where the first cycle is a 3-cycle of the form $(1 \quad x \quad y)$ such that $y \neq 2$. To have image permutations with 3-cycles of the form $(1 \quad x \quad 2)$ we go to the third case.

Case III. When the first cycle of π has one element and it is 1, then it is certain that π is of the form $\pi = (1)(2 \quad a_1)$. We now apply the correspondence

$$\pi = (1)(2 \quad a_1) \cdots \rightarrow (1 \quad a_1 \quad 2) \in D,$$

so the resulting permutation will have a 3-cycle of the form $(1 \quad x \quad 2)$, as we wanted.

Case IV. In this case we map the π permutations where 1 is a fixed point and the second cycle has at least three elements as follows:

$$\pi = (1)(2 \quad a_1 \quad a_2 \cdots \quad a_j) \cdots \rightarrow (1 \quad a_1)(2 \quad a_2 \quad \cdots \quad a_j) \cdots \in D.$$

This mapping hits all derangements beginning with a cycle $(1 \quad x)$ where $x \neq 2$.

Considering all four cases, we see that it remains to describe those bijections between F and D which act on a π beginning with $(1 \quad 2)$. To describe our bijection in this case, we apply the following definition. A permutation is said to be of type k if it begins like

$$(1 \quad 2)(3 \quad 4) \cdots (2k-1 \quad 2k) \cdots,$$

and the next cycle is *not* of the form $(2k+1 \quad 2k+2)$. Let us introduce the notation

$$\sigma_k = (1 \quad 2)(3 \quad 4) \cdots (2k-1 \quad 2k).$$

The bijection acts as before where the elements 1 and 2 are replaced with $2k+1$ and $2k+2$, respectively. That is, the bijection acts as follows, considering cases I-IV:

$$\sigma_k (2k+1 \quad a_1 \quad a_2 \quad \cdots \quad a_j) \cdots (\alpha) \cdots \rightarrow$$
$$\sigma_k (2k+1 \quad \alpha \quad a_1 \quad a_2 \quad \cdots \quad a_j) \cdots,$$
$$\sigma_k (2k+1 \quad a_1) \cdots (\alpha) \cdots \rightarrow \sigma_k (2k+1 \quad \alpha \quad a_1) \cdots,$$
$$\sigma_k (2k+1)(2k+2 \quad a_1) \cdots \rightarrow \sigma_k (2k+1 \quad a_1 \quad 2k+2) \cdots,$$
$$\sigma_k (2k+1)(2k+2 \quad a_1 \quad a_2 \quad \cdots \quad a_j) \cdots \rightarrow$$
$$\sigma_k (2k+1 \quad a_1)(2k+2 \quad a_2 \quad \cdots \quad a_j) \cdots.$$

Notice that when n is even, there is only one permutation of type $n/2-1$, which is the permutation $\sigma_{n/2-1} (n-1)(n)$. This permutation has two fixed points so it does not belong to F or D; it is out of consideration of our mapping. Also, still when n is even, there is one permutation of type $n/2$, namely $\sigma_{n/2}$ itself. This permutation belongs to D but not F. Therefore, when n is even, the bijection is only an "almost" bijection between F and D: it is a bijection from F onto $D \setminus \{\sigma_{n/2}\}$. In this case, therefore, $D_n = nD_{n-1} + 1$.

When n is odd, there is only one permutation which is of type $(n-1)/2$: $\sigma_{(n-1)/2}(n)$. This is an element of F but not of D. So our map, for odd n, is a bijection between $F \setminus \{\sigma_{(n-1)/2}(n)\}$ and D_n. Hence, $D_n = nD_{n-1} - 1$. The proof is now complete.

10.3 Universal Stirling numbers of the second kind

During studying the restricted and associated Stirling numbers, we saw that the exponential generating functions reflected in a very clear way what kind of restrictions were applied. For example,

$$\sum_{n=mk}^{\infty} \left\{ {n \atop k} \right\}_{\geq m} \frac{x^n}{n!} = \frac{1}{k!} \left(e^x - 1 - \frac{x^1}{1!} - \frac{x^2}{2!} - \frac{x^3}{3!} - \cdots - \frac{x^{m-1}}{(m-1)!} \right)^k$$

shows that from the partitions belonging to $\left\{ {n \atop k} \right\}_{\geq m}$ we subtract the blocks having $1, 2, \ldots, m-1$ elements. To the contrary,

$$\sum_{n=k}^{mk} \left[{n \atop k} \right]_{\leq m} \frac{x^n}{n!} = \frac{1}{k!} \left(x + \frac{x^2}{2} + \frac{x^3}{3} + \cdots + \frac{x^m}{m} \right)^k$$

belongs to numbers counting permutations. Note that the numbers in the denominators (with or without factorial) reflect the fact that in one structure (partition) the elements in an individual block are of arbitrary order, while in the other structure (permutation) the order is fixed, up to order-preserving shifting.

These observations suggest a far reaching generalization of the Stirling numbers. We can decide whether we permit the existence of j pieces of blocks of size i. If we permit we set $a_{i,j} = 1$, otherwise we set $a_{i,j} = 0$. Clearly, j can be 0 (it makes sense to avoid blocks of size 0) but $i > 0$ is necessary (it makes no sense to talk about 0 sized blocks). We collect these $a_{i,j}$ numbers in an (infinite) matrix.

$$\mathcal{A} = \begin{pmatrix} a_{1,0} & a_{1,1} & a_{1,2} & \cdots \\ a_{2,0} & a_{2,1} & a_{2,2} & \cdots \\ a_{3,0} & a_{3,1} & a_{3,2} & \cdots \\ \vdots & \vdots & \vdots & \vdots \end{pmatrix}$$

We also introduce the sets

$$K_i = \{ j \mid a_{i,j} = 1 \}.$$

This simply contains the *permitted* number of the blocks of size i.

We are ready to introduce the universal Stirling numbers.

Definition 10.3.1. *Given a matrix $\mathcal{A}$, the* universal Stirling number of the second kind *is denoted by $\left\{{n \atop k}\right\}_{\mathcal{A}}$ and gives that how many k-partitions there are on n elements such that the number of the blocks of size i is contained in the set K_i.*

Definition 10.3.2. *The*

$$B_{n,\mathcal{A}} = \sum_{k=0}^{n} \left\{{n \atop k}\right\}_{\mathcal{A}}$$

numbers are the universal Bell numbers, *and the polynomials*

$$B_{n,\mathcal{A}}(x) = \sum_{k=0}^{n} \left\{{n \atop k}\right\}_{\mathcal{A}} x^k$$

are the universal Bell polynomials.

Examples

If all the elements of $\mathcal{A}$ are 1 (this matrix will be denoted by $\mathcal{A}_1$), then we have no restriction at all: $\left\{{n \atop k}\right\}_{\mathcal{A}_1} = \left\{{n \atop k}\right\}$.

How can we get back already known cases like, for example, the m-restricted Stirling numbers? All the blocks contain maximum m elements, so blocks of size $m+1, m+2, \ldots$ are not permitted. It does not matter how many blocks there are as far as there is zero number of oversized blocks. We can transfer this restriction to the $\mathcal{A}$ matrix if the K_i sets are defined as follows.

$$\begin{aligned} K_i &= \{0,1,2,\ldots\} \quad (i \le m) \\ K_i &= \{0\} \quad\quad\quad\; (i > m) \end{aligned}$$

The corresponding matrix is

$$\mathcal{A}_{\le m} = \begin{pmatrix} 1 & 1 & 1 & 1 & \cdots \\ 1 & 1 & 1 & 1 & \cdots \\ & \vdots & & & \\ 1 & 1 & 1 & 1 & \cdots \\ 1 & 0 & 0 & 0 & \cdots \\ 1 & 0 & 0 & 0 & \cdots \\ & \vdots & & & \end{pmatrix}.$$

The first row in which zero appears is the $(m+1)$-th. Hence,

$$\left\{{n \atop k}\right\}_{\mathcal{A}_{\le m}} = \left\{{n \atop k}\right\}_{\le m}.$$

Now it is clear that the m-associated Stirling numbers belong to the set

$$K_i = \{0\} \qquad (i < m)$$
$$K_i = \{0, 1, 2, \ldots\} \quad (i \geq m),$$

and the

$$\mathcal{A}_{\geq m} = \begin{pmatrix} 1 & 0 & 0 & 0 & \cdots \\ 1 & 0 & 0 & 0 & \cdots \\ \vdots & & & & \\ 1 & 0 & 0 & 0 & \cdots \\ 1 & 1 & 1 & 1 & \cdots \\ 1 & 1 & 1 & 1 & \cdots \\ \vdots & & & & \end{pmatrix} \qquad (10.17)$$

matrix:

$$\left\{ n \atop k \right\}_{\mathcal{A}_{\geq m}} = \left\{ n \atop k \right\}_{\geq m}.$$

Note that $a_{i,0} = 1$, because from every block we can have 0 piece. What is more, the number of small blocks *must be* zero in the associated case.

At last we determine the matrix of partitions which have only blocks of even size. Clearly enough,

$$K_i = \{0, 1, 2, \ldots\}, \quad \text{if } i \text{ is even}$$
$$K_i = \{0\}, \qquad\qquad \text{if } i \text{ is odd}.$$

From where the following matrix arises:

$$\mathcal{A}_e = \begin{pmatrix} 1 & 0 & 0 & 0 & \cdots \\ 1 & 1 & 1 & 1 & \cdots \\ 1 & 0 & 0 & 0 & \cdots \\ 1 & 1 & 1 & 1 & \cdots \\ 1 & 0 & 0 & 0 & \cdots \\ 1 & 1 & 1 & 1 & \cdots \\ \vdots & & & & \end{pmatrix}. \qquad (10.18)$$

More examples can be found on p. 226 in Comtet's book [158].

The universal generating functions

We are going to determine the generating functions of the universal Stirling numbers relying on (2.30). It can be transformed into a formula which is more useful for the present purpose:

$$\sum_{n=0}^{\infty} B_{n,\mathcal{A}}(x) \frac{y^n}{n!} =$$

$$\sum_{n=0}^{\infty}\left(\sum_{k=0}^{n}\sum_{\substack{j_1+2j_2+\cdots+nj_n=n\\j_1+j_2+\cdots+j_n=k\\j_i\in K_i}}\frac{x^{j_1}x^{j_2}\cdots x^{j_n}}{j_1!j_2!\cdots j_n!}\left(\frac{y^1}{1!}\right)^{j_1}\left(\frac{y^2}{2!}\right)^{j_2}\cdots\left(\frac{y^n}{n!}\right)^{j_n}\right)$$

The assumption that $j_i\in K_i$ can be eliminated if we apply the elements of $\mathcal{A}$. If some j_is do not belong to K_i, then those members – regardless of the values of the other j indices – do not appear in the sum. This cancellation is achieved by putting the product of a_{i,j_i} in the nominator:

$$=\sum_{n=0}^{\infty}\left(\sum_{k=0}^{n}\sum_{\substack{j_1+2j_2+\cdots+nj_n=n\\j_1+j_2+\cdots+j_n=k}}\frac{a_{i,j_1}\cdots a_{i,j_n}x^{j_1+\cdots+j_n}}{j_1!j_2!\cdots j_n!}\left(\frac{y^1}{1!}\right)^{j_1}\cdots\left(\frac{y^n}{n!}\right)^{j_n}\right).$$

To shorten the formula, we use the $\prod$ symbol inside the sum, and the summation over the condition $j_1+j_2+\cdots+j_n=k$ can be relaxed: $j_1+j_2+\cdots+j_n\geq 0$, and so

$$=\sum_{n=0}^{\infty}\left(\sum_{\substack{j_1+2j_2+\cdots+nj_n=n\\j_1+j_2+\cdots+j_n\geq 0}}\prod_{i=1}^{n}\frac{a_{i,j_i}x^{j_i}}{j_i!}\left(\frac{y^i}{i!}\right)^{j_i}\right).$$

The equality

$$\prod_{i=1}^{\infty}\sum_{k_i=0}^{\infty}\frac{a_{i,k_i}x^{k_i}}{k_i!}\left(\frac{y^i}{i!}\right)^{k_i}=\prod_{i=1}^{\infty}\sum_{j=0}^{\infty}\frac{a_{i,j}x^j}{j!}\left(\frac{y^i}{i!}\right)^{j}$$

holds, whence

$$\sum_{n=0}^{\infty}B_{n,\mathcal{A}}(x)\frac{y^n}{n!}=\prod_{i=1}^{\infty}\sum_{j=0}^{\infty}\frac{a_{i,j}x^j}{j!}\left(\frac{y^i}{i!}\right)^{j}. \tag{10.19}$$

This is the exponential generating function of the universal Bell polynomials. In particular,

$$\sum_{n=0}^{\infty}B_{n,\mathcal{A}}\frac{y^n}{n!}=\prod_{i=1}^{\infty}\sum_{j=0}^{\infty}\frac{a_{i,j}}{j!}\left(\frac{y^i}{i!}\right)^{j}.$$

Application of the universal generating functions to the former examples

Let us see how the universal generating functions work on the former examples. First, consider the associated Bell polynomials $B_{n,\mathcal{A}_{\geq m}}(x)=B_{n,\geq m}(x)$. Applying (10.19) to the matrix (10.17):

$$\sum_{n=0}^{\infty}B_{n,\mathcal{A}_{\geq m}}(x)\frac{y^n}{n!}=\prod_{i=m}^{\infty}\exp\left(\frac{xy^i}{i!}\right). \tag{10.20}$$

By the functional identity $e^x e^y = e^{x+y}$

$$\prod_{i=m}^{\infty} \exp\left(\frac{xy^i}{i!}\right) = \exp\left[x\left(e^y - 1 - y - \frac{y^2}{2} - \cdots - \frac{y^{m-1}}{(m-1)!}\right)\right].$$

Where the exponential generating function of $\left\{{n \atop k}\right\}_{\geq m}$ also follows (after setting $x = 1$ and renaming the variable y to x):

$$\sum_{n=mk}^{\infty} \left\{{n \atop k}\right\}_{\geq m} \frac{x^n}{n!} = \frac{1}{k!}\left(e^x - 1 - \frac{x^1}{1!} - \frac{x^2}{2!} - \frac{x^3}{3!} - \cdots - \frac{x^{m-1}}{(m-1)!}\right)^k.$$

The proof of (9.8) is very similar.

Partitions with blocks of even size

Another interesting application of the universal generating functions is the case when we permit only blocks of even size. The matrix is already calculated, see (10.18), whence

$$\sum_{n=0}^{\infty} B_{n,\mathcal{A}_p}(x)\frac{y^n}{n!} = \prod_{\substack{i=1 \\ 2|i}}^{\infty} \exp\left(\frac{xy^i}{i!}\right) =$$

$$\exp\left[x\left(\frac{y^2}{2!} + \frac{y^4}{4!} + \frac{y^6}{6!} + \cdots\right)\right].$$

The inner sum is the cosh (*hyperbolic cosine function*) (more exactly $\cosh(y) - 1$):

$$\cosh(y) = \sum_{i=0}^{\infty} \frac{y^{2i}}{(2i)!}$$

It is equally correct to say that cosh is the exponential generating function of the sequence

$$a_n = \begin{cases} 0, & \text{if } n \text{ is odd}; \\ 1, & \text{if } n \text{ is even}. \end{cases}$$

(Compare this to Exercise 1 of Chapter 2.) The cosh function helps to simplify the result:

$$\sum_{n=0}^{\infty} B_{n,\mathcal{A}_p}(x)\frac{y^n}{n!} = \exp\left[x\left(\cosh(y) - 1\right)\right].$$

We show how this equality leads to a recursion for the $B_{n,\mathcal{A}_p} = B_{n,\mathcal{A}_p}(1)$ numbers. Substitute $x = 1$ and take the derivative of both sides ($\cosh' = \sinh$, where sinh is the *hyperbolic sine function*):

$$\sum_{n=1}^{\infty} B_{n,\mathcal{A}_p} \frac{y^{n-1}}{(n-1)!} = \exp(\cosh(y) - 1)\sinh(y).$$

The hyperbolic sine is the exponential generating function of

$$a_n = \begin{cases} 1, & \text{if } n \text{ is odd;} \\ 0, & \text{if } n \text{ is even.} \end{cases}$$

Thus,

$$\sum_{n=1}^{\infty} B_{n,\mathcal{A}_p} \frac{y^{n-1}}{(n-1)!} = \left(\sum_{n=0}^{\infty} B_{n,\mathcal{A}_p} \frac{y^n}{n!} \right) \left(\sum_{n=1}^{\infty} \frac{y^{2n-1}}{(2n-1)!} \right) =$$

$$\sum_{n=1}^{\infty} \left(\sum_{k=1}^{[n/2]} \binom{n-1}{2k-1} B_{n-2k,\mathcal{A}_p} \right) \frac{y^{n-1}}{(n-1)!}.$$

Comparing the coefficients, the wanted recursion follows:

$$B_{n,\mathcal{A}_p} = \sum_{k=1}^{[n/2]} \binom{n-1}{2k-1} B_{n-2k,\mathcal{A}_p} \tag{10.21}$$

The initial values are $B_{0,\mathcal{A}_p} = 1$, $B_{1,\mathcal{A}_p} = 0$.

10.4 Universal Stirling numbers of the first kind

We know already very well that the only essential difference in the generating functions when we turn from partitions to permutations is the $k! \to k$ change in the denominators. Therefore, the universal Stirling numbers of the first kind do not give much surprise in the formalism[5].

The set K_i and the matrix $\mathcal{A}$ are similarly defined as in the previous section.

Definition 10.4.1. *The* universal Stirling number of the first kind *is denoted by $\begin{bmatrix} n \\ k \end{bmatrix}_{\mathcal{A}}$ and it gives that how many permutations there are on n elements which contain k cycles and the number of cycles of length i is in the set K_i.*

Definition 10.4.2. *The*

$$A_{n,\mathcal{A}} = \sum_{k=0}^{n} \begin{bmatrix} n \\ k \end{bmatrix}_{\mathcal{A}}$$

numbers are called universal factorials, *while the polynomials*

$$A_{n,\mathcal{A}}(x) = \sum_{k=0}^{n} \begin{bmatrix} n \\ k \end{bmatrix}_{\mathcal{A}} x^k$$

are the universal factorial polynomials.

---

[5]This statement does not mean, however, that the underlying number sequences and polynomials are similar. It is enough to remember that the horizontal generating function of the $\begin{bmatrix} n \\ k \end{bmatrix}$ numbers is the rising factorial, while of $\begin{Bmatrix} n \\ k \end{Bmatrix}$ is the Bell polynomial, and these are by no means similar to each other.

The exponential generating function of these polynomials can be deduced as we did in the second kind case:

$$\sum_{n=0}^{\infty} A_{n,\mathcal{A}}(x)\frac{y^n}{n!} = \prod_{i=1}^{\infty}\sum_{j=0}^{\infty}\frac{a_{i,j}x^j}{j!}\left(\frac{y^i}{i}\right)^j. \tag{10.22}$$

More on the universal Stirling numbers can be found in [158, p. 225], [235, Chapter 2], and [495, p. 74 and p. 80-89]. Particularly for permutations, see Section 3.4 in [83].

Exercises

1. How many options do we have if we want to give 7 gifts to 3 friends such that we want everyone to get at least two gifts?

2. Give a combinatorial interpretation for the (10.1) binomial sum.

3. How many surjective functions are there from an n-set to a k-set such that each image is taken at least m times?

4. Give a combinatorial interpretation for (10.3) and (10.10).

5. Generalize (10.3) and (10.10) identities by following the pattern of Exercises 6 and 7 in Chapter 9.

6. Show that
$$\left\{ {mk \atop k} \right\}_{\geq m} = \frac{1}{k!} \left({mk \atop \underbrace{m, m, \ldots, m}_{k}} \right)$$
(see the Appendix for the definition of the multinomial coefficient).

7. Prove the validity of
$$\left[{mk \atop k} \right]_{\geq m} = \frac{(mk)!}{k! m^k}.$$
(It is fun to note that this yields that $(mk)!$ is divisible by $k! m^k$ for any m. Try to prove this latter fact without using combinatorics.)

8. Prove the following identity:
$$\left\{ {n \atop k} \right\}_{\geq m} = \frac{1}{k} \sum_{k=m}^{n-(k-1)m} \binom{n}{k} \left\{ {n-k \atop k-1} \right\}_{\geq m} \quad (n \geq km).$$

9. Find the analogue of the identity in the previous exercise with respect to the associated Stirling numbers of the first kind.

10. Show that the 2-associated Stirling numbers take the following values:
$$\left\{ {n \atop 2} \right\}_{\geq 2} = \frac{1}{2}(2^n - 2n - 2),$$
$$\left\{ {n \atop 3} \right\}_{\geq 2} = \frac{1}{6}(3^n - 3 \cdot 2^n) - \frac{1}{2}n(2^{n-1} - 1) + \frac{1}{2}(n^2 + 1),$$
$$\left\{ {n \atop 4} \right\}_{\geq 2} = \frac{4^n}{24} - \frac{3^n}{18}(n+3) - \frac{1}{6}(n^3 + 2n + 1) + \frac{2^n}{16}(n^2 + 3n + 4).$$

Here $n \geq 4, 6, 8$, respectively. (Hint: the first can easily be proven combinatorially. The other two can be established by using the first together with the previous exercise.)

11. Prove that the above special values are valid:

$$\left\{ {n \atop n-1} \right\}_{\leq m} = \binom{n}{2} \quad (n \geq 2, \ m \geq 2),$$

$$\left\{ {n \atop n-2} \right\}_{\leq m} = \begin{cases} \frac{3n-5}{4}\binom{n}{3} & (n \geq 4, m \geq 3); \\ 3\binom{n}{4} & (n \geq 4, m = 2), \end{cases}$$

$$\left\{ {n \atop n-3} \right\}_{\leq m} = \begin{cases} \binom{n}{4}\binom{n-2}{2} & (n \geq 4, m \geq 4); \\ 15\binom{n}{6} + 10\binom{n}{5} & (n \geq 4, m = 3); \\ 15\binom{n}{6} & (n \geq 4, m = 2). \end{cases}$$

(See [340].)

12. Provide a proof for the below special values

$$\left[{n \atop n-1} \right]_{\geq m} = \begin{cases} \binom{n}{2} & (m = 1, m = n = 2); \\ 0 & (otherwise), \end{cases}$$

$$\left[{n \atop n-2} \right]_{\geq m} = \begin{cases} \frac{3n-1}{4}\binom{n}{3} & (m = 1); \\ 3 & (m = 2, n = 4); \\ 2 & (m = 2, 3; n = 3); \\ 0 & (otherwise), \end{cases}$$

$$\left[{n \atop n-3} \right]_{\geq m} = \begin{cases} \binom{n}{4}\binom{n}{2} & (m = 1); \\ 6 & (2 \leq m \leq 4, n = 4); \\ 20 & (m - 2, n = 5); \\ 15 & (m = 2, n = 6); \\ 0 & (otherwise). \end{cases}$$

(See [341].)

13. Show that

$$A_{n, \geq 2}(-1) = 1 - n.$$

(See [158, p. 256].)

14. Show that

$$n! B_n = \sum_{m=0}^{n} \binom{n}{m} m^n D_{m-n} \quad (n \geq 0).$$

(Hint: check (2.21).)

15. Let the number of partitions into odd number of blocks be $B_{n,\mathcal{A}_{\mathrm{odd}}}$. Prove that this satisfies the recursion

$$B_{n,\mathcal{A}_{\mathrm{odd}}} = \sum_{k=0}^{[(n-1)/2]} \binom{n-1}{2k-1} B_{n-2k-1,\mathcal{A}_{\mathrm{odd}}}$$

with $B_{0,\mathcal{A}_{\mathrm{odd}}} = 1$ initial value (this is the pair of (10.21)). Describe the corresponding matrix $\mathcal{A}_{\mathrm{odd}}$ and the exponential generating function of $B_{n,\mathcal{A}_{\mathrm{odd}}}$.

16. Prove that the universal factorial's exponential generating function is

$$\exp\left(\frac{y^2}{1-y^2}\right)$$

in the case when we permit only even number of cycles.

17. Use the generating function interpretation of sinh and cosh given in the text to prove that

$$\sinh(x) = \frac{1}{2}(e^x - e^{-x}) \quad \text{and} \quad \cosh(x) = \frac{1}{2}(e^x + e^{-x}).$$

18. Prove that
$$B_{n,\geq 2} \equiv 1 \pmod{p}$$
for any prime p. (See [87, Corollary 3.])

19. Let $h(n)$ be the number of fixed point free involutions on $2n$ elements (clearly, there are no such permutations on odd number of elements). Show that the exponential generating function of $h(n)$ is $e^{x^2/2}$, and also that
$$h(n) = \frac{(2n)!}{2^n n!}.$$

(This is Example 3.54 in [83].)

20. Let S be a subset of the positive integers. Moreover, let K_i be the set of positive integers for $i \in S$, and $K_i = \{0\}$ otherwise. Show that for this setting

$$\begin{bmatrix} n+1 \\ k \end{bmatrix}_{\mathcal{A}} = \sum_{s \in S} \frac{n!}{(n-j+1)!} \begin{bmatrix} n-s+1 \\ k-1 \end{bmatrix}_{\mathcal{A}},$$

$$\left\{ \begin{matrix} n+1 \\ k \end{matrix} \right\}_{\mathcal{A}} = \sum_{s \in S} \binom{n}{s-1} \left\{ \begin{matrix} n-s+1 \\ k-1 \end{matrix} \right\}_{\mathcal{A}}.$$

(See [578] for these results and more on $\begin{bmatrix} n \\ k \end{bmatrix}_{\mathcal{A}}, \left\{\begin{matrix} n \\ k \end{matrix}\right\}_{\mathcal{A}}, A_{n,\mathcal{A}}, B_{n,\mathcal{A}}$ for this particular $\mathcal{A}$.)

21. Prove that the $(n+1)$st order Hankel determinant of the derangement numbers is

$$
\det
\begin{pmatrix}
D_0 & D_1 & \cdots & D_n \\
D_1 & D_2 & \cdots & D_{n+1} \\
\vdots & & \ddots & \\
D_n & D_{n+1} & \cdots & D_{2n}
\end{pmatrix}
= \left(\prod_{k=0}^{n} k! \right)^2 .
$$

(This is actually Radoux's theorem, see [478] and also [211] for a nice combinatorial proof.) Hint: use the tool we developed in Subsection 2.10.2. Note that the derivatives of $e^{-y}/(1-y)$ are more complex to calculate than those of the derangement polynomials, so try to find the Hankel determinants of the $A_{n,\geq 2}(x)$ polynomials first. The exponential generating function of this sequence is $e^{-y}/(1-xy)$. The derivatives of this function with respect to x are much simpler.

22. Show that the Hankel determinant of all the polynomials $B_{n,\leq 2}(x)$, $B_{n,\geq 2}(x)$ and $B_{n,=2}(x)$ is

$$
x^{\binom{n+1}{2}} \prod_{k=0}^{n} k!.
$$

Here $B_{n,=2}(x)$ is the polynomial for the $\left\{{n \atop k}\right\}_{=2}$ numbers, where the block size can only be 2. Note that $B_{n,=2}(1) = h(n)$ of Exercise 19. (This was proven by Ehrenborg [211], although for $B_{n,\leq 2}(x)$ it was known earlier by Radoux [479] as well.)

23. Prove the following integral representation for the derangements [39, p. 855]:

$$
D_n = \int_0^\infty e^{-t}(t-1)^n dt.
$$

24. Let $D_{n,k}$ denote the number of permutations with exactly k fixed points (so $D_n = D_{n,0}$). Apply the inclusion-exclusion principle to show that

$$
D_{n,k} = \frac{n!}{k!} \sum_{j=0}^{n-k} \frac{(-1)^j}{j!}.
$$

The $D_{n,k}$ numbers are called *rencontres numbers*.

25. From (10.15) it follows that

$$
D_n = (n-1)(D_{n-1} + D_{n-2}) \quad (n \geq 2).
$$

Find a combinatorial proof of this simple recursion. (The initial values are $D_0 = 1$ and $D_1 = 0$.)

26. A *signed permutation* is an ordinary permutation with optional bars on each of the elements (equally, as if we painted all the elements black or white, independently). Show that the number of signed derangements, which is denoted in the literature by D_n^B, can be calculated as

$$D_n^B = n! \sum_{k=0}^{n} \frac{(-1)^k}{k!} 2^{n-k}.$$

Note that, for example, $\begin{pmatrix} 1 & 2 & 3 & 4 \\ 1 & 3 & 4 & 2 \end{pmatrix}$ is not a signed derangement as one is a fixed point, while $\begin{pmatrix} 1 & 2 & 3 & 4 \\ \bar{1} & 3 & 4 & 2 \end{pmatrix}$ is, because one is not a fixed point anymore, it is mapped to its barred counterpart.

27. Give a proof for the derivative formula

$$A'_{n,\geq 2}(x) = -nA_{n-1,\geq 2}(x) + n! \sum_{k=0}^{n-1} \frac{1}{k!} \frac{A_{k,\geq 2}(x)}{n-k} \quad (n \geq 1).$$

In particular,

$$A'_{n,\geq 2}(1) = -nD_{n-1} + n! \sum_{k=0}^{n-1} \frac{1}{k!} \frac{D_k}{n-k} \quad (n \geq 1).$$

(Hint: it comes from (10.14) that the exponential generating function of $A_{n,\geq 2}(x)$ is $\left(\frac{e^{-y}}{1-y} \right)^x$. Taking the derivative with respect to x, we get the exponential generating function of $A'_{n,\geq 2}(x)$. The rest is a simple calculation.)

28. Show that for any $n \geq 0$

$$xB'_{n,\geq 2}(x) = B_{n+1,\geq 2}(x) - nxB_{n-1,\geq 2}(x),$$

in particular,

$$B'_{n,\geq 2}(1) = B_{n+1,\geq 2} - nB_{n-1,\geq 2}.$$

29. Prove the log-convexity of the $A_{n,\leq m}$ and $B_{n,\leq m}$ sequences (for fixed m). (Hint: use the Bender-Canfield theorem of Subsection 4.5.2. $A_{n,\leq m}$ and $B_{n,\leq m}$ were the examples in [59] to show how the theorem can be applied.)

Outlook

1. Historical remarks on the Stirling, Lah, r-Stirling, and associated Stirling numbers can be found on p. 319-321 of the book of Charalambides [134].

2. A very recent treatise on the particular version of the universal Stirling numbers when the permitted block sizes belong to a fixed subset of the positive integers (but the number of the blocks of a given size is arbitrary) is presented in [69].

3. We met with the Bernoulli and Cauchy numbers in Chapter 5. These numbers can be generalized many ways. Two of these generalizations connect to the restricted and associated Stirling numbers. See the paper [341] for the definition and basic properties of the restricted and associated Cauchy numbers, and [340] for the restricted and associated (poly-)Bernoulli numbers. The latter paper gives a relation between these poly-Bernoulli numbers, the Lambert W function and the *Riemann zeta function*.

4. The real zero property of the associated factorial polynomials $A_{n,\geq m}(x)$ (see (10.13)) was studied by several authors. Canfield [114], Tricomi [567] (in this latter paper $A_{n,\geq 2}(x) = l_n(-x)$), and [120] in the $m = 2$ case and Brenti [95] for general m showed that $A_{n,\geq m}(x)$ has only real zeros. Tricomi also showed that the negative zeros are smaller than -1. See the paper of Temme [560] about the asymptotics of $l_n(x)$.

   Compare these polynomials with the (2.41) horizontal generating function of the $\begin{bmatrix} n \\ k \end{bmatrix}$ numbers. Latter has zeros at the negative integers and at zero. One might ask whether this simple property generalizes somehow for $A_{n,\geq m}(x)$ once we know that the roots of them are real. The answer is due to M. Bóna [85] and it says that for any given m there exists a positive integer N such that if $n > N$, then one zero of $A_{n,\geq m}(x)$ will be close to -1, one other will be close to -2, and so on, and one zero will be close to $-n$.

5. The real zero property for the associated Bell polynomials $B_{n,\geq 2}(x)$ were proven in [87], and for $B_{n,\geq 3}(x)$ in [559]. It is not known in general whether $B_{n,\geq m}(x)$ has only real zeros or not. However, it can be checked easily that $A_{9,\leq 3}(x)$ *has* complex zeros as well as $B_{9,\leq 3}(x)$, so it is not true in general that the restricted Bell and factorial polynomials have only real zeros.

6. The divisibility properties of restricted and associated Fubini numbers were probably first studied in [416]. See [106, 442] for other results regarding these sequences and polynomials.

7. The asymptotics (10.16) of the derangement or associated factorial numbers can be generalized for arbitrary m. It is true that

$$\frac{A_{n,\geq m}}{n!} \to \frac{1}{e^{H_{m-1}}} \quad (n \to \infty).$$

Here H_{m-1} is the $m - 1$st harmonic number. See [235, p. 261].

8. The asymptotic behavior of the associated Bell numbers and the associated first and second kind Stirling numbers was studied in [585]. The asymptotics of the restricted Bell numbers, factorials, and first and second kind Stirling numbers was considered in [586]. However, there is an incorrect statement in [585] which was subsequently corrected in [179], where even more general statements can be found. We had known before the paper [586] that $A_{n,\leq 2}$, the number of involutions, behave asymptotically like

$$A_{n,\leq 2} \sim \frac{n!}{\sqrt{2\pi}} \frac{1}{\sqrt{2n}} \frac{\exp\left(\frac{n}{2} + \sqrt{n} - \frac{1}{4}\right)}{n^{n/2}} \qquad \text{as} \quad n \to \infty$$

(cf. [151, Theorem 8], [597, p. 187], [235, pp. 558-560]). Recently, G. Louchard studied the case when the number of singletons is given and fixed [381]. In [586] it was proven for general m that

$$A_{n,\leq m} \sim \frac{n!}{\sqrt{2\pi}} \frac{1}{\sqrt{mn}} \frac{\exp\left(n/m + O(n^{1-1/m})\right)}{n^{n/m}},$$

see also (the even more detailed) [17]. Similar statements hold for the restricted Bell numbers:

$$B_{n,\leq m} \sim \frac{n!}{\sqrt{2\pi}} \frac{1}{\sqrt{mn}} \frac{\exp\left(n/m + O(n^{1-1/m})\right)}{(n(m-1)!)^{n/m}}.$$

9. The behavior of the associated Stirling numbers [585] does not differ from that of the classical Stirling numbers:

$$\left[\begin{matrix} n \\ k \end{matrix}\right]_{\geq m} \sim \left[\begin{matrix} n \\ k \end{matrix}\right], \quad \left\{\begin{matrix} n \\ k \end{matrix}\right\}_{\geq m} \sim \left\{\begin{matrix} n \\ k \end{matrix}\right\}.$$

This also means, by the idea introduced in Section 7.3, that the maximizing index of the associated Stirling numbers of both kinds behaves similarly as the maximizing index of the classical Stirling numbers. This, however, does not imply that the sub-dominant terms are also the same. It would be an interesting problem to study the m-dependence of the maximizing indices.

The "restricted version" makes sense only if the lower parameter tends to infinity together with the upper parameter (otherwise, the restricted

Stirlings eventually become zero). In [586] it is proven that for fixed positive integers m and a, there holds

$$\lim_{n \to \infty} \left[\begin{matrix} n \\ n-a \end{matrix} \right]_{\leq m} = \frac{1}{2^a a!} n^{2a} \left(1 + \frac{a - 2a^2}{n} + O\left(\frac{1}{n^2}\right) \right),$$

$$\lim_{n \to \infty} \left\{ \begin{matrix} n \\ n-a \end{matrix} \right\}_{\leq m} = \frac{1}{2^a a!} n^{2a} \left(1 + \frac{a - 2a^2}{n} + O\left(\frac{1}{n^2}\right) \right).$$

For a more profound analysis with respect to permutations, see the recent PhD thesis of R. Petuchovas [461].

10. An extensive set of results (combinatorial identities, Hankel transform, log-convexity, arithmetical properties) was obtained for the restricted and associated Bell and factorial numbers in [442].

11. In [76] one can find some finite sum results with respect to the derangement numbers. For example,

$$\sum_{k=1}^{n} \frac{D_{k+2}^2}{(k+1)!} = \frac{D_{n+2} D_{n+3}}{(n+2)!} - 1.$$

See also [332, 397].

12. The derangement numbers (associated factorials) can be generalized such that we consider the restriction we had in the r-Stirling number case, i.e., some elements are forced to be in different cycles. If we consider permutations without fixed points such that the first r elements are in separate cycles, we get the r-derangement numbers. These numbers were studied in detail [587].

13. Restricted and associated Lah numbers and their horizontal sums were recently defined and studied in [55, 215]. In [215] a combinatorial interpretation is given for the absolute values of the elements in the inverse matrix of $\left\{ \begin{matrix} n \\ k \end{matrix} \right\}_{\mathcal{A}}$, where $\mathcal{A}$ is as in Exercise 20.

14. The generalized Bell polynomials of Section 3.3 are not including the r-Stirling numbers (the reader should check this by comparing the generating functions). Mihoubi and Rahmani generalized these Bell polynomials to include the r-Stirlings, r-Lah, r-Whitney numbers (for the latter, see the Outlook of Chapter 8) and even the restricted and associated versions of all of these numbers. See [435].

15. In [143] the common generalization of the r-Stirling and restricted Stirling numbers is defined and studied by the exponential Riordan array method.

16. Some relations between $\left\{ \begin{matrix} n \\ k \end{matrix} \right\}_{\geq 2}$ and the so-called *Eulerian numbers of the second kind* [262] are deduced through the Lambert W function in [314].

17. The Lambert W function has been arising a couple of times in the book. Here is another connection to W. The 3-associated Stirling numbers of the first and second kind appear in some Taylor series representations of the different branches of the Lambert W function [301].

18. The two-dimensional derangement problem was studied by Fisk [232], see also [184].

19. The "r-version" of the universal Stirling numbers was introduced in [68].

20. One can study the *relative derangements*. A permutation, given in word representation (see p. 23) as $\pi_1 \pi_2 \cdots \pi_n$, is a relative derangement if $\pi_{i+1} \neq \pi_i + 1$ for all $1 \leq i \leq n-1$. Let Q_n denote the number of relative derangements on $\{1, \ldots, n\}$. The following relation holds between this sequence and the sequence of derangement numbers (see [103, Theorem 6.5.1], and [23, Example 6.11]:

$$Q_n = D_n + D_{n-1}.$$

In [136] one finds a combinatorial proof which employs skew derangements.

Relative derangements can be defined for signed permutations just as standard derangements can be (see Exercise 26). Denoting the number of signed relative derangements by Q_n^B, it can be shown [138] that

$$Q_n^B = D_n^B + D_{n-1}^B.$$

Signed permutations are also, and often, called *type B permutations*. Accordingly, signed derangements are also called *type B derangements*. These were originally introduced in 2006 by C.-O. Chow [150].

Part III

Number theoretical properties

Chapter 11

Congruences

The counting sequences often have very interesting number theoretical properties. In the third part of this book, we will learn some of these and we develop the tools to study such questions.

It is interesting to know that, for example, which Bell numbers are even, or under what circumstances the Stirling numbers are divisible by a given prime p. Such questions will be studied in the first part of the chapter. In the second part, we apply the basic theory of divisibility to deduce the non-integer property of harmonic numbers and some of its generalizations.

11.1   The notion of congruence

We introduce some notations first. The fact that the integer a divides the integer b is denoted as

$$a \mid b.$$

In other words, a divides b means that there exists an integer c such that

$$ac = b.$$

This fact[1] equivalently means that b is an *integer multiple* of a.

If this is not the case (in symbols, $a \nmid b$), then we can look for the remainder of the division b/a. In any circumstances, b can be expressed as

$$b = ak + r,$$

k is a non-negative integer, and r ($0 \le r < a$) is the *remainder* or *residue*.

If two numbers, say b_1 and b_2, have the same residue if we divide them by a, then we say that b_1 and b_2 are *congruent modulo a*. This fact is denoted as follows:

$$b_1 \equiv b_2 \pmod{a}.$$

The number a is called *modulus*. For instance, 5 and 8 are congruent with respect to the modulus 3, because the residue is 2 in both cases. In symbols,

$$5 \equiv 8 \pmod{3}.$$

---

[1] That a divides b is often also expressed as a is a factor of b.

It is particularly important when the modulus is a prime number.

$$b \equiv 0 \pmod{p},$$

means that b is an integer multiple of p.

Congrunces of fractions

It is often useful to generalize the congruence notation to rational numbers. The rational numbers will be written in reduced form, that is, the nominator and denominator have no factor in common. For fractions, the congruence is understood to be taken on the *nominators*. For example,

$$\frac{3}{7} \equiv \frac{8}{7} \pmod 5.$$

Taking another example,

$$\frac{a}{b} \equiv 0 \pmod{c^2}$$

means that a is divisible by the second power of c (and, by our reduced form agreement, b is not divisible by c).

11.2   The parity of the binomial coefficients

The theorem on the parity (modulo 2 behavior) of the binomial coefficients is interesting in itself, and it is good to know about it because it can be applied in many situations. This question was first studied in detail by Glaisher[2]. The following statement can be proven for non-negative integers n and k:

$$\binom{n}{k} \equiv \begin{cases} 0 \pmod 2, & \text{if } n \text{ is even and } k \text{ is odd;} \\ \binom{\lfloor \frac{n}{2} \rfloor}{\lfloor \frac{k}{2} \rfloor} \pmod 2, & \text{otherwise.} \end{cases} \tag{11.1}$$

The proof is elementary, but needs to be broken down into some subcases. We could not do better than Jonathan Gross in his book [266, Chapter 4], so we do not repeat the proof here. Divisibility with other primes and more sophisticated congruences will be the topic of the next chapter.

What we still mention is an interesting corollary of (11.1): the number of odd binomial coefficients in the row n of the Pascal triangle is $a(n) = 2^w$, where w is the number of 1-digits in the binary representation of n. The proof is again easy, see [266, Proposition 4.1.11].

---

[2]James Whitbread Lee Glaisher (1848-1928), English mathematician and astronomer.

Thus, for example, there are 8 odd numbers among $\binom{14}{0}, \binom{14}{1}, \ldots, \binom{14}{14}$, because

$$14 = 2^3 + 2^2 + 2^1 = 1110_2,$$

thus $w = 3$ and so $a(14) = 2^w = 8$.

11.2.1 The Stolarsky-Harborth constant

There is an interesting function $f(n)$ counting the number of odd elements in the first n rows of the Pascal triangle:

$$f(n) = \sum_{i=0}^{n-1} a(i). \tag{11.2}$$

We have that $f(0) = 0$, $f(1) = 1$, and for $n \geq 2$

$$f(n) = \begin{cases} 3f(m), & \text{if } n = 2m; \\ 2f(m) + f(m+1), & \text{if } n = 2m+1. \end{cases} \tag{11.3}$$

The function $f(n)$ can be approximated by using the constant $\theta = \log(3)/\log(2)$ such that

$$\liminf_{n\to\infty} \frac{f(n)}{n^\theta} = \lambda,$$

$$\limsup_{n\to\infty} \frac{f(n)}{n^\theta} = 1.$$

The constant λ is known as the *Stolarsky-Harborth constant*, and its numeric value is

$$\lambda = 0.8125565590\ldots.$$

More on the function f and the liminf-limsup relations, see [278, 357, 540]. See also [230, Section 2.16].

11.3 Lucas congruence for the binomial coefficients

A strikingly powerful modulo p reduction is possible by the congruence of Lucas. First let us see an example, how it can be applied before we give its proof.

Let $n = 637\,899$, and $k = 77\,681$. Let us determine $\binom{n}{k}$ modulo p. Note that $\binom{637\,899}{77\,681}$ has $102\,625$ digits, so some trick and theorem is necessary, indeed.

The method of Lucas is as follows. First let us write n and k in base p:

$$n = \mathbf{3} + \mathbf{2} \cdot 7 + \mathbf{5} \cdot 7^2 + \mathbf{4} \cdot 7^3 + \mathbf{6} \cdot 7^4 + \mathbf{2} \cdot 7^5 + \mathbf{5} \cdot 7^6,$$

$$k = \mathbf{2} + \mathbf{2} \cdot 7 + \mathbf{3} \cdot 7^2 + \mathbf{2} \cdot 7^3 + \mathbf{4} \cdot 7^4 + \mathbf{4} \cdot 7^5.$$

Then we reduce the binomial coefficient modulo p according to the *digits* of n and k (typeset in bold) as follows:

$$\binom{637\,899}{77\,681} \equiv \binom{3}{2}\binom{2}{2}\binom{5}{3}\binom{4}{2}\binom{6}{4}\binom{2}{4}\binom{5}{0} \pmod{7}.$$

The penultimate binomial coefficient, $\binom{2}{4}$ is zero, so we have that $\binom{637\,899}{77\,681}$ is divisible by 7:

$$\binom{637\,899}{77\,681} \equiv 0 \pmod{7}.$$

The proof of Lucas's theorem

The below very simple proof comes from [231]. The binomial theorem says that

$$\sum_{k=0}^{n} \binom{n}{k}x^k = (1+x)^n.$$

Taking

$$n = n_0 + n_1 p + n_2 p^2 + \cdots + n_a p^a,$$

we can write the right-hand side of the above as

$$(1+x)^n = (1+x)^{n_0}(1+x)^{n_1 p}\cdots(1+x)^{n_a p^a} = \prod_{i=0}^{a}((1+x)^{p^i})^{n_i}.$$

To proceed, we need the following statement with respect to the binomial coefficients. If $0 < k < p$, then $\binom{p}{k}$ is divisible by p, that is,

$$\binom{p}{k} \equiv 0 \pmod{p}. \tag{11.4}$$

This can be seen easily, because

$$\binom{p}{k} = \frac{1}{k!}(p-k+1)(p-k+2)\cdots p$$

and p is not divisible by any numbers among $1, 2, \ldots, p-1$. Hence, in particular, none of the factors of $k!$ are divisible by p. We cannot cancel p from the nominator.

Hence, in particular, we have

$$(1+x)^p = \sum_{i=0}^{p} \binom{p}{i}x^i \equiv 1 + x^p \pmod{p}.$$

Whence, by induction, it follows that

$$(1+x)^{p^i} \equiv 1 + x^{p^i} \pmod{p}.$$

Continuing our original calculation, we get

$$\sum_{k=0}^{n}\binom{n}{k}x^k = \prod_{i=0}^{a}(1+x^{p^i})^{n_i} \pmod{p}.$$

Applying the binomial theorem in each term on the right-hand side,

$$\sum_{k=0}^{n}\binom{n}{k}x^k \equiv \prod_{i=0}^{a}\sum_{s_i=0}^{n_i}\binom{n_i}{s_i}x^{p^i s_i}. \tag{11.5}$$

To match the coefficients of the powers of x on both sides, take a fix index k on the left, and write it as

$$k = k_0 + k_1 p + k_2 p^2 + \cdots + k_a p^a.$$

On the right-hand side we have x^k only if

$$s_0 p^0 + s_1 p^1 + \cdots + s_a p^a = k.$$

But, as $0 \le s_i < p$, this can happen only if for every i it holds that $s_i = k_i$. There is only one such s_i in each inner sum in (11.5). Therefore, modulo p,

$$\binom{n}{k} \equiv \prod_{i=0}^{a}\binom{n_i}{s_i} = \prod_{i=0}^{a}\binom{n_i}{k_i}.$$

In its final form, Lucas's theorem therefore reads as

$$\binom{n}{k} \equiv \binom{n_0}{k_0}\binom{n_1}{k_1}\cdots\binom{n_a}{k_a} \pmod{p}.$$

We will see that theorems similar to Lucas's can be proven with respect to the Stirling numbers, but first we need some additional results.

11.4   The parity of the Stirling numbers

First, we study the parity of the Stirling numbers. It will turn out that the test whether a given Stirling number (let it be of the first or second kind) is even or odd can be traced back to the parity check of the binomial coefficients.

Parity of the Stirling numbers of the first kind

In the first kind case, we apply the horizontal generating function (2.44):

$$\sum_{k=0}^{n}\begin{bmatrix}n\\k\end{bmatrix}x^k = x(x+1)(x+2)\cdots(x+n-1). \tag{11.6}$$

On the right-hand side, we clearly have this congruence:

$$x(x+1)(x+2)\cdots(x+n-1) \equiv x(x+1)x(x+1)\cdots \pmod 2,$$

moreover,

$$x(x+1)x(x+1)\cdots = x^{\lceil\frac{n}{2}\rceil}(x+1)^{\lfloor\frac{n}{2}\rfloor}.$$

Now we apply the binomial theorem:

$$x^{\lceil\frac{n}{2}\rceil}(x+1)^{\lfloor\frac{n}{2}\rfloor} = \sum_{m=0}^{\lfloor\frac{n}{2}\rfloor}\binom{\lfloor\frac{n}{2}\rfloor}{m}x^{m+\lceil\frac{n}{2}\rceil}.$$

To compare the coefficients of the powers of x here and in (11.6), we have to substitute $m = k - \lceil\frac{n}{2}\rceil$. Since the congruence has been preserved in all the steps, we have that

$$\begin{bmatrix}n\\k\end{bmatrix} \equiv \binom{\lfloor\frac{n}{2}\rfloor}{k-\lceil\frac{n}{2}\rceil} \pmod 2.$$

In other words, the quantities on the two sides are even or odd simultaneously. If we are only interested in the parity of $\begin{bmatrix}n\\k\end{bmatrix}$, then it is easier to calculate the binomial coefficient and its parity.

Parity of the Stirling numbers of the second kind

A same result can be proven with respect to the second kind Stirling numbers, but in this case the (2.46) horizontal generating function cannot be applied. It is better to consider (2.22):

$$\sum_{n=0}^{\infty}\begin{Bmatrix}n\\k\end{Bmatrix}x^n = \frac{x^k}{(1-x)(1-2x)\cdots(1-kx)}.$$

The factors in the denominator of the form $1-2x$, $1-4x$, etc. are all congruent to one. Therefore,

$$\sum_{n=0}^{\infty}\begin{Bmatrix}n\\k\end{Bmatrix}x^n \equiv \frac{x^k}{(1-x)^{\lceil\frac{k}{2}\rceil}} \pmod 2.$$

By using Example 23 on page 81 we have that

$$\frac{x^k}{(1-x)^{\lceil\frac{k}{2}\rceil}} = x^k\sum_{m=0}^{\infty}\begin{bmatrix}k\\2\end{bmatrix}^{\overline{m}}\frac{x^m}{m!} = \sum_{m=0}^{\infty}\binom{\lceil\frac{k}{2}\rceil+m-1}{m}x^{m+k}.$$

Comparing the coefficients on the left- and right-hand sides, we infer

$$\begin{Bmatrix}n\\k\end{Bmatrix} \equiv \binom{\lceil\frac{k}{2}\rceil+n-k-1}{n-k} \pmod 2.$$

The number of odd elements in a given row of the first kind Stirling triangle is the A060632 sequence, while the same for the second kind is the sequence A007306 in OEIS. Some other interesting results can be found there.

11.5 Stirling numbers with prime parameters

There is a nice property of both kinds of Stirling numbers: if the upper parameter is a prime number p, then all the Stirling numbers in this line of the triangle are divisible by p except the leftmost and rightmost elements:

$$\begin{bmatrix} p \\ k \end{bmatrix} = \begin{Bmatrix} p \\ k \end{Bmatrix} \equiv 0 \pmod{p} \quad (1 < k < p). \tag{11.7}$$

The proof we give is combinatorial and can be carried out in both kinds of Stirling numbers at the same time. We have to count the k-partition (or k-cycle decomposition) of the set $\{1, 2, \ldots, p\}$. We pick up an arbitrary partition (permutation) to start the process. If a block or cycle decomposition is given, then we map every element to the subsequent element, and to p we map 1. In the next step, we do the same mapping to our new partition (or permutation), and so on. For example, if $p = 5$, then we have a possible sequence of partitions (permutations) like

$$1, 2, 3|4, 5 \to 2, 3, 4|5, 1 \to 3, 4, 5|1, 2 \to 4, 5, 1|2, 3 \to 5, 1, 2|3, 4 \to 1, 2, 3|4, 5$$

We have a sequence of p different partitions (permutations), and in every step the resulting partition (permutation) is different.

Next, we pick a new partition (permutation) of $\{1, 2, \ldots, p\}$ which we have not covered so far, and repeat the algorithm. We will get another p partition (permutation). We follow this process until we have partition (permutation) to choose from. At the end of the process, we will have ap partitions (permutations). This means that the total number of k-partitions or (k-permutations) is a multiple of p. This is what we wanted to prove. Note that this algorithm cannot be applied when $k = 1$ or $k = n$, because in this case the algorithm does not result in new partitions (permutations).

Hence, for instance, in the seventh line of the Stirling matrices of both kinds, every element except the first and last one is divisible by 7.

Let us consider the recursion of the first kind Stirlings:

$$\begin{bmatrix} p+1 \\ k \end{bmatrix} = p \begin{bmatrix} p \\ k \end{bmatrix} + \begin{bmatrix} p \\ k-1 \end{bmatrix}.$$

The first term on the right-hand side is automatically divisible by p. By our previous theorem, the same is true for the second term at least if $1 < k-1 < p$. The same argument applies for the second kind Stirlings. Hence,

$$\begin{bmatrix} p+1 \\ k \end{bmatrix} = \begin{Bmatrix} p+1 \\ k \end{Bmatrix} \equiv 0 \pmod{p} \quad (2 < k \le p). \tag{11.8}$$

The same reasoning gives that

$$\begin{bmatrix} p+2 \\ k \end{bmatrix} = \begin{Bmatrix} p+2 \\ k \end{Bmatrix} \equiv 0 \pmod{p} \quad (3 < k \le p) \tag{11.9}$$

also holds. Finally, after $p - 2$ steps k must be equal to p and the upper parameter is $p + p - 2$:

$$\begin{bmatrix} 2p - 2 \\ p \end{bmatrix} \equiv \begin{Bmatrix} 2p - 2 \\ p \end{Bmatrix} \equiv 0 \pmod{p}.$$

Applying the basic recursions of the Stirling numbers of the first kind, we get divisibility properties for the upper parameter of the form $p - 1$:

$$\begin{bmatrix} p - 1 \\ k \end{bmatrix} \equiv 1 \pmod{p}. \tag{11.10}$$

Let us see how to prove this. We have that

$$\begin{bmatrix} p \\ k \end{bmatrix} = (p - 1)\begin{bmatrix} p - 1 \\ k \end{bmatrix} + \begin{bmatrix} p - 1 \\ k - 1 \end{bmatrix},$$

in which the left-hand side and $p\begin{bmatrix} p-1 \\ k \end{bmatrix}$ are divisible by p, so

$$0 \equiv -\begin{bmatrix} p - 1 \\ k \end{bmatrix} + \begin{bmatrix} p - 1 \\ k - 1 \end{bmatrix} \pmod{p}.$$

Rewriting this

$$\begin{bmatrix} p - 1 \\ k \end{bmatrix} \equiv \begin{bmatrix} p - 1 \\ k - 1 \end{bmatrix} \pmod{p}. \tag{11.11}$$

This nice congruence, which holds for all $1 \le k \le p - 1$, says that all the elements in a line which precedes a prime line have the same remainder if we divide them by p. We can go further. If we take $k = p - 1$, then $\begin{bmatrix} p-1 \\ p-1 \end{bmatrix} = 1$, so $\begin{bmatrix} p-1 \\ 1 \end{bmatrix} \equiv 1 \pmod{p}$. By congruence (11.11) this value modulo p inherits all k. Therefore,

$$\begin{bmatrix} p - 1 \\ k \end{bmatrix} \equiv 1 \pmod{p} \quad (1 \le k \le p - 1). \tag{11.12}$$

11.5.1   Wilson's theorem

Wilson's[3] *theorem* [305, p. 70] states that

$$(p - 1)! \equiv -1 \pmod{p}. \tag{11.13}$$

This statement is very easy to prove once we have (11.12) at hand. Recall (2.41):

$$\sum_{k=0}^{n} \begin{bmatrix} n \\ k \end{bmatrix} x^k = x(x + 1)(x + 2) \cdots (x + n - 1).$$

Substituting $x = 1$ and $n = p - 1$, we have

$$(p - 1)! = \sum_{k=0}^{p-1} \begin{bmatrix} p - 1 \\ k \end{bmatrix} \overset{(11.12)}{\equiv} \sum_{k=1}^{p-1} 1 = p - 1 \equiv -1 \pmod{p}.$$

---

[3] John Wilson (1741-1793), English mathematician. Wilson's theorem is named after him; however, this result was reportedly known around 1000 by Ibn al-Haytham (c. 965-1040), Arab mathematician.

11.5.2 Wolstenholme's theorem

We have seen that $\begin{bmatrix} p \\ k \end{bmatrix}$ is divisible by p for $1 < k < p$. In the particular case when $k = 2$, even more is true. The theorem of Wolstenholme[4] [15, 160, 602] says that if $p > 3$, then

$$\begin{bmatrix} p \\ 2 \end{bmatrix} \equiv 0 \pmod{p^2}.$$

To prove this nice divisibility property, we recall (2.42):

$$\sum_{k=0}^{n} \overline{\begin{bmatrix} n \\ k \end{bmatrix}} x^k = x(x-1)(x-2)\cdots(x-n+1).$$

Taking $x = p$ and $n = p$

$$\sum_{k=0}^{p} \overline{\begin{bmatrix} p \\ k \end{bmatrix}} p^k = p!.$$

Writing this out in details,

$$p\begin{bmatrix} p \\ 1 \end{bmatrix} - p^2\begin{bmatrix} p \\ 2 \end{bmatrix} + p^3\begin{bmatrix} p \\ 3 \end{bmatrix} - \cdots + p^{p-2}\begin{bmatrix} p \\ p-2 \end{bmatrix} - p^{p-1}\begin{bmatrix} p \\ p-1 \end{bmatrix} + p^p\begin{bmatrix} p \\ p \end{bmatrix} = p!$$

Since $\begin{bmatrix} p \\ 1 \end{bmatrix} = (p-1)!$, $p!$ cancels. Rearranging and dividing by p^2:

$$p\begin{bmatrix} p \\ 3 \end{bmatrix} - \cdots + p^{p-4}\begin{bmatrix} p \\ p-2 \end{bmatrix} - p^{p-3}\begin{bmatrix} p \\ p-1 \end{bmatrix} + p^{p-2}\begin{bmatrix} p \\ p \end{bmatrix} = \begin{bmatrix} p \\ 2 \end{bmatrix} \qquad (11.14)$$

Applying (11.7) for $\begin{bmatrix} p \\ 3 \end{bmatrix}$ we get that the left-hand side is a multiple of p^2. This is the content of the Wolstenholme theorem.

11.5.3 Wolstenholme's theorem for the harmonic numbers

Recalling (1.10), we know that the first kind Stirlings and the harmonic numbers are related in a simple way. Fixing $n = p$ to be a prime, we have that

$$H_{p-1} = \frac{1}{(p-1)!}\begin{bmatrix} p \\ 2 \end{bmatrix}.$$

We have just seen that $\begin{bmatrix} p \\ 2 \end{bmatrix} \equiv 0 \pmod{p^2}$ (for primes greater than three). What consequence does this have on the H_{p-1} harmonic number? Since $(p-1)!$ is not divisible by p, we have that H_{p-1} can always be written in the form

$$H_{p-1} = p^2 \frac{a}{b} \quad (p > 3), \qquad (11.15)$$

where b is not divisible by p. Equivalently, by using the extended congruence definition for fractions,

$$H_{p-1} \equiv 0 \pmod{p^2} \quad (p > 3). \qquad (11.16)$$

---

[4] Joseph Wolstenholme (1829-1891), English mathematician.

For example,

$$H_{19-1} = H_{18} = \frac{14\,274\,301}{4\,084\,080} = 19^2 \frac{39\,541}{2^4 \cdot 3 \cdot 5 \cdot 7 \cdot 11 \cdot 13 \cdot 17}.$$

Wolstenholme proved his theorem in 1862 [602]. The weaker result that $H_{p-1} = p\frac{a}{b}$ for odd primes was known already by Babbage[5] [41] in 1819.

11.5.4   Wolstenholme's primes

We have to note that $\left[\begin{smallmatrix} p \\ 2 \end{smallmatrix}\right] \equiv 0 \pmod{p^2}$ is not always strict, it can happen that $\left[\begin{smallmatrix} p \\ 2 \end{smallmatrix}\right] \equiv 0 \pmod{p^3}$ is still true. Indeed,

$$\begin{bmatrix} 16\,843 \\ 2 \end{bmatrix} \equiv 0 \pmod{16\,843^3}$$

and

$$\begin{bmatrix} 2\,124\,679 \\ 2 \end{bmatrix} \equiv 0 \pmod{2\,124\,679^3},$$

thus, for these two primes,

$$H_{p-1} = p^3 \frac{a}{b},$$

for some a and b which are not divisible by the respective primes. There are no more known primes with this cubic divisibility property.

The primes for which the above cubic divisibility holds are called *Wolstenholme primes*. There are no more known Wolstenholme primes up to 10^9 [400], although it is believed that there are infinitely many such primes [399].

11.5.5   Wolstenholme's theorem for $H_{p-1,2}$

The generalized harmonic numbers

$$H_{n,k} = \sum_{j=1}^{n} \frac{1}{j^k}$$

were introduced in Section 2.7.1 to express some special values of the Stirling numbers of the first kind. In his paper [602], Wolstenholme proved another statement together with (11.15). Namely,

$$H_{p-1,2} = p\frac{a}{b}$$

---

[5]Charles Babbage (1791-1871), English mathematician, engineer, inventor, and philosopher. He worked on cryptography also, and invented but never finished his differential machine which was supposed to calculate polynomial values.

for all prime $p > 3$ for some integers a and b, depending on p but not divisible by p.

This statement follows from (11.15); let us see how. Consider the product

$$\prod_{j=1}^{n}(\lambda - x_j).$$

This is a polynomial of degree n in λ, so there must exist some coefficients $e_1, e_2, \ldots e_n$ depending on $x_1, \ldots, x_n$ such that

$$\prod_{j=1}^{n}(\lambda-x_j) = e_0(x_1,\ldots,x_n)\lambda^n - e_1(x_1,\ldots,x_n)\lambda^{n-1} + e_2(x_1,\ldots,x_n)\lambda^{n-2} + \cdots +$$

$$(-1)^n e_n(x_1,\ldots,x_n).$$

The constant coefficient in the product is $(-1)^n x_1 x_2 \cdots x_n$, thus

$$e_n(x_1,\ldots,x_n) = x_1 x_2 \cdots x_n.$$

The coefficient of λ^n is one; therefore,

$$e_0(x_1, x_2, \ldots, x_n) = 1. \tag{11.17}$$

Similarly,

$$e_1(x_1, x_2, \ldots, x_n) = x_1 + x_2 + \cdots + x_n, \tag{11.18}$$

$$e_2(x_1, x_2, \ldots, x_n) = x_1 x_2 + x_1 x_3 + \cdots + x_1 x_n + x_2 x_3 + x_2 x_4 + \cdots + x_{n-1} x_n. \tag{11.19}$$

These functions are called *elementary symmetric polynomials* [386].

Now consider the following expression:

$$\binom{\lambda - 1}{n} = \frac{1}{n!}(\lambda - 1)(\lambda - 2) \cdots (\lambda - n) = (\lambda - 1)\left(\frac{\lambda}{2} - 1\right) \cdots \left(\frac{\lambda}{n} - 1\right) =$$

$$\prod_{j=1}^{n}\left(\frac{\lambda}{j} - 1\right) = (-1)^n \lambda^n \prod_{j=1}^{n}\left(\frac{1}{\lambda} - \frac{1}{j}\right).$$

We apply the symmetric polynomial decomposition for the last expression (with $\lambda \rightsquigarrow \frac{1}{\lambda}$ and $x_j = 1/j$ $(j = 1, \ldots, n)$) to get

$$\binom{\lambda - 1}{n} = (-1)^n \lambda^n \prod_{j=1}^{n}\left(\frac{1}{\lambda} - \frac{1}{j}\right) =$$

$$(-1)^n \lambda^n \left[\lambda^{-n} - \lambda^{-n+1} e_1\left(\frac{1}{1}, \frac{1}{2}, \ldots, \frac{1}{n}\right) + \cdots + (-1)^n e_n\left(\frac{1}{1}, \frac{1}{2}, \ldots, \frac{1}{n}\right)\right] =$$

$$(-1)^n \left[1 - \lambda e_1 \left(\frac{1}{1}, \frac{1}{2}, \ldots, \frac{1}{n} \right) + \cdots + (-1)^n \lambda^n e_n \left(\frac{1}{1}, \frac{1}{2}, \ldots, \frac{1}{n} \right) \right].$$

Setting $\lambda = p > 2$ and $n = p - 1$ we get, in particular,

$$1 = 1 - pe_1 \left(\frac{1}{1}, \frac{1}{2}, \ldots, \frac{1}{p-1} \right) + \cdots + (-1)^{p-1} p^{p-1} e_{p-1} \left(\frac{1}{1}, \frac{1}{2}, \ldots, \frac{1}{p-1} \right).$$

We will consider this expression modulo p^3. One can easily see that in the denominator of $e_k \left(\frac{1}{1}, \frac{1}{2}, \ldots, \frac{1}{p-1} \right)$ the prime p does not appear, thus

$$1 \equiv 1 - pe_1 \left(\frac{1}{1}, \frac{1}{2}, \ldots, \frac{1}{p-1} \right) + p^2 e_2 \left(\frac{1}{1}, \frac{1}{2}, \ldots, \frac{1}{p-1} \right) \quad (\bmod \ p^3). \quad (11.20)$$

Let us express these symmetric functions in familiar terms. Recalling (11.18)-(11.19),

$$e_1 \left(\frac{1}{1}, \frac{1}{2}, \ldots, \frac{1}{p-1} \right) = H_{p-1},$$

while

$$e_2 \left(\frac{1}{1}, \frac{1}{2}, \ldots, \frac{1}{p-1} \right) = \sum_{i=1}^{p-1} \sum_{j=i+1}^{p-1} \frac{1}{ij}.$$

The reader may easily verify that

$$\sum_{i=1}^{p-1} \sum_{j=i+1}^{p-1} \frac{1}{ij} = \frac{1}{2} \left(H_n^2 - H_{n,2} \right).$$

Hence, from (11.20) we get

$$pH_{p-1} \equiv \frac{p^2}{2} H_{p-1}^2 - \frac{p^2}{2} H_{p-1,2} \quad (\bmod \ p^3). \quad (11.21)$$

(Actually, the more general statement

$$\binom{p-1}{n} \equiv (-1)^n \left[1 - pH_n + \frac{p^2}{2} \left(H_n^2 - H_{n,2} \right) \right] \quad (\bmod \ p^3).$$

is also true for $n < p$.) For $p > 3$ Wolstenholme's theorem (11.16) yields that the left-hand side of (11.21) is congruent to zero modulo p^3, as well as $\frac{p^2}{2} H_{p-1}^2$. (Note that this latter is zero even modulo p^4 by Wolstenholme's theorem.) Therefore,

$$\frac{p^2}{2} H_{p-1,2} \equiv 0 \quad (\bmod \ p^3).$$

This means that the nominator of $H_{p-1,2}$ must contain a factor of p, i.e.,

$$H_{p-1,2} \equiv 0 \quad (\bmod \ p)$$

for primes greater than three. This is the other theorem of Wolstenholme we wanted to prove.

We note that the idea of the above proof comes from Emma Lehmer's[6] paper [361, p. 360], where (11.21) is presented, but not mentioned that it is valid only for $n < p$. The statement and its proof can also be found in [551].

11.6 Lucas congruence for the Stirling numbers of both kinds

The Stirling numbers are known since 1730, and Lucas's theorem for the binomial coefficients was born in 1878. However, it was not earlier than 1985 when Lucas's theorem was transferred to $\begin{bmatrix} n \\ k \end{bmatrix}$ ([502], later [288, 457]), and 2000 when it was presented for $\begin{Bmatrix} n \\ k \end{Bmatrix}$ [508].

11.6.1 The first kind Stirling number case

The idea of Fine can easily be transferred from the binomial coefficients to the first kind Stirling numbers. For the first, we started from

$$\sum_{k=0}^{n} \binom{n}{k} x^k = (1 + x)^n,$$

and for the second, we will depart from

$$\sum_{k=0}^{n} \begin{bmatrix} n \\ k \end{bmatrix} x^k = x(x + 1) \cdots (x + n - 1) = x^{\overline{n}}.$$

Let us fix a prime p and write n as

$$n = n' + n_0$$

such that $n' = 0$ or it is divisible by p, and $0 \le n_0 < p$. Then

$$\sum_{k=0}^{n} \begin{bmatrix} n \\ k \end{bmatrix} x^k = \prod_{a=0}^{n'-1} (x + ap)(x + ap + 1) \cdots (x + ap + p - 1) \prod_{b=0}^{n_0-1} (x + n'p + b) \equiv$$

$$\prod_{a=0}^{n'-1} x(x + 1) \cdots (x + p - 1) \prod_{b=0}^{n_0-1} (x + b) = (x^{\overline{p}})^{n'} x^{\overline{n_0}}.$$

---

[6]Emma Lehmer (1906-2007), mathematician with Russian origin. She mainly worked on number theory, special primes and congruences.

We can simplify $x^{\overline{p}}$ modulo p by (11.7), and by Wilson's theorem (11.13). That is, modulo p,

$$x^{\overline{p}} = \sum_{k=0}^{p} \begin{bmatrix} p \\ k \end{bmatrix} x^k \overset{(11.7)}{\equiv} \begin{bmatrix} p \\ 1 \end{bmatrix} x + \begin{bmatrix} p \\ p \end{bmatrix} x^p = (p-1)!x + x^p \overset{(11.13)}{\equiv} x^p - x.$$

Hence, we have that

$$\sum_{k=0}^{n} \begin{bmatrix} n \\ k \end{bmatrix} x^k \equiv (x^p - x)^{n'} x^{\overline{n_0}} \pmod{p}.$$

Expanding the polynomials on the right-hand side,

$$\sum_{k=0}^{n} \begin{bmatrix} n \\ k \end{bmatrix} x^k \equiv \left(\sum_{j=0}^{n'} \binom{n'}{j} x^{pj} (-x)^{n'-j} \right) \left(\sum_{i=0}^{n_0} \begin{bmatrix} n_0 \\ i \end{bmatrix} x^i \right) \pmod{p}. \quad (11.22)$$

We have nothing else to do but comparing the coefficients of both sides. This is more convenient if we write $k - n'$ as

$$k - n' = j(p-1) + i \quad (11.23)$$

for some i and j (such that if $n_0 = 0$, then $0 \le i < p-1$ and if $n_0 > 0$, then $0 < i \le p-1$). Then we get our result:

$$\begin{bmatrix} n \\ k \end{bmatrix} \equiv (-1)^{n'-j} \binom{n'}{j} \begin{bmatrix} n_0 \\ i \end{bmatrix} \pmod{p}. \quad (11.24)$$

A corollary of (11.24)

There is an interesting corollary of (11.24), noted by Peele et al. [457]: for any fixed k, the set A of those n for which $\begin{bmatrix} n \\ k \end{bmatrix}$ is not divisible by p is finite. The greatest element of A is pk, the smallest is k.

Indeed, if $n > pk$, then $j < 0$ in (11.23); therefore, $\begin{bmatrix} n \\ k \end{bmatrix} \equiv 0 \pmod{p}$. That is, A cannot have any element greater than pk.

If $n = pk$, then $n' = n_0 = i = j = 0$ and

$$\begin{bmatrix} pk \\ k \end{bmatrix} \equiv (-1)^k \pmod{k}. \quad (11.25)$$

Thus, the maximal element of the set A is pk, as we stated.

If $n = k$, then $\begin{bmatrix} k \\ k \end{bmatrix} = 1 \equiv 1 \pmod{p}$. If, finally, $n < k$, then $\begin{bmatrix} n \\ k \end{bmatrix} = 0 \equiv 0 \pmod{p}$. So, the smallest element in A is k indeed.

11.6.2 The second kind Stirling number case

The easiness of the proof of the Lucas congruence both for the binomial co-efficients and for the first kind Stirling numbers came from the fact that their horizontal generating functions are simple. In the case of the second kind Stirling numbers, the horizontal generating functions (the Bell polynomials) have no simple product form (we do not have closed-form description of the roots of $B_n(x)$ in general). Thus, we cannot expect a simple Lucas-type congruence. The following result was found by Roberto Sánchez-Peregrino [508]. Let

$$n = n_0 + n_1 p + \cdots + n_m p^m,$$
$$k = k_0 + k_1 p + \cdots + k_m p^m.$$

Then we have

$$\left\{ {n \atop k} \right\} \equiv \sum_{n_{10}+n_{11}=n_1} \sum_{n_{20}+n_{21}+n_{22}=n_2} \cdots \sum_{n_{m0}+n_{m1}+\cdots+n_{mm}=n_m}$$

$$\binom{n_1}{n_{10}, n_{11}} \binom{n_2}{n_{20}, n_{21}, n_{22}} \cdots \binom{n_m}{n_{m0}, n_{m1}, \ldots, n_{mm}} \left\{ {N \atop K} \right\} \pmod{p},$$

where

$$N = n_0 + n_{10} + n_{20} + \cdots + n_{m0},$$

$$K = k_0 + (k_1 - n_{m1} - \cdots - n_{21} - n_{11})p + (k_2 - n_{m2} - \cdots - n_{22})p^2 + \cdots + (k_m - n_{mm})p^m.$$

The proof is not much less complicated than the expression itself, so we do not give a discussion of it here, we refer to the original paper [508]. We note, however, the following two propositions which were used in the proof and are interesting in themselves:

$$\left\{ {n+p \atop k} \right\} \equiv \left\{ {n+1 \atop k} \right\} + \left\{ {n \atop k-p} \right\} \pmod{p}.$$

From this it comes that

$$\left\{ {n_0 + tp \atop k_0 + tp} \right\} \equiv \sum_{i=0}^{t} \binom{t}{i} \left\{ {n_0 + i \atop k_0 + ip} \right\} \pmod{p}.$$

11.7 Divisibility properties of the Bell numbers

11.7.1 Theorems about B_p and B_{p+1}

Let p be a fixed but arbitrary prime as before. By (11.7) every term of the sum

$$B_p = \sum_{k=0}^{p} \left\{ {p \atop k} \right\}$$

are divisible by p, except for $k = 1, p$. We have our first divisibility result for the Bell numbers.

$$B_p \equiv \left\{ {p \atop 1} \right\} + \left\{ {p \atop p} \right\} = 2 \pmod{p}.$$

For B_{p+1} it is also easy to find something, but now we use (1.1):

$$B_{p+1} = \sum_{k=0}^{p} \binom{p}{k} B_k.$$

(The reader can verify that the Stirling number sum representation also offers something like $B_{p+1} \equiv \left\{ {p+1 \atop 1} \right\} + \left\{ {p+1 \atop 2} \right\} \pmod{p}$. The details are among the exercises.)

Applying our (11.4) divisibility for the binomial coefficients we have that

$$B_{p+1} \equiv \binom{p}{0} + \binom{p}{p} B_p = 1 + B_p \equiv 3 \pmod{p},$$

knowing that B_p congruent to 2 modulo p.

Some interesting remarks about B_{p-1} modulo p will be given in Section 11.9.

11.7.2   Touchard's congruence

We can prove even more with respect to the Bell numbers. Touchard[7] proved in 1933 that, in general [563],

$$B_{n+p} \equiv B_{n+1} + B_n \pmod{p}, \tag{11.26}$$

where p is an arbitrary prime and $n \geq 0$.

This results in our former congruences $B_p \equiv 2 \pmod{p}$ and $B_{p+1} \equiv 3 \pmod{p}$ when we choose $n = 0$ and $n = 1$, respectively.

The congruence of Touchard can be proven with a minor effort if we apply a nice identity of M. Z. Spivey[8] from 2008 [531]:

$$B_{n+m} = \sum_{k=0}^{n} \sum_{j=0}^{m} j^{n-k} \left\{ {m \atop j} \right\} \binom{n}{k} B_k \tag{11.27}$$

(See [58, Theorem 1.] and [260] for the Bell polynomial extension, and Exercise 19.) The proof is as follows. All the partitions of an $(n+m)$-element set (which are counted by B_{n+m} on the left) can be constructed as follows. We put m elements into j blocks in $\left\{ {m \atop j} \right\}$ ways and from the other n elements we select k and form an arbitrary partition with them in $\binom{n}{k} B_k$ ways. There are $n - k$

---

[7]Jacques Touchard (1885-1968), French mathematician.
[8]Michael Z. Spivey (1960-), American mathematician.

elements not selected before from that of n. We put these into the j blocks of the m elements one-by-one. This can be done, of course, in j^{n-k} ways. Summing over j and k, we get all the partitions on $n + m$ elements.

Going back to the congruence of Touchard, we take $m = p$ in Spivey's formula. Since $\left\{{p \atop j}\right\} \equiv 0 \pmod{p}$ if $1 < j < p$, so

$$B_{n+p} = \sum_{k=0}^{n} \sum_{j=0}^{p} j^{n-k} \left\{{p \atop j}\right\} \binom{n}{k} B_k \equiv$$

$$\left\{{p \atop 1}\right\} \sum_{k=0}^{n} \binom{n}{k} B_k + \left\{{p \atop p}\right\} \sum_{k=0}^{n} \binom{n}{k} B_k p^{n-k} \pmod{p}.$$

In the last sum all the terms $k = 0, \ldots, n - 1$ are divisible by p so only B_n does not cancel. The special values $\left\{{p \atop 1}\right\} = \left\{{p \atop p}\right\} = 1$ and the recursion $\sum_{k=0}^{n} \binom{n}{k} B_k = B_{n+1}$ now result in the congruence of Touchard.

A generalization of Touchard's congruence is presented among the exercises. See the Outlook of this chapter for more information.

11.8 Divisibility properties of the Fubini numbers

11.8.1 Elementary congruences

Thanks to the $k!$ in their definition, the Fubini numbers

$$F_n = \sum_{k=0}^{n} k! \left\{{n \atop k}\right\},$$

satisfy some simple congruence relations. For example,

$$F_n = 1! \left\{{n \atop 1}\right\} + 2! \left\{{n \atop 2}\right\} + \cdots \equiv 1 \left\{{n \atop 1}\right\} = 1 \pmod{2}, \tag{11.28}$$

and therefore *all* the Fubini numbers are odd. Actually, even more is true. The Fubini numbers have the same remainder after dividing not only by 2 but also by $2^5 = 32$. This is Theorem 4 in [469]:

$$F_{n+1} \equiv F_n \pmod{32}$$

for $n \geq 5$.

It is easy to prove that

$$F_p \equiv 1 \pmod{p}$$

for any prime p. Indeed,

$$F_p = 1! \begin{Bmatrix} p \\ 1 \end{Bmatrix} + 2! \begin{Bmatrix} p \\ 2 \end{Bmatrix} + \cdots + p! \begin{Bmatrix} p \\ p \end{Bmatrix} \equiv 1 \pmod{p},$$

since all the non-extremal terms in the sum are divisible by p as we have shown on p. 303. The last term is divisible by p for any p prime, so we get the statement.

11.8.2   A Touchard-like congruence

The Touchard congruence for Bell numbers says that

$$B_{n+p} = B_{n+1} + B_n.$$

It is not hard (actually, even easier) to prove the corresponding congruence for the Fubini numbers:

$$F_{n+p} \equiv F_{n+1} \pmod{p}.$$

To present the one-line proof, we need Fermat's[9] little theorem. It states that for any integer a and any prime p

$$a^p \equiv a \pmod{p},$$

or, equivalently, if $a \not\equiv 0 \pmod{p}$,

$$a^{p-1} \equiv 1 \pmod{p}. \tag{11.29}$$

This theorem can be found in any introductory textbook on number theory, see for example [305]. Once we know this theorem of Fermat, we can use the ordered Dobiński formula (6.6). Let p be an odd prime. Then

$$F_{n+p} = \sum_{k=0}^{\infty} \frac{k^{n+p}}{2^{k+1}} = \sum_{k=0}^{\infty} \frac{k^n}{2^{k+1}} k^p \equiv \sum_{k=0}^{\infty} \frac{k^{n+1}}{2^{k+1}} = F_{n+1} \pmod{p}.$$

This is what we wanted to prove. The $p = 2$ case was excluded, because the denominator would cause a problem. Fortunately, we do not need Fermat's theorem when $p = 2$ because we have already seen that

$$F_{n+2} \equiv F_{n+1} \pmod{2},$$

(see (11.28)).

It costs no additional effort to prove a more general form:

$$F_{n+mp} \equiv F_{n+m} \pmod{p}. \tag{11.30}$$

All the steps can be repeated verbatim after noting that $k^{mp} = (k^p)^m \equiv k^m \pmod{p}$. The $p = 2$ case must be treated separately, but it gives the same congruence.

---

[9] Pierre de Fermat (1607-1665), French mathematician and lawyer, worked also on physical optics.

11.8.3 The Gross-Kaufman-Poonen congruences

A very nice congruence property of the Fubini numbers was noted by Gross in 1962 [267]:

$$F_{n+4} \equiv F_n \pmod{10}.$$

The proof of Gross is short but uses finite differences. It might be more apparent what is going on if we use only the definition of the Stirling numbers and their special values. The proof presented here is from [416].

Let $n > 4$. By the definition,

$$F_{n+4} - F_n = \sum_{k=0}^{n+4} k! \left\{ {n+4 \atop k} \right\} - \sum_{k=0}^{n} k! \left\{ {n \atop k} \right\} =$$

$$\sum_{k=5}^{n+4} k! \left\{ {n+4 \atop k} \right\} - \sum_{k=5}^{n} k! \left\{ {n \atop k} \right\} + \sum_{k=0}^{4} k! \left\{ {n+4 \atop k} \right\} - \sum_{k=0}^{4} k! \left\{ {n \atop k} \right\} \equiv$$

$$\left\{ {n+4 \atop 0} \right\} + \left\{ {n+4 \atop 1} \right\} + 2 \left\{ {n+4 \atop 2} \right\} + 6 \left\{ {n+4 \atop 3} \right\} + 24 \left\{ {n+4 \atop 4} \right\}$$

$$- \left\{ {n \atop 0} \right\} - \left\{ {n \atop 1} \right\} - 2 \left\{ {n \atop 2} \right\} - 6 \left\{ {n \atop 3} \right\} - 24 \left\{ {n \atop 4} \right\} \pmod{10}.$$

Because of the special values $\left\{ {n \atop 0} \right\} = 0$, $\left\{ {n \atop 1} \right\} = 1$, the first two members cancel. The remaining terms are divisible by two, so it is enough to prove that

$$5 \left| \left(\left\{ {n+4 \atop 2} \right\} + 3 \left\{ {n+4 \atop 3} \right\} + 12 \left\{ {n+4 \atop 4} \right\} - \left\{ {n \atop 2} \right\} - 3 \left\{ {n \atop 3} \right\} - 12 \left\{ {n \atop 4} \right\} \right) \right. .$$

We need the special values

$$\left\{ {n \atop 2} \right\} = 2^{n-1} - 1, \qquad \left\{ {n \atop 3} \right\} = \frac{1}{2} \left(3^{n-1} - 2^n + 1 \right),$$

$$\left\{ {n \atop 4} \right\} = \frac{1}{6} 4^{n-1} - \frac{1}{2} 3^{n-1} + 2^{n-2} - \frac{1}{6},$$

from which the first was proven on p. 8, and the others can be seen by using (1.5). We then have that

$$\left\{ {n+4 \atop 2} \right\} - \left\{ {n \atop 2} \right\} = 2^{n+3} - 1 - (2^{n-1} - 1) = 15 \cdot 2^{n-1},$$

$$3 \left\{ {n+4 \atop 3} \right\} - 3 \left\{ {n \atop 3} \right\} = \frac{3}{2} \left(3^{n-1}(3^4 - 1) - 2^n(2^4 - 1) \right),$$

$$12 \left\{ {n+4 \atop 4} \right\} - 12 \left\{ {n \atop 4} \right\} = \frac{12}{2} \left(\frac{1}{3} 4^{n-1}(4^4 - 1) - 3^{n-1}(3^4 - 1) + 2^{n-2}(2^4 - 1) \right).$$

All of these numbers – independently of n – are divisible by 5, so we proved Gross's result.

Gross's congruence states that the *last digits* of the members of the sequence F_n form a periodic sequence of length four. This can readily be checked for small n, see the table for F_n at the end of the book.

Generalized Gross congruence

Kauffman generalized Gross's congruence and showed that such periodicity property holds for the last two, three and four digits [328]! Her results read as

$$F_{n+20} \equiv F_n \pmod{100} \quad (n > 1),$$
$$F_{n+100} \equiv F_n \pmod{1\,000} \quad (n > 2),$$
$$F_{n+500} \equiv F_n \pmod{10\,000} \quad (n > 3).$$

This, together with Gross's congruence,

$$F_{n+4} \equiv F_n \pmod{10}$$

suggests that there might be some general statement saying that the last k digits of Fubini numbers form periodic sequences for all k. Surprisingly, this statement is indeed true. This is a result of Poonen from 1986; it states that

$$F_{n+m(k)} \equiv F_n \pmod{10^k}$$

for all $k \geq 1$, where the function $m(k)$ is

$$m(k) = 4 \cdot 5^{k-1} \quad (k = 1, 2, 3, 4),$$

and

$$m(k) = 2^{k-4} 5^{k-1} \quad (k > 4).$$

In other words, the last k digits of the sequence F_n form a recurring sequence with period $m(k)$. It is worth it to note that even more is true: the last digits form a period sequence *in any number base*. Poonen's result was originally phrased in this general form.

11.9   Kurepa's conjecture

To close the chapter of the elementary congruences, we show an interesting consequence of some of our results so far.

Kurepa[10] conjectured that the so-called left factorial function

$$!n = 0! + 1! + \cdots + (n-1)!$$

---

[10]Đuro Kurepa (1907-1993), Yugoslav mathematician.

is never congruent to zero for primes greater than two. That is,

$$!p \not\equiv 0 \pmod{p}$$

when $p > 2$ is a prime. This conjecture is still unsolved[11].

We will prove[12] now that

$$!p \equiv D_{p-1} \equiv B_{p-1} - 1 \pmod{p}. \tag{11.31}$$

The first part: $!p \equiv D_{p-1} \pmod{p}$

By (10.15) we can write that

$$D_{p-1} = (p-1)! \sum_{k=0}^{p-1} \frac{(-1)^k}{k!} = \sum_{k=0}^{p-1} \binom{p-1}{k} (-1)^k (p-1-k)! =$$

$$\sum_{k=0}^{p-1} \binom{p-1}{k} (-1)^{p-1-k} k! = \sum_{k=0}^{p-1} \binom{p-1}{k} (-1)^k k! \equiv$$

$$\sum_{k=0}^{p-1} (-1)^k (-1)^k k! = \sum_{k=0}^{p-1} k! = !p \pmod{p}.$$

In the penultimate step, we used the congruence

$$\binom{p-1}{k} \equiv (-1)^k \pmod{p},$$

which is a consequence of (11.21).

The second part: $!p \equiv B_{p-1} - 1 \pmod{p}$

By the first part, it is enough to show that $B_{p-1} \equiv D_{p-1} + 1 \pmod{p}$. To this end we recall (2.21) which, in particular, says that

$$(p-1)! B_{p-1} = \sum_{m=1}^{p-1} \binom{p-1}{m} m^{p-1} D_{p-1-m}.$$

Fermat's little theorem (11.29) and Wilson's theorem (11.13) offer a way of simplification:

$$(-1) B_{p-1} \equiv \sum_{m=0}^{p-1} \binom{p-1}{m} D_{p-1-m} - D_{p-1}.$$

---

[11] For a while it was believed that a solution was found, see the attempt in [44] and the falsification in [45]. Kurepa himself also announced a proof, but he never published it. See the historical remarks in [297].

[12] Our proof is completely elementary. The first proof of these congruences was given in [248] which uses p-adic analysis.

(Here we added the term $m = 0$ and compensate this addition by subtracting D_{p-1}.) Appealing to the symmetry of the binomial coefficients, and multiplying by minus one, we get that

$$B_{p-1} \equiv -\sum_{m=0}^{p-1} \binom{p-1}{m} D_m + D_{p-1}.$$

It is an easy exercise to verify that the derangement number sequence is the inverse binomial transform of the sequence $k!$; therefore,

$$\sum_{m=0}^{p-1} \binom{p-1}{m} D_m = (p-1)!.$$

(See the exercises for a hint.) Wilson's theorem is handy again to simplify this factorial. We have now what we wanted to prove:

$$B_{p-1} \equiv 1 + D_{p-1}. \tag{11.32}$$

Alternative formulations of Kurepa's conjecture

Congruences (11.31) give another way to formulate Kurepa's conjecture: there is no odd prime p for which $B_{p-1} - 1$ or D_{p-1} would be divisible by p.

11.10   The non-integral property of the harmonic and hyperharmonic numbers

We are now going to prove that the harmonic numbers and some of their sums are non-integer numbers. To shorten our formulas, we will introduce some notations. Let u be a rational number. After fixing a prime p, u can always be written in the form

$$u = p^\alpha \frac{a}{b},$$

where a and b are not divisible by p and α is an integer. α is called the *p-adic valuation* of u, and we often use the notation $\nu_p(u) = \alpha$.

For example, if $u = \frac{352}{475}$, and we fix, say, $p = 5$, then

$$u = 5^{-2} \frac{352}{19}.$$

That is, the 5-adic valuation of $u = \frac{352}{475}$ is -2:

$$\nu_5\left(\frac{352}{475}\right) = -2.$$

It is an easy exercise to see that

$$\nu_p(ab) = \nu_p(a) + \nu_p(b), \tag{11.33}$$
$$\nu_p(a+b) \geq \min\{\nu_p(a), \nu_p(b)\}. \tag{11.34}$$

In the latter, we have equality if $\nu_p(a) \neq \nu_p(b)$.

The *p-adic absolute value* or *p-adic norm* of a rational number u is denoted and defined as

$$|u|_p = p^{-\alpha} = p^{-\nu_p(u)}.$$

Thus,

$$\left| \frac{352}{475} \right|_5 = 5^2 = 25.$$

From (11.33)-(11.34) it follows that

$$|ab|_p = |a|_p|b|_p, \tag{11.35}$$
$$|a+b|_p \leq \max\{|a|_p, |p|_p\}. \tag{11.36}$$

The latter of these two properties is called *strong triangle inequality*.

We will mainly use these because of notational conveniences, but the theory of p-adic valuations is a non-trivial, beautiful, and profound area of mathematics. See the book of F. Gouvêa [261] or of Koblitz [337] for an introduction, and Robert [497] for a rich and detailed material both on the algebraic and analytic aspects of the theory.

11.10.1 H_n is not integer when $n > 1$

As a first and simplest application of the p-adic machinery, we will study the non-integer property of the harmonic numbers and then of the hyperharmonics. One way to prove the non-integerness of H_n (when $n > 1$) is to show that (when it is written as a reduced fraction) its denominator is always divisible by some positive power of 2. The exact statement that we will prove reads as follows:

$$H_n = \frac{1}{2^{\lfloor \log_2 n \rfloor}} \frac{a_n}{b_n},$$

where a_n and b_n are odd integers, and $\log_2$ is the logarithm of base 2. Having the p-adic norm introduced, the statement can be rewritten in an equivalent form:

$$|H_n|_2 = 2^{\lfloor \log_2 n \rfloor}. \tag{11.37}$$

First, let n be even. Since $|x|_2 = |-x|_2$ for all rational x, by the strong triangle inequality we get

$$\max\{|H_n|_2, |1|_2\} = \max\left\{ |H_n|_2, |1|_2, \left|\frac{1}{3}\right|_2, \left|\frac{1}{5}\right|_2, \ldots, \left|\frac{1}{n-1}\right|_2 \right\} \geq$$

$$\left| H_n - 1 - \frac{1}{3} - \frac{1}{5} - \cdots - \frac{1}{n-1} \right|_2 =$$

$$\left|\frac{1}{2} + \frac{1}{4} + \cdots + \frac{1}{n-2} + \frac{1}{n}\right|_2 =$$

$$\left|\frac{1}{2}\right|_2 \left|1 + \frac{1}{2} + \frac{1}{3} + \cdots + \frac{1}{n/2}\right|_2 = 2\left|H_{n/2}\right|_2.$$

If n is odd, the situation is the same, as the reader can check:

$$\max\left\{|H_n|_2, |1|_2\right\} \geq 2\left|H_{(n-1)/2}\right|_2.$$

So we get that the 2-adic norm of the harmonic numbers is monotone increasing. Since $|H_2|_2 = \left|\frac{3}{2}\right|_2 = 2$, the 2-adic norm of all the harmonic numbers are greater than 1. As a corollary, this means that the harmonic numbers are not integers, which is the statement that we wanted to show.

We can continue the calculations on $H_{n/2}$ (or on $H_{(n-1)/2}$, depending on the parity of n) instead of H_n. For instance, let us consider that $n/2$ is even. Then the method described above gives that

$$|H_{n/2}|_2 \geq \left|\frac{1}{2}\right|_2 |H_{n/4}|_2 = 2|H_{n/4}|_2.$$

This and the previous estimation implies that

$$|H_n|_2 \geq \left|\frac{1}{2}\right|_2 |H_{n/2}|_2 \geq \left|\frac{1}{2}\right|_2 \left|\frac{1}{2}\right|_2 |H_{n/4}|_2 = 4|H_{n/4}|_2,$$

and so on. If $n/2$ is odd, then we choose $(n/2 - 1)/2$ instead of $n/4$. We can perform these steps exactly $\lfloor \log_2(n) \rfloor$ times. (Note that the greatest power of two occurring between 1 and n is $\lfloor \log_2(n) \rfloor$.)

After all, we shall have the following:

$$|H_n|_2 \geq \left|\frac{1}{2^{\lfloor \log_2(n) \rfloor}}\right|_2 |H_1|_2 = 2^{\lfloor \log_2(n) \rfloor}.$$

On the other hand,

$$|H_n|_2 \leq \max\left\{|1|_2, \left|\frac{1}{2}\right|_2, \cdots, \left|\frac{1}{n}\right|_2\right\} = \left|\frac{1}{2^{\lfloor \log_2(n) \rfloor}}\right|_2 = 2^{\lfloor \log_2(n) \rfloor}.$$

The inequalities detailed above give the statement.

Another approach can be found in [101] and in [262] among the exercises of Chapter 6.

Finally, note that if we want to prove only the non-integer property (and not the stronger statement about the exact 2-adic evaluation), less effort is enough. Since

$$H_n = \sum_{k=0}^{n} \binom{n}{k} \frac{(-1)^{k-1}}{k},$$

(this is an exercise at the end of this chapter), and a sequence is integer if and only if its binomial transform is integer, it is immediate that H_n is not an integer.

11.10.2   The 2-adic norm of the hyperharmonic numbers

We are going to prove that the sums of harmonic numbers are never integers either. To this end we shall need a statement which is interesting in itself. This statement gives the exact valuation of the factorials for an arbitrary prime p.

The exact valuation of the factorials

For all $n \geq 1$

$$|n!|_p = p^{-\frac{n - A_p(n)}{p-1}}. \tag{11.38}$$

Here $A_p(n)$ is the sum of the digits of n in base p. By using valuations instead of the p-adic absolute value, the statement reads as

$$\nu_p(n!) = \frac{n - A_p(n)}{p-1}.$$

Before proving this theorem, we give an example. Let $p = 2$ and $n = 11$. We determine what is the highest power of $p = 2$ dividing $n!$. To this end, it basically suffices to calculate $A_2(11)$. Since

$$11 = 8 + 2 + 1 = 1 \cdot 2^3 + 12^1 + 12^0,$$

it comes that

$$A_2(11) = 1 + 1 + 1 = 3.$$

Now

$$\nu_2(11!) = \frac{11 - 3}{2 - 1} = 8.$$

That is, 2^8 divides $11!$, and no higher power of two divides $11!$:

$$|11!|_2 = 2^{-8}.$$

Let us prove now (11.38). The proof goes by induction. For $n = 1$, the statement is clear. We suppose that (11.38) holds for $n - 1$. Let us write

$$n - 1 = a_0 + a_1 p + a_2 p^2 + \cdots + a_m p^m.$$

We want to determine $A_p(n)$. If $a_0 < p - 1$ then

$$A_p(n) = A_p(n-1) + 1.$$

If, in turn, $a_0 = p - 1$, then adding one to $n - 1$, involve carries. Supposing that

$$a_0 = a_1 = \cdots = a_{t-1} = p - 1, \quad \text{but} \quad a_t < p - 1, \tag{11.39}$$

it is then clear that

$$A_p(n) = A_p(n-1) - (p-1)t + 1.$$

It is therefore necessary to distinguish two cases: (1) $a_0 < p - 1$, and (2) $a_0 = p - 1$.

In the first case n is not divisible by p and thus $|n|_p = 1$; whence

$$|n|_p = |n(n-1)!|_p = |(n-1)!|_p.$$

By the induction hypothesis,

$$|(n-1)!|_p = p^{-\frac{n-1-A_p(n-1)}{p-1}} = p^{-\frac{n-(A_p(n-1)+1)}{p-1}} = p^{-\frac{n-A_p(n)}{p-1}}.$$

In the second case, recalling (11.39), $|n|_p = p^{-t}$. Again by the induction hypothesis, and by (11.35)

$$|n|_p = |n(n-1)!|_p = |n|_p |(n-1)!|_p = p^{-t} p^{-\frac{n-1-A_p(n-1)}{p-1}} =$$

$$p^{-\frac{n-(A_p(n-1)-t(p-1)+1)}{p-1}} = p^{-\frac{n-A_p(n)}{p-1}},$$

and our proof is complete.

The exact valuation of the hyperharmonic numbers

To shorten the formulas a bit, we introduce the notation $l_2(n) = \lfloor \log_2(n) \rfloor$. To prove that the consecutive sums of harmonic numbers is never integer, we first determine the 2-adic absolute value of the hyperharmonic numbers in general.

If $l_2(n + r - 1) > l_2(r - 1)$, then

$$|H_n^{(r)}|_2 = 2^{A_2(n+r-1)-A_2(n)-A_2(r-1)+l_2(n+r-1)},$$

otherwise

$$|H_n^{(r)}|_2 = 2^{A_2(n+r-1)-A_2(n)-A_2(r-1)+\max\{|\frac{1}{r}|_2, |\frac{1}{r+1}|_2, \dots, |\frac{1}{n+r-1}|_2\}}.$$

The proof is as follows.

$$\left| H_n^{(r)} \right|_2 = \left| \binom{n+r-1}{r-1}(H_{n+r-1} - H_{r-1}) \right|_2 =$$

$$\left| \binom{n+r-1}{r-1} \right|_2 \left| \frac{a}{2^{l_2(n+r-1)}b} - \frac{c}{2^{l_2(r-1)}d} \right|_2 =$$

$$\left| \binom{n+r-1}{r-1} \right|_2 \left| \frac{2^{l_2(r-1)}ad - 2^{l_2(n+r-1)}bc}{2^{l_2(n+r-1)+l_2(r-1)}bd} \right|_2.$$

Because of the condition $l_2(n + r - 1) > l_2(r - 1)$ we get

$$\left| H_n^{(r)} \right|_2 = \left| \binom{n+r-1}{r-1} \right|_2 \left| \frac{ad - 2^{l_2(n+r-1)-l_2(r-1)}bc}{2^{l_2(n+r-1)}bd} \right|_2.$$

Since the nominator is odd, we get the following:

$$\left| \frac{ad - 2^{l_2(n+r-1)-l_2(r-1)}bc}{2^{l_2(n+r-1)}bd} \right|_2 = 2^{l_2(n+r-1)}.$$

To compute the 2-adic norm of the binomial coefficient, we use (11.38).

$$\left| \binom{n+r-1}{r-1} \right|_2 = \left| \frac{(n+r-1)!}{(r-1)!n!} \right|_2 =$$

$$\frac{2^{A_2(n+r-1)-n-r+1}}{2^{A_2(r-1)-r+1}2^{A_2(n)-n}} = 2^{A_2(n+r-1)-A_2(n)-A_2(r-1)}.$$

This and the previous equality give the result with respect to the condition $l_2(n + r - 1) > l_2(r - 1)$.

Let us fix an arbitrary n for which $l_2(n + r - 1) = l_2(r - 1)$.

$$H_{n+r-1} - H_{r-1} = \frac{1}{r} + \frac{1}{r+1} + \cdots + \frac{1}{n+r-1}.$$

Let us subtract all of the fractions with odd denominators. Then we can take $\frac{1}{2}$ out of the remainder and continue the recursive method just the same way as we did during proving (11.37). We can make such subtraction steps

$$\max\left\{ \left|\frac{1}{r}\right|_2, \left|\frac{1}{r+1}\right|_2, \ldots, \left|\frac{1}{n+r-1}\right|_2 \right\}$$

times. The result:

$$|H_{n+r-1} - H_{r-1}|_2 \geq \max\left\{ \left|\frac{1}{r}\right|_2, \left|\frac{1}{r+1}\right|_2, \ldots, \left|\frac{1}{n+r-1}\right|_2 \right\}.$$

On the other hand, by the strong triangle inequality

$$|H_{n+r-1} - H_{r-1}|_2 \leq \max\left\{ \left|\frac{1}{r}\right|_2, \left|\frac{1}{r+1}\right|_2, \ldots, \left|\frac{1}{n+r-1}\right|_2 \right\}.$$

The proof is therefore complete.

11.10.3 $H_1 + H_2 + \cdots + H_n$ is not integer when $n > 1$

We are ready to prove that $H_1 + H_2 + \cdots + H_n$ is not integer.

By definition, $H_1 + H_2 + \cdots + H_n = H_n^{(2)}$. The condition with respect to the order of n and r holds because $l_2(n + 2 - 1) > l_2(2 - 1) = 0$ for all $n \geq 1$. Furthermore,

$$\left| H_n^{(2)} \right|_2 = 2^{A_2(n+1)-A_2(n)-A_2(1)+l_2(n+1)}.$$

Let $m = l_2(n+1)$. Our goal is to minimize the power of 2. $l_2(n+1) = m$ implies that $n + 1 < 2^{m+1}$; therefore, $1 \leq A_2(n + 1) \leq m + 1$ and $1 \leq A_2(n) \leq m$.

The minimum in the power is taken when $A_2(n) = m$ and $A_2(n+1) = 1$. It is possible if and only if $n = 2^m - 1$. In this case,

$$A_2(n+1) - A_2(n) - A_2(1) + l_2(n+1) = 1 - m - 1 + m = 0.$$

We get that if $n \neq 2^m - 1$ for some m, then $|H_n^{(2)}|_2 > 1$, that is, $H_n^{(2)} \notin \mathbb{N}$. On the other hand, let us assume that n has the form $2^m - 1$. This implies that

$$H_n^{(2)} = \binom{n+2-1}{2-1}(H_{n+2-1} - H_{2-1}) =$$

$$(n+1)(H_{n+1} - 1) = 2^m \left(\frac{a}{2^{l_2(n+1)}b} - 1 \right) = \frac{a}{b} - 2^m \notin \mathbb{N}.$$

And this is what we wanted to prove.

See the Outlook of Chapter 8 for references on generalizations.

Exercises

1. Is $\left[\begin{smallmatrix} 180 \\ 70 \end{smallmatrix}\right]$ even or odd?

2. Is $\left\{\begin{smallmatrix} 180 \\ 70 \end{smallmatrix}\right\}$ even or odd?

3. Determine $\left[\begin{smallmatrix} 1567 \\ 789 \end{smallmatrix}\right]$ modulo 7.

4. Let n and k be positive integers such that $n + k$ is odd. Then prove that
   (a) $\left[\begin{smallmatrix} n \\ k \end{smallmatrix}\right]$ is divisible by the odd part of $n - 1$,
   (b) $\left\{\begin{smallmatrix} n \\ k \end{smallmatrix}\right\}$ is divisible by the odd part of k.

   (The odd part of a number n is n divided by the maximal power of two which still divides n. These statements were proven in [455]. See the Outlook of this chapter for sharper statements.)

5. Let $n > lp$ for some integer l and prime p. Show that

$$\begin{bmatrix} n \\ k \end{bmatrix} \equiv 0 \pmod{p}$$

   for all $k \leq l$. Note that, because of (11.25), this result is the best possible. (Hint: use the horizontal generating function modulo p.)

6. Generalize the previous exercise: if $n > klp$, then

$$\begin{bmatrix} n \\ k \end{bmatrix} \equiv 0 \pmod{p^r}.$$

   (This is (41) in [502].)

7. Confirm the following mod p^2 congruences where k is odd and belongs to the respective intervals:

$$\begin{bmatrix} p+1 \\ k \end{bmatrix} \equiv 0 \pmod{p^2} \quad (3 \leq k \leq p - 2),$$

$$\begin{bmatrix} 2p \\ k \end{bmatrix} \equiv 0 \pmod{p^2} \quad (3 \leq k \leq p).$$

   (These results were proven in [247].)

8. Show the validity of the recursion (11.3).

9. Deduce that $f(2^r + x) = f(2^r)$ $(0 \leq x \leq 2^r)$, and from here, prove that $f(2^n) = 3^n$. Here $f(n)$ is the function which we defined by (11.2). (See [278].)

10. Show that if we write the $2p$-th harmonic number as $H_{2p} = \frac{a}{b}$, then the maximal power of p which divides b is p^2. (Hint: use the previous exercise with $n = 2p + 1$ and $k = 2$.)

11. Prove that
$$B_{p+2} \equiv 7 \pmod{p}.$$

12. Show that the congruence (11.32) generalizes to the r-Bell numbers as follows:
$$B_{p-1,r} \equiv 1 + (-1)^{r_0} D_{p-1-r_0} \pmod{p},$$
where r_0 is the modulo p reduction of $-r$.

13. Applying
$$\binom{p}{k} \equiv 0 \pmod{p} \quad (0 < k < p)$$
prove that
$$\binom{p-1}{k} \equiv (-1)^k \pmod{p} \quad (0 \le k < p).$$
(Use the recursion $\binom{n}{k} = \binom{n-1}{k} + \binom{n-1}{k-1}$.)

14. Show the following parity connection between the binomial coefficients and Eulerian numbers:
$$\left\langle {n \atop k} \right\rangle \equiv \binom{n-1}{k} \pmod{2} \quad (n > 0).$$

15. Confirm that
$$p \left| \left\langle {p-1 \atop k} \right\rangle - 1 \right. \quad (0 \le k < p).$$

16. Generalize the (11.26) Touchard congruence as follows:
$$B_{n+kp} \equiv \sum_{i=0}^{k} \binom{k}{i} B_{n+i} \pmod{p} \quad (k \ge 0).$$

Hint: use induction based on (11.26).

17. Prove the Touchard congruence in the more general form
$$B_{n+p^m} \equiv mB_n + B_{n+1} \pmod{p},$$
where n is a non-negative integer, m is a positive integer, and p is an arbitrary prime. In particular,
$$B_{p^m} \equiv m + 1 \pmod{p}.$$

Hint: use induction based on (11.26), or see the paper [293] for a proof. The particular $n = 0$ form was proven by Touchard [564], and rediscovered by Tsumura [568].

18. Prove the dual of Spivey's formula (11.27) for the first kind Stirling numbers

$$(n+m)! = \sum_{k=0}^{n} \sum_{j=0}^{m} m^{\overline{n-k}} \begin{bmatrix} m \\ j \end{bmatrix} \binom{n}{k} k! \qquad (11.40)$$

(What kind of divisibility statement comes from this for the factorials? Follow the steps that we followed before to prove Touchard's congruence.)

19. Prove Spivey's formula for the Bell polynomials:

$$B_{n+m}(x) = \sum_{k=0}^{n} \sum_{j=0}^{m} \binom{n}{k} x^j B_k(x) \begin{Bmatrix} m \\ j \end{Bmatrix} j^{n-k}.$$

(See [605].)

20. Prove that the special value

$$\begin{Bmatrix} n \\ 3 \end{Bmatrix} = \frac{1}{2} \left(3^{n-1} - 2^n + 1 \right),$$

which we required in the proof of Gross's congruence. Hint: use (1.5).

21. Show that

$$F_{p+1} \equiv 3 \pmod{p}$$

for all prime $p > 3$.

22. Show that

$$F_{p+2} \equiv 13 \pmod{p}$$

for all prime $p > 13$. Hint: use (11.8) and (11.9), respectively.

23. Show that

$$F_{n+ap^k} \equiv F_{n+a} \pmod{p} \quad (k, a \geq 0).$$

(You can successively apply (11.30).)

24. With the aid of (2.57) show that

$$L_{m+p} \equiv L_m \pmod{p},$$

and deduce its consequence

$$L_{mp} \equiv 1 \pmod{p}$$

for any $m \geq 0$.

25. Show that $D_p \equiv -1 \pmod{p}$.

26. Show that the derangements satisfy the following congruence:
$$D_{n+m} \equiv D_m D_n \pmod{n}.$$

Thus, in particular (by using the previous exercise), $D_{n+p} \equiv -D_n$ (mod p). (This was shown by group actions in [252].)

27. Deduce from the previous two exercises that
$$D_{n+m} \equiv (-1)^m D_n \pmod{m}.$$

28. Present a proof for the identity
$$B_{n,r} = \frac{1}{n!} \sum_{m=0}^{n} \binom{n}{m} (m+r)^n D_{n-m}.$$

(Hint: see the proof of (2.21).)

29. Show that the last digits of L_n form a periodic sequence of period 5. In other words,
$$L_{n+5} \equiv L_n \pmod{10}.$$

(See the table of L_n for numerical evidence. This result was proven in [430].)

30. Prove that
$$H_n = \sum_{k=0}^{n} \binom{n}{k} \frac{(-1)^{k-1}}{k}.$$

(Use Euler's (2.14) theorem on the binomial transform.)

31. Prove that H_n^3 is not integer when $n > 1$ (see [415]).

32. Prove that p is a Wolstenholme prime if and only if
$$p \left| \sum_{k=1}^{p-1} \frac{H_{k-1}}{k(p-k)} \right..$$

(Hint: in view of (11.14), p is a Wolstenholme prime if and only if p^2 divides $\left[\begin{smallmatrix}p\\3\end{smallmatrix}\right]$. Now recall (2.25): $\left[\begin{smallmatrix}p\\3\end{smallmatrix}\right] = \frac{p!}{3} \sum_{k=1}^{p-1} \frac{H_{k-1}}{k(p-k)}$. The $p!$ factor is responsible only for one power of p, and thus it is necessary that p divides the sum, too.)

33. Show that the p-th Lah polynomial (p is a prime), if it has rational zero, it must belong to the set
$$\{-p, -2p, \ldots, -(p-1)!p\}.$$

Hint: Use the rational root theorem to show that any rational root must be a divisor of $p!$. Then fix $d > 0$ such that $p \nmid d$ and prove that $L_p(-d) \equiv d \not\equiv 0 \pmod{p}$, so such a $-d$ cannot be a zero of $L_p(x)$.

Outlook

1. The parity of the binomial coefficients can be represented in a very interesting way, via cellular automata. If we go through the Pascal triangle, and substitute 1 wherever we have an odd coefficient, and 0 wherever there is an even one, we get the following diagram:

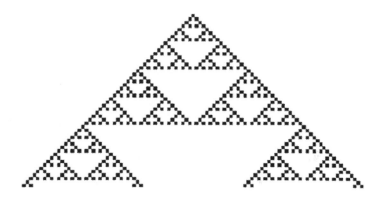

This diagram is self-similar, as we see. Cellular automata is an algorithm where the evolution of black and white dots in a row, or on a plane or in higher dimension is governed by some rules. Elementary cellular automata operates on dots in one dimension, and there are only 256 possible elementary automaton[13]. These are identified by a number called Rule. The above parity diagram of the binomial coefficients can be constructed by the Rules 18, 22, 26, 82, 90, 146, 154, 210 and 218. (It is not possible, however, to represent the modulo 2 first or second kind Stirling triangle with cellular automata. Do you see why?) More on this fascinating topic can be found in [601].

It is also interesting to note that the infinite fractal version of the above self-similar diagram is called Sierpiński triangle [601].

2. An instance of the few multinomial-coefficient congruences is due to Rota and Sagan [499]:

$$\binom{p^m}{i_1, \dots, i_n} \equiv \binom{p^{m-1}}{i_1/p, \dots, i_n/p} \pmod{p^m}.$$

(Here the convention is used which says that a multinomial coefficient having a fraction is zero.)

---

[13]The plural form of automata is automaton.

3. Lucas's theorem of Section 11.3 can be generalized to arbitrary powers of p. This was worked out by A. Granville [263], and – in a slightly restricted form – by K. S. Davis and W. A. Webb [188], see also [189, 523]. In his work, he gave connections to cellular automata, mentioned in the previous point. One can read more on the long and rich history of the divisibility properties of the binomial coefficients in [409].

4. The parity of the Stirling numbers of the second kind was thoroughly studied by L. Carlitz [122, 123, 124]. If we define $\theta(n)$ to be the number of odd $\left\{{n+1 \atop 2r+1}\right\}$ $(0 \leq 2n < r)$, then the generating function of the sequence $\theta(n)$ is

$$\sum_{n=0}^{\infty} \theta(n)x^n = \prod_{n=0}^{\infty} (1 + x^{2^n} + x^{2^{n+1}}).$$

$\theta(n)$ is also the number of odd binomial coefficients of the form $\binom{a}{b}$, where $a + b = n$. A number of relations for $\theta(n)$ are discovered in [123]. The sequence $\theta(n)$ is strongly related to the so-called Stern sequence [492, 493].

The paper [124, 125] contains congruences modulo primes in general.

Bondarenko et al. [88] study the distribution of the Stirling and Euler numbers modulo primes.

5. Divisibility of the Stirling numbers by small primes was studied by a number of authors [19, 153, 362, 363, 613]. Divisibility via algebraic methods was studied by Sagan [502]. See also [343] on the p-adic valuation of $\left[{n \atop k}\right]$ for general primes. A whole chapter is devoted to this question in [441]. The congruence

$$\left[{n \atop k}\right] \equiv \left\{{p-k \atop p-n}\right\} \pmod{p} \quad (1 \leq k \leq n < p)$$

was discovered by Hoffman via multiple harmonic sums [284]. A generalization to prime powers can be found in [248]:

$$\left[{p^\nu - j \atop p^\nu - i}\right] = \left\{{i \atop j}\right\} \pmod{p^\nu}$$

when $1 \leq i, j \leq p - 1$, and for all $\nu \geq 1$.

6. The modulo 4 description of the Stirling numbers of the second kind was described in [132].

7. If n and k are positive integers such that $n + k$ is odd, then

$$\left[{n \atop k}\right] \equiv 0 \pmod{\binom{n}{2}},$$

$$\left\{{n \atop k}\right\} \equiv 0 \pmod{\binom{k+1}{2}}.$$

These results are of Howard [288]. In the same paper, it is also proved (with different notations) that for any $h > 0$, and $0 \le m < p$

$$\begin{bmatrix} hp + m \\ k \end{bmatrix}_{\ge 2} \equiv (-1)^h \begin{bmatrix} m \\ k - h \end{bmatrix}_{\ge 2} \pmod{p},$$

$$\left\{ \begin{matrix} hp + m \\ k \end{matrix} \right\}_{\ge 2} \equiv \sum_{r=0}^{h} \binom{h}{t} \left\{ \begin{matrix} m + h - r \\ k - r \end{matrix} \right\}_{\ge 2} \pmod{p}.$$

8. Higher prime power congruences were considered by a number of mathematicians. Rota and Sagan [499] used group theory to prove that if $k = \lfloor a/p \rfloor$, then

$$\left\{ \begin{matrix} p^m \\ a \end{matrix} \right\} \equiv \left\{ \begin{matrix} p^{m-1} \\ a \end{matrix} \right\} \pmod{p^{m-k}}.$$

Note that this gives the following particular for $1 < a < p$:

$$\left\{ \begin{matrix} p^m \\ a \end{matrix} \right\} \equiv 0 \pmod{p} \quad (j \ge 1).$$

9. It is known that if $p^j \le k < p^{j+1}$, then

$$\left\{ \begin{matrix} n + p^{r+j}(p-1) \\ k \end{matrix} \right\} \equiv \left\{ \begin{matrix} n \\ k \end{matrix} \right\} \pmod{p^r}.$$

This is a result of Carlitz [121], but also was proven by Tsumura [568], and for $r = 1$ Becker and Riordan [53].

10. The last digits of the Lah number sum L_n form a periodic sequence of length ten. More generally, it holds true [430] that

$$L_{n+10^k} \equiv L_n \pmod{10^k}.$$

Better can be said when $k = 1$:

$$L_{n+5} \equiv L_n \pmod{10}.$$

It can happen that the above periodicity results can be improved. It is conjectured in [430] that the *minimal* period mod m is

$$m(k) = \begin{cases} \frac{k}{2}, & \text{if } k \equiv 2 \pmod 4; \\ k, & \text{otherwise.} \end{cases}$$

11. Fubini numbers satisfy many other congruences which are not presented in the main text. See Diagana and Maiga [193].

12. Touchard's congruence can be extended not only as it is given in Exercise 17, but also to polynomials. See [249, 308, 475, 476] for the details. Touchard's congruence was rediscovered right after Touchard by Hall in 1934 [275].

13. Another congruence, due to Comtet, says that for any prime p

$$B_{np} \equiv B_{n+1} \pmod{p}.$$

This was later extended to the following statement [249] (solving the conjecture of Zuber): If ν is the highest power of $p \neq 2$ that divides n, then

$$B_{np} \equiv B_{n+1} \pmod{p^{\nu+1}}.$$

If $p = 2$ then

$$B_{np} \equiv B_{n+1} \pmod{p^{\nu}}.$$

A more complex congruence, due to Sagan [502], says that the three-fold sum

$$\sum_{i=0}^{r}(-1)^i \binom{r}{i} \sum_{j=0}^{i} \left\{ {i \atop j} \right\} (-p(p-1))^{i-j} \sum_{l=0}^{j} \binom{j-1+\delta_{ij}}{l} B_{n+(r-i)p+l}$$

is divisible by p^r. Here $\delta_{ij} = 1$ if $i = j$, otherwise it is zero.

Similar divisibility properties are given for both kinds of Stirling numbers in the same paper. Both

$$\sum_{i=0}^{r}(-1)^i \binom{r}{i} \sum_{j=0}^{i} \left\{ {i \atop j} \right\} (-p(p-1))^{i-j} \sum_{l=0}^{j} \binom{j-1+\delta_{ij}}{l} \left\{ {n+(r-i)p+l \atop k-(j-l)p} \right\},$$

and

$$\sum_{i=0}^{r}(-1)^i \binom{r}{i} \sum_{j=0}^{i} \left[{i \atop j} \right] (-p(p-1))^{i-j} \sum_{l=0}^{j} \binom{j-1+\delta_{ij}}{l} (p-1)^l \left[{n+(r-i)p \atop k-(j-l)p-l} \right]$$

are divisible by p^r.

14. Congruences with respect to the Stirling numbers of both kinds and composite modulus can be found in [502].

15. A congruence for B_n involving p^2 is due to Touchard [564, p. 314]:

$$B_{2p} - 2B_{p+1} - 2B_p + p + 5 \equiv 0 \pmod{p^2}.$$

Another one was deduced in [384, 502]:

$$[(E^p - E - 1)^2 - p(p-1)]B_n \equiv 0 \pmod{p^2}.$$

Here $Ea_n = a_{n+1}$ is the right shift operator.

An interesting higher power congruence is due to Junod [309]:

$$B_{m+np^s} \equiv \sum_{j=0}^{n} \binom{n}{j} s^{n-j} B_{m+j} \pmod{p^{k+1}}.$$

Junod also studied finite field relations with respect to the Bell polynomials [310].

16. We saw two reformulations of Kurepa's conjecture in the text. There is another equivalent statement [309] connected to the Bell numbers: if p does not divide n, then p does not divide $B_{p^n-1} - 1$ for all odd primes p and $n \geq 1$.

There is yet another form of Kurepa's conjecture: Kurepa asked whether $(!n, n!) = 2$ for all integers $n > 1$. The relation among this formulation, the other in our text and yet other equivalents can be found in [297].

No odd prime p is found up to $p < 2^{34}$ which divides $!p$ [24].

17. The alternating left factorial function

$$A(n) = \sum_{i=1}^{n} (-1)^{n-i} i!$$

was studied in [387].

18. For sums with Stirling numbers and arithmetic functions, like

$$\sum_{d|n} \mu(d) \left\{ \begin{matrix} \frac{n}{d} \\ a \end{matrix} \right\} \equiv 0 \pmod{n},$$

see [499, 502].

19. A large number of results is known for $k!\left\{ \begin{matrix} n \\ k \end{matrix} \right\}$, see the citations on pp. 496-499 in [306].

20. Howard [289] studied congruences of $\left[\begin{matrix} n \\ k \end{matrix} \right]$ in terms of Bernoulli numbers. He got, among others, that

$$\left[\begin{matrix} n \\ n-2r \end{matrix} \right] \equiv -\frac{n}{2r} \binom{n-1}{2r} B_{2r} \pmod{p^2 t},$$

where p is an odd prime, t is the highest power of p dividing n, and $0 < 2r < 2p - 2$ plus $1 < 2r + 1 < 2p - 2$.

Using Howard's results, Adelberg [4] proved additional congruences for the first kind Stirling numbers, like the following one:

$$\left[\begin{matrix} n \\ n-p \end{matrix} \right] \equiv \binom{n-1}{p} \frac{n^2}{2} \pmod{pn^2},$$

where n is divisible by p, and p is odd.

21. For each positive n, there are integers $a_0, a_1, \ldots, a_{n-1}$ such that for all $m \geq 0$

$$B_{m+n} + a_{n-1}B_{m+n-1} + \cdots + a_0 B_m \equiv 0 \quad (\text{mod } n!).$$

In other words, it is possible to form a linear combination of $n + 1$ consecutive Bell numbers such that this linear combination is a multiple of $n!$. This is a result of Gessel [250]. Minimal length linear combinations modulo an integer were studied in [384].

22. Z.-W. Sun and D. Zagier [552] gave the interesting rational congruence

$$\sum_{k=1}^{p-1} \frac{B_k}{(-m)^k} \equiv (-1)^{m-1} D_{m-1} \quad (\text{mod } p)$$

for any non-zero integer m and prime p not dividing m. Here D_{m-1} is the derangement number. Sun and Zagier presented this result in a polynomial version in the same paper, and later it was extended to an even more general form [548]. The r-Bell polynomial extension reads as

$$(-x)^m \sum_{k=1}^{p-1} \frac{B_{k,r}(x)}{(-m)^k} \equiv (-x)^{p-r} D_{m+r-1} \quad (\text{mod } p\mathbb{Z}_p[x]).$$

This was presented in [427]. Here $\mathbb{Z}_p[x]$ is the polynomial ring over the field of p elements.

23. The Spivey congruence for polynomials was proven by several authors [58, 605, 260] (see also Exercise 19.), as we have already mentioned in the text. In [605] an extension can be found for the generalized Stirling numbers of L. C. Hsu, and P. J.-Sh. Shiue [293]. See [171, 323] for q-extension, and also [385, 395] for other generalizations. Interestingly, the proof in [171] uses rook placements on specific shapes.

24. The overwhelming majority of the Bell numbers seems to be a composite number. The only known prime Bell numbers are

$$B_2 = 2, \quad B_3 = 5, \quad B_7 = 877, \quad B_{13} = 27\,644\,437,$$

$$B_{42} = 35\,742\,549\,198\,872\,617\,291\,353\,508\,656\,626\,642\,567,$$

$$B_{55} = 359\,334\,085\,968\,622\,831\,041\,960\,188\,598\,043\,661\,065\,388\,726\,959\,079\,837,$$

and B_{2841}, which has approximately 6539 digits (to determine the approximative number of digits, we made use of (7.10)). To find the evidence that $B_{2\,841}$ is a prime, it took 17 months, and the calculation was carried out in 2003-2004 by Ignacio Larrosa Cañestro (see the remarks on OEIS: A051130).

It was known until 2006 that there are no other prime Bell numbers below $B_{6\,000}$. On the 23rd of April, 2006, this limit was pushed forward by E. W. Weisstein [594] who proved that there are no new primes among the Bell numbers at least until $B_{30\,447}$.

25. Chowla, Herstein and Moore [151], and independently Sagan [502] proved three congruences with respect to the involutions I_n:

$$I_{n+p^r} \equiv I_n \pmod{p^r} \quad (p \geq 3),$$
$$I_{n+m} \equiv I_n \pmod{m} \quad (m \text{ is odd}),$$
$$I_n \equiv 0 \pmod{2^r} \quad (n \geq 4r - 2).$$

26. Some congruential properties of the Eulerian numbers can be found in [334]. See [488] for generalizations to the r-Whitney number case we mentioned in the Outlook of Chapter 8.

27. An analogue of Wolstenholme's 1862 theorem says that the congruence

$$\sum_{\substack{k=1 \\ (k,n)=1}}^{n} \frac{1}{i} \equiv 0 \pmod{n^2}$$

holds for all number n not divisible by six. This was proven in 1888 by Leudesdorf [371].

For historical remarks, generalizations, extensions and equivalents, see [410]. For congruences involving harmonic numbers, see [551].

28. Another analogue of Wolstenholme's theorem is when, in place of H_{p-1}, we sum reciprocals congruent to a given integer. The following interesting statement is due to Zhi-Wei Sun [550]:

$$\sum_{\substack{k=1 \\ k \equiv r \,(\text{mod } m)}}^{p-1} \frac{1}{k} \equiv \frac{1}{m} \left(B_{p-1}\left(\left\{\frac{r}{m}\right\}\right) - B_{p-1}\left(\left\{\frac{r-p}{m}\right\}\right) \right) \pmod{p}$$

for positive integers m not divisible by the odd prime p, while r can be an arbitrary integer. The $\{\cdot\}$ denotes the fractional part.

29. We have seen that the numerator of H_{p-1} is always divisible by p (and also by p^2). An interesting question is the following. Given a prime p, determine the set of the positive integers n for which p divides the nominator of H_n. This set is denoted by $J(n)$ in the literature. We know from Wolstenholme's result that $p - 1 \in J(p)$ for $p > 3$. The study of the set $J(p)$ was started in [222], where the authors showed that $p^2 - 1$ and $p^2 - p$ are also in $J(p)$ for all $p > 3$. There are primes for which $J(p)$ contains only the above three numbers. These numbers are called *harmonic*. For example, 5, 13, 17, 23, 67 are harmonic primes.

It is believed that $J(p)$ is finite for all primes, and that there are infinitely many harmonic primes. We will see in Chapter 11 that $p = 2$ is always a divisor of the denominator if $n > 1$, so $J(2) = \{0\}$. It takes more effort to show that $J(3) = \{0, 2, 7, 22\}$ [222]. For more on the set $J(p)$, apart of [222], see [94, 507, 603, 604] for estimations on the cardinality of $J(p)$, and further statements. The set $J(p)$ can be generalized to hyperharmonic numbers [258], as well as Wolstenholme's theorem.

30. Wilson's theorem states that $(p - 1)! \equiv -1 \pmod{p}$, i.e., $(p - 1)! + 1$ is divisible by p. It is an interesting question what the remainder is of this division. To this end, one defines

$$w_p = \frac{(p - 1)! + 1}{p}.$$

These numbers are called *Wilson quotients*.

Similarly, the little Fermat theorem says that $a^{p-1} \equiv 1 \pmod{p}$, that is,

$$q_p(a) = \frac{a^{p-1} - 1}{p}$$

is an integer number, called *Fermat quotient*. The Wilson and Fermat quotients are interesting quantities. The latter is strongly connected to special harmonic number sums. We give a remarkable example of this: Eisenstein and, in a slightly different form, Sylvester [554] proved that

$$H_{\frac{p-1}{2}} \equiv -2q_p(2) \pmod{p}.$$

Lehmer [361] gave the stronger modulo p^2 relation

$$H_{\frac{p-1}{2}} \equiv -2q_p(2) + 2q_p(2)^2 \pmod{p^2}.$$

This was later extended to modulo p^3 by Zhi-Hong Sun [544]:

$$H_{\frac{p-1}{2}} \equiv -2q_p(2) + 2q_p(2)^2 - \frac{2}{3}p^2 q_p(2)^3 - \frac{7}{12}p^2 B_{p-3} \pmod{p^3}.$$

Another direction of research is the shortening of the sum $H_{\frac{p-1}{2}}$ to express the Fermat quotient $q_p(2)$ modulo a prime. It is known [254] that

$$-3q_p(2) \equiv H_{\lfloor p/4 \rfloor} \pmod{p},$$

and [196, 361]

$$2q_p(2) \equiv \sum_{k=\lfloor p/6 \rfloor + 1}^{\lfloor p/3 \rfloor} \frac{1}{k} \pmod{p}.$$

For similar relations with respect to more general Fermat quotients, see

[197, 201]. Such formulas were studied in [524] in the framework of p-adic numbers. See the book [494] of Ribenboim, too.

Since the Fermat quotient $q_p(a)$ is an integer, it can be asked which is the smallest a such that $q_p(a) \equiv 0 \pmod{p}$. This was studied in [91] and in the references of this paper. Numerical results are contained in [221]. The value set of the Fermat quotients is also interesting: one can ask, which is the smallest number L such that $\{q_p(1), q_p(2), \ldots, q_p(L)\}$ represents all the residues modulo p. See [519] and its references for the details and bounds on L.

It is Glaisher's [255] result that

$$g_p(2) \equiv -\frac{1}{2} \sum_{j=1}^{p-1} \frac{2^j}{j} \pmod{p}.$$

This was later (much later) generalized to [264]

$$g_p(2)^2 \equiv -\sum_{j=1}^{p-1} \frac{2^j}{j^2} \pmod{p},$$

and to the cubic formula [198, 411]

$$g_p(2)^3 \equiv -3 \sum_{j=1}^{p-1} \frac{2^j}{j^3} + \frac{7}{4} \sum_{j=1}^{p-1} \frac{(-1)^j}{j^3} \pmod{p}.$$

Quartic formula, although it is more complicated, still exists [9].

The Fermat and Wilson quotients are connected via Lerch's formula [370, 527, 528]:

$$\sum_{j=1}^{p-1} q_p(j) = w_p \pmod{p}.$$

(See [361, 437] for similar congruences.) In his paper, Sondow introduced and studied the *Lerch quotient*

$$\ell_p = \frac{1}{p} \left(\sum_{j=1}^{p-1} q_p(j) - w_p \right),$$

which, thanks to Lerch's formula, is an integer number.

Connections of these quotients and the Euler numbers were found in [298].

The primes p for which $q_p(a) \equiv 0 \pmod{p}$ are called *Wieferich primes* (of base a). 1093 and 3511 are the only known base 2 Wieferich primes, although it is conjectured that there are infinitely many [178]. This notion can be extended, and one can define *Wieferich numbers* in base a.

The integer m is a Wieferich number in base a if $\frac{a^{\phi(m)}-1}{m}$ is divisible by m. Here $\phi(m)$ is the *Euler totient function*. It gives that how many co-prime positive integers are there to m in the set $\{1,\ldots,m-1\}$. For the newest reference to Wieferich numbers, see [13].

The primes p for which $w_p \equiv 0 \pmod{p}$ are called *Wilson primes* (see [8] and the references therein). The only known Wilson primes are 5, 13, 563. It is conjectured that there are infinitely many Wilson primes.

Similarly, the primes p for which the Lerch number ℓ_p is divisible by p are called *Lerch primes*. Up to three million, there are four Lerch primes: $3; 103; 839; 2\,237$ [527]. The generalized Lerch primes were introduced and studied in [272].

The websites "Prime Pages" (`http://primes.utm.edu`) and the "Prime Curios!" (`http://www.primecurios.com/`) are rich and entertaining sources of facts about primes. See the printed compilation [108], too.

On the Prime Pages you can learn, among others, that $2\,237$ is not only the greatest known Lerch prime up to now, but it is also the next prime year in which Pluto is closer to the Sun than Neptune.

31. It is a result of Kürschák from 1918 that $H_n - H_m$ are not integers [347]. This was later extended by Nagell [448] and Erdős [219]: $\sum_{j=0}^{k-1} \frac{1}{md+j} \notin \mathbb{N}$ if m and d have no common prime factors. Erdős's result was further extended by Belbachir and Khelladi [57]: $\sum_{j=0}^{k-1} \frac{1}{(md+j)^{\alpha_j}} \notin \mathbb{N}$ for arbitrary $\alpha_1,\ldots,\alpha_{k-1}$ positive integers, and co-prime m and d. See [520, 610] for the most recent results, and [462] for extensions to multiple harmonic sums. For the non-integerness results with respect to the hyperharmonic numbers, see the footnote on p. 227 in this book.

For more on the p-adic properties of H_n, see the short writing of K. See Conrad [160].

32. The 2-adic valuation of the binomial coefficients from an asymptotic distribution point of view was recently studied in [530].

33. The 2-adic valuation of the Bell numbers was studied by Ambdeberhan, de Angelis and Moll [18], and of the r-Bell numbers in [427].

34. The difference operator Δ_c is defined on an sequence a_n by

$$\Delta_c a_n = a_{n+c} - a_n,$$

where c is a fixed non-negative integer. The powers Δ_c^k of Δ_c are defined by using the composition of operators. P. T. Young [611] studied congruences of the Bernoulli, Euler and Stirling and r-Stirling numbers of the second kind (the latter he calls "weighted Stirling number") under the effect of the $D\Delta_c^k$ operators.

35. The nested sum

$$H(n,k) = \sum_{1 \le i_1 \le \cdots \le i_k \le n} \frac{1}{i_1 \cdots i_k},$$

which is a generalization of the harmonic numbers is integer only for $(n,k) \in \{(1,1),(3,2)\}$ [139]. The p-adic properties of these numbers were studied in [369].

36. (The Lengyel-Wannemacker formula.) It was conjectured by T. Lengyel in 1994 [363] that

$$\nu_2\left(\left\{ {2^h \atop k} \right\}\right) = S_2(k) - 1 \quad (1 \le k \le 2^h),$$

where $S_2(k)$ is the sum of the digits of k written in base two. This conjecture was proven to be true by S. de Wannemacker in 2005 [588]. In [588] it is also proved that

$$\nu_2\left(\left\{ {n \atop k} \right\}\right) \ge S_2(k) - S_2(n).$$

Later Lengyel, by using Wannemacker's proof, strenghtened the statement above and proved that [364]

$$\nu_2\left(\left\{ {c2^h \atop k} \right\}\right) = S_2(k) - 1 \quad (1 \le k \le 2^h)$$

for all integer $c \ge 1$.

A related result, due to Hong et al. [286] is that

$$\nu_2\left(\left\{ {2^h + 1 \atop k+1} \right\}\right) = S_2(h) - 1.$$

The 2-adic valuation of the Stirling numbers of the second kind is rather subtle, only partial description is known. See the already cited [286] for more results, and also [19].

37. The p-adic valuation of the Stirling numbers of the first kind was recently studied in details by T. Lengyel [366]. A remarkable result among the many in [366] is that for *any* prime p and $k \ge 1$

$$\lim_{n \to \infty} \nu_p\left(\left[{n \atop k} \right]\right) = \infty.$$

In other words, as n grows $\left[{n \atop k} \right]$ contains more and more primes (in limit, all the primes appear) and on higher and higher powers. (This is, by the way, not some curious property, the $a_n = n!$ sequence has the very same property, as the reader sees.) Other works on the topic are [73, 253]. Simplifications of some proofs of Lengyel and other results can be found in [5]. See also the most recent work [439].

38. The theory p-adic numbers has a large number of applications in number theory, analysis, and even in physics. For applications in number theory, see [155, 156], and for an application in physics, see [208]. It is believed by some physicists that physics should be done "locally" by measuring distances p-adically (via the p-adic norm as a distance), and "globally," as we are used to: using the standard absolute value.

Chapter 12

Congruences via finite field methods

The result of this and the following chapter is mainly collections of the material scattered in the literature rather than a systematic treatise on the topic. The latter cannot be expected, since the literature is rather hollow. Some properties that are known for a given counting sequence might be completely unknown for another. Still there is a lot of research to be done here.

12.1 An application of the Hankel matrices

The modulo p properties of the counting sequences can be studied in the framework of finite fields. It turns out that our counting sequences satisfy linear congruences modulo p, and the linear congruences over finite fields have a well-developed theory.

In this chapter, some preliminary knowledge is supposed. Familiarity with the arithmetic in finite fields, polynomials and splitting fields will be necessary. The field of p elements will be denoted by $\mathbb{F}_p$.

The Hankel matrices were defined in Section 2.10. We now show an interesting application of these matrices in the modulo p theory of counting sequences. Let us imagine that we are looking for a congruence in the form

$$A_{n+k} \equiv a_0 A_n + a_1 A_{n+1} + \cdots a_{k-1} A_{n+k-1} \pmod{p}. \qquad (12.1)$$

This is a recurrence relation[1] of *order* k. We have already seen some special cases of such congruences. Taking $A_n = B_n$, the Bell number sequence, we know that

$$B_{n+p} \equiv 1 \cdot B_n + 1 \cdot B_{n+1};$$

this is just Touchard's congruence, see (11.26). In this case $a_0 = a_1 = 1$, and all the other a_is are zero. For the Fubini numbers, $F_{n+p} \equiv F_{n+1} \pmod{p}$, so $a_1 = 1$, and the other coefficients are zero. Thus, it makes sense to look for a general approach to prove such congruences. Consider again the general Equation (12.1), but now take a collection of these, say, for $n = 0$, $n = 1$, until

---

[1] More exactly, it is a homogeneous linear recurrence relation, because there is no constant additive term and the sequence members are all on the first power.

$n = p - 1$:

$$A_k \equiv a_0 A_0 + a_1 A_1 + \cdots a_{k-1} A_{k-1} \pmod{p}$$
$$A_{k+1} \equiv a_0 A_1 + a_1 A_2 + \cdots a_{k-1} A_k \pmod{p}$$
$$\vdots$$
$$A_{k-1+k} \equiv a_0 A_{k-1} + a_1 A_{k-1+1} + \cdots a_{k-1} A_{k-1+k-1} \pmod{p}.$$

This is a linear system of equations, which can conveniently be written in matrix form:

$$\begin{pmatrix} A_0 & A_1 & \cdots & A_{k-1} \\ A_1 & A_2 & \cdots & A_k \\ \vdots & \vdots & \vdots & \vdots \\ A_{k-1} & A_k & \cdots & A_{2k-2} \end{pmatrix} \begin{pmatrix} a_0 \\ a_1 \\ \vdots \\ a_{k-1} \end{pmatrix} = \begin{pmatrix} A_k \\ A_{k+1} \\ \vdots \\ A_{2k-1} \end{pmatrix},$$

where the coefficient matrix is just a $k \times k$ Hankel matrix! We are looking for the a_i coefficients, which are given uniquely by

$$\begin{pmatrix} a_0 \\ a_1 \\ \vdots \\ a_{k-1} \end{pmatrix} = \begin{pmatrix} A_0 & A_1 & \cdots & A_{k-1} \\ A_1 & A_2 & \cdots & A_k \\ \vdots & \vdots & \vdots & \vdots \\ A_{k-1} & A_k & \cdots & A_{2k-2} \end{pmatrix}^{-1} \begin{pmatrix} A_k \\ A_{k+1} \\ \vdots \\ A_{2k-1} \end{pmatrix}, \tag{12.2}$$

if the matrix is invertible. We are working modulo p, but this does not cause any difference: a square matrix is invertible modulo p if and only if its determinant is not zero modulo p.

The determinants of Hankel matrices are just the Hankel determinants, that we have calculated for many sequences. We know take some examples, applying our discovery that Hankel determinants and modulo p recurrences are related.

The Touchard congruence – once again

If we would like to find the coefficients $a_0, a_1, \ldots, a_{k-1}$ such that

$$B_{n+k} \equiv a_0 B_n + a_1 B_{n+1} + \cdots a_{k-1} B_{n+k-1} \pmod{p},$$

we can recall that the $k \times k$ Hankel determinant (i.e., the Hankel transform) of the Bell numbers is

$$\begin{vmatrix} B_0 & B_1 & \cdots & B_{k-1} \\ B_1 & B_2 & \cdots & B_k \\ \vdots & \vdots & \vdots & \vdots \\ B_{k-1} & B_k & \cdots & B_{2k-2} \end{vmatrix} = \prod_{i=0}^{k-1} i!,$$

which was a nice result of Chapter 2 (see (2.62)). Thanks to the fact that p is a prime, this determinant is not zero modulo p for $k = p$, so the coefficients can be found, and they are unique[2].

Taking $k = p = 3$, the Hankel matrix of the Bell numbers is

$$\begin{pmatrix} B_0 & B_1 & B_2 \\ B_1 & B_2 & B_3 \\ B_2 & B_3 & B_4 \end{pmatrix} = \begin{pmatrix} 1 & 1 & 2 \\ 1 & 2 & 5 \\ 2 & 5 & 15 \end{pmatrix} \equiv \begin{pmatrix} 1 & 1 & 2 \\ 1 & 2 & 2 \\ 2 & 2 & 0 \end{pmatrix} \quad (\text{mod } 3).$$

The inverse of the matrix on the right-hand side is

$$\begin{pmatrix} 1 & -1 & \frac{1}{2} \\ -1 & -1 & 0 \\ \frac{1}{2} & 0 & -\frac{1}{4} \end{pmatrix} \equiv \begin{pmatrix} 1 & 2 & 2 \\ 2 & 2 & 0 \\ 2 & 0 & 2 \end{pmatrix} \quad (\text{mod } 3).$$

To complete Equation (12.2), we still calculate

$$\begin{pmatrix} B_3 \\ B_4 \\ B_5 \end{pmatrix} = \begin{pmatrix} 5 \\ 15 \\ 52 \end{pmatrix} \equiv \begin{pmatrix} 2 \\ 0 \\ 1 \end{pmatrix} \quad (\text{mod } 3).$$

Hence, we have

$$\begin{pmatrix} a_0 \\ a_1 \\ a_2 \end{pmatrix} = \begin{pmatrix} 1 & 2 & 2 \\ 2 & 2 & 0 \\ 2 & 0 & 2 \end{pmatrix} \begin{pmatrix} 2 \\ 0 \\ 1 \end{pmatrix},$$

where it comes that

$$\begin{pmatrix} a_0 \\ a_1 \\ a_2 \end{pmatrix} = \begin{pmatrix} 4 \\ 4 \\ 6 \end{pmatrix} \equiv \begin{pmatrix} 1 \\ 1 \\ 0 \end{pmatrix} \quad (\text{mod } p),$$

resulting in

$$B_{n+3} \equiv B_n + B_{n+1} \quad (\text{mod } 3) \quad (n \geq 0),$$

which we recognize as a particular case of Touchard's congruence.

If the reader has a new sequence to study, the above approach is of great help. One can perform the above calculations for some small primes, and if there is a common pattern in the resulting recurrences, a good conjecture can be formed for general primes.

A simple congruence for the involutions

For another example, we take the I_n involutions of Section 3.4. Let

---

[2] Note that the $(p+1) \times (p+1)$-size Hankel matrix would have zero determinant modulo p, so the highest order recurrence we can find is of order p in the case of Bell numbers.

$k = p = 5$, and the Hankel matrix of this order modulo 5 is

$$\begin{pmatrix} 1 & 1 & 2 & 4 & 0 \\ 1 & 2 & 4 & 0 & 1 \\ 2 & 4 & 0 & 1 & 1 \\ 4 & 0 & 1 & 1 & 2 \\ 0 & 1 & 1 & 2 & 4 \end{pmatrix},$$

and the vector on the right-hand side of our equation is

$$\begin{pmatrix} I_5 \\ I_6 \\ I_7 \\ I_8 \\ I_9 \end{pmatrix} = \begin{pmatrix} 1 \\ 1 \\ 2 \\ 4 \\ 0 \end{pmatrix} \quad (\text{mod } 5).$$

In this case, we do not even need to solve the equation, because we can note that this vector is the same as the first column of the coefficient matrix. By *Cramer's*[3] *rule*[4], all the determinants we should calculate are zero (having two equal columns), except the very first one when we substitute our vector into the first row. In this case the determinants in the nominator and denominator are equal, resulting in the solution $a_0 = 1$, $a_2 = \cdots = a_4 = 0$. Hence, we get the congruence

$$I_{n+5} \equiv I_n \quad (\text{mod } 5).$$

This congruence is actually a particular case of

$$I_{n+m} \equiv I_n \quad (\text{mod } m)$$

for odd integers m and $n \geq 0$, see [151]. Note that the longest possible recurrence for I_n is of length $p-1$, because the higher order Hankel determinants are not regular, just as in the case of the Bell numbers (Exercise 17 of Chapter 3).

12.2   The characteristic polynomials

The characteristic polynomials play a prominent role in the modulo p theory of recurrence sequences. Now we define and study these polynomials in details.

---

[3] Gabriel Cramer (1704-1752), Swiss mathematician.

[4] This rule says that an equation $Ax = b$ having as many variables as equations, and the coefficient matrix A is regular, then the ith solution x_i can be given as follows: change the ith column of A by the vector b, and divide the determinant of this new matrix by the original determinant of A.

Definition 12.2.1. *Let us suppose that we are given a sequence satisfying a modulo p recurrence of the form* (12.1) *with* $a_0, \ldots, a_{k-1}$ *coefficients being in* $\mathbb{F}_p$. *Then the* characteristic polynomial *of the sequence* A_n *is*

$$f(x) = x^k - a_{k-1}x^{k-1} - a_{k-2}x^{k-2} - \cdots - a_1 x - a_0.$$

For example, the characteristic polynomial of the modulo p Bell number sequence is

$$f(x) = x^p - x - 1,$$

due to the Touchard congruence.

There are some good reasons why these polynomials are so useful. These reasons will be explained in the below subsections.

12.2.1 Representation of mod p sequences via the zeros of their characteristic polynomials

Let A_n be a linear recurrence of order k in $\mathbb{F}_p$ (that is, modulo p). Moreover, we suppose that its characteristic polynomial is $f(x)$ such that all the $\alpha_1, \ldots, \alpha_k$ zeros of $f(x)$ are distinct. Then we can express the terms of our sequence as follows:

$$A_n \equiv \sum_{j=1}^{k} \beta_j \alpha_j^n \pmod{\mathbb{F}_p / f(x)} \quad (n \geq 0), \tag{12.3}$$

where the β_j numbers are uniquely determined by the sequence and belong to the splitting field of $f(x)$.

The proof of (12.3)

First we show that the β_j coefficients are uniquely determined by A_n. In fact, β_j can be calculated from the system of equations

$$\sum_{j=1}^{k} \beta_j \alpha_j^n = A_n \quad (n = 0, 1, \ldots, k-1).$$

Note that the coefficient matrix is of Vandermonde type, and it is invertible because of the imposed conditions on the zeros of $f(x)$. Moreover, the determinant of this matrix is the product of factors of the form $(\alpha_i - \alpha_j)$, it follows that the β_j coefficients (the solutions) are found in the splitting field of $f(x)$.

It still remains to see that (12.3) indeed holds. It suffices to show that the right-hand side satisfies the same recurrence as A_n, namely

$$A_{n+k} = a_0 A_n + a_1 A_{n+1} + \cdots a_{k-1} A_{n+k-1} \quad (n \geq 0).$$

But

$$\sum_{j=1}^{k} \beta_j \alpha_j^{n+k} - a_{k-1} \sum_{j=1}^{k} \beta_j \alpha_j^{n+k-1} - a_{k-2} \sum_{j=1}^{k} \beta_j \alpha_j^{n+k-2} - \cdots - a_0 \sum_{j=1}^{k} \beta_j \alpha_j^n =$$

$$\sum_{j=1}^{k} \beta_j f(\alpha_j) \alpha_j^n = 0,$$

since α_js are the zeros of $f(x)$. The proof is therefore complete.

12.2.2 The Bell numbers modulo 2 and 3

Let us apply our theorem to the Bell numbers, giving them a new representation mod 2 and mod 3. As we said before, their characteristic polynomial mod p is $x^p - x - 1$. Its zeros are all different. If this were not the case, $x^p - x - 1$ and its derivative $px^p - 1 \equiv -1 \pmod{p}$ would have common zeros which is impossible, because the derivative is simply -1 which has no zeros at all.

The Bell numbers mod 2

Taking $p = 2$ first, the splitting field of $x^2 - x - 1$ is

$$\mathbb{F}_2/(x^2 - x - 1) = \{0, 1, x, 1 + x\}.$$

One zero of an irreducible polynomial in its splitting field is always x. As 0 and 1 are not zeros of $x^2 - x - 1$, the only remaining possibility is $1 + x$. (Indeed, $(1 + x)^2 - (1 + x) - 1 = 0$ in $\mathbb{F}_2/(x^2 - x - 1)$.) Our theorem above therefore says that there are two constants β_1 and β_2 such that

$$B_n \equiv \beta_1 x^n + \beta_2 (1 + x)^n \pmod{2}.$$

To determine these coefficients take $n = 0$ and $n = 1$. Then, respectively,

$$1 = \beta_1 + \beta_2$$
$$1 = \beta_1 x + \beta_2 (1 + x)$$

It comes readily that $\beta_1 = x$, $\beta_2 = 1 + x$ thus

$$B_n \equiv x^{n+1} + (1 + x)^{n+1} \pmod{\mathbb{F}_2/(x^2 - x - 1)}. \tag{12.4}$$

This seems to be a bit complicated. It is not even obvious at first sight that the polynomial on the right-hand side simplifies to a *number*. However, (12.4) can considerably be simplified. Below we use the "=" sign while doing calculations in our splitting field. Let us consider the first term x^{n+1} as n runs from $n = 0$ in the field $\mathbb{F}_2/(x^2 - x - 1)$:

$$x^{0+1} = x,$$
$$x^{1+1} = x^2 = x + 1,$$
$$x^{2+1} = x^3 = x^2 \cdot x = (x + 1)x = x^2 + x = 1,$$
$$x^{3+1} = x^4 = x^3 \cdot x = x,$$
$$\vdots$$

We conclude at the following fact in our field:

$$x^{n+1} = \begin{cases} x, & \text{if } n \equiv 0 \pmod 3; \\ x + 1, & \text{if } n \equiv 1 \pmod 3; \\ 1, & \text{if } n \equiv 2 \pmod 3. \end{cases}$$

Considering the first term, similar calculation results in

$$(1 + x)^{n+1} = \begin{cases} x + 1, & \text{if } n \equiv 0 \pmod 3; \\ x, & \text{if } n \equiv 1 \pmod 3; \\ 1, & \text{if } n \equiv 2 \pmod 3. \end{cases}$$

Since $x + (x + 1) = 1$, and $1 + 1 = 0$, we have the following conclusion modulo 2:

$$B_n \equiv \begin{cases} 1, & \text{if } n \equiv 0 \pmod 3; \\ 1, & \text{if } n \equiv 1 \pmod 3; \\ 0, & \text{if } n \equiv 2 \pmod 3. \end{cases}$$

That is, the nth Bell number B_n is even if and only if n is of the form $3k + 2$ ($k = 0, 1, \dots$).

This can be put in another form: the Bell numbers modulo 2 form a periodic sequence of length three.

A slightly weaker version,

$$B_n + B_{n+1} + B_{n+2} \equiv 0 \pmod 2$$

was discovered by Williams [599].

The Bell numbers mod 3

Now we consider the modulo 3 case, because some new aspects of our theory show themselves more clearly for higher primes. So let $p = 3$. Then the characteristic polynomial of B_n (mod 3) is $x^3 - x - 1$. In its splitting field $\mathbb{F}_3/(x^3 - x - 1)$ one of its zeros is x. If we recall an important fact from the theory of finite fields, we do not need to work hard to find the other zeros.

This theorem says that if f is an irreducible polynomial of degree m in $\mathbb{F}_q$, then all the roots of f are simple and are given by the m distinct elements $\alpha, \alpha^q, \alpha^{q^2}, \ldots, \alpha^{q^{m-1}}$ of its splitting field with q^m elements. (See [373, Theorem 2.14].)

Hence, the roots of the *irreducible* characteristic polynomial $x^3 - x - 1$ are x, x^3, x^9 in $\mathbb{F}_3/(x^3 - x - 1)$. These can be reduced in this field, resulting in

$$\alpha_1 = x,$$
$$\alpha_2 = x^3 = x + 1,$$
$$\alpha_3 = x^9 = (x^3)^3 = (x+1)^3 = 1 + x^3 = 1 + (x+1) = 2 + x.$$

The β_j coefficients of (12.3) are still missing. Taking $n = 0, 1, 2$ in (12.3) we respectively get

$$1 = \beta_1 + \beta_2 + \beta_3,$$
$$1 = \beta_1\alpha_1 + \beta_2\alpha_2 + \beta_3\alpha_3,$$
$$2 = \beta_1\alpha_1^2 + \beta_2\alpha_2^2 + \beta_3\alpha_3^2.$$

In matrix form

$$\begin{pmatrix} 1 & 1 & 1 \\ \alpha_1 & \alpha_2 & \alpha_3 \\ \alpha_1^2 & \alpha_2^2 & \alpha_3^2 \end{pmatrix} \begin{pmatrix} \beta_1 \\ \beta_2 \\ \beta_3 \end{pmatrix} = \begin{pmatrix} 1 \\ 1 \\ 2 \end{pmatrix}.$$

We can recognize the matrix as a *Vandermonde matrix*. There exists a relatively simple expression for the inverse of Vandermonde matrices [569] by using LU-decomposition. Even without using this expression, one can easily find that

$$\beta_1 = 2x^2 + 2x + 2,$$
$$\beta_2 = 2x^2,$$
$$\beta_3 = 2x^2 + x + 2,$$

whence

$$B_n \equiv (2x^2 + 2x + 2)x^n + 2x^2 x^{3n} + (2x^2 + x + 2)x^{9n}$$

in the field $\mathbb{F}_3/(x^3 - x - 1)$. Our observation above is that with respect to the values of α_i we can reduce the powers:

$$B_n \equiv (2x^2 + 2x + 2)x^n + 2x^2(1 + x)^n + (2x^2 + x + 2)(2 + x)^n.$$

12.2.3 General periodicity modulo p

Can we say something similar as we said in the case $p = 2$, namely, that there is some periodicity modulo 3 of B_n? Let us put the question in its most general form: knowing that representation of the form (12.3) exists, what can we say about the period k such that

$$A_{n+k} \equiv A_n \pmod{p}? \tag{12.5}$$

A straightforward observation is that if k is a period, then multiples of k are periods also. Under period we, from now on, mean *shortest* period.

We saw, for example, that $B_{n+3} \equiv B_n \pmod 2$, which is another recurrence for the Bell numbers apart of the Touchard recurrence (and it is even simpler, involving two terms in place of three!). Our above question with respect to periodicity can then be put into another form: is it possible to find the *simplest* recurrence of the form (12.5) having only two terms?

If we have a characteristic polynomial, we can find such a simple recurrence. It is so, because in

$$A_n = \sum_{j=1}^{k} \beta_j \alpha_j^n$$

α_j's are elements of a finite field, so the sequence α_j^n can take only finitely many values, and for some k_j

$$\alpha_j^{n+k_j} = \alpha_j^n.$$

One might conclude that then we just take the *least common multiple* of all the k_j's and we get the k in (12.5). The problem at this level is actually even simpler: we can express all the α_js with α_1:

$$\alpha_j^n = \left[(x^p)^{j-1} \right]^n = (x^n)^{p(j-1)} = (\alpha_1^n)^{p(j-1)}.$$

Thus,

$$\alpha_j^{n+k} = (\alpha_1^{n+k})^{p(j-1)} = (\alpha_1^n)^{p(j-1)} = \alpha_j^n,$$

where k is the order of the periodicity of $\alpha_1^n = x^n$ modulo the characteristic polynomial.

Hence, the periodicity k of a sequence A_n satisfying a linear homogeneous recurrence modulo a prime p, having irreducible characteristic polynomial f with pairwise distinct roots, equals to the periodicity of the sequence $(x^n)_{n \geq 0}$ modulo f.

The periodicity k can be expressed in another way, involving only the characteristic polynomial. We need that $x^{n+k} \equiv x^n \pmod{f(x)}$ for all $n \geq 0$. It is sufficient, however, to claim only that the $n = 0$ case satisfies:

$$x^k \equiv 1 \pmod{f(x)}, \tag{12.6}$$

the rest comes by multiplying with x^n. Congruence (12.6) satisfies if and only if

$$x^k - 1 \equiv 0 \pmod{f(x)},$$

that is, when

$$x^k - 1 = c(x)f(x)$$

with some polynomial $c(x)$. We therefore have the following statement: the (shortest) period of the sequence A_n is equal to the smallest integer k such that the characteristic polynomial $f(x)$ divides $x^k - 1$. This k is called the *order* of $f(x)$.

If $x^{n+k} \equiv x^n \pmod{f(x)}$ satisfies only from a given n on, the sequence is periodic from that n onward. The elements until $n-1$ are from the *pre-period*. For example, the sequence $3, 2, 1, 0, 1, 0, 1, 0, \ldots$ is periodic from $n = 2$, and the part $3, 2$ is the pre-period. Note also that the pre-period is zero when $f(0) \neq 0$. Moreover, by the reasons said above, the pre-period is of length h if $f(x)$ is of the form $x^h g(x)$, when $g(x) \neq 0$.

12.2.4   Shortest period of the Fubini numbers

We saw in Subsection 11.8.2 that for all prime p

$$F_{n+p} \equiv F_{n+1} \pmod{p},$$

and thus the characteristic polynomial of the Fubini sequence modulo p is

$$f(x) = x^p - x = x(x^{p-1} - 1).$$

Taking into account what we have said about the pre-periods above, we see also from the form of $f(x)$ that F_n modulo p has a pre-period of length one. It readily comes that the period of F_n modulo p is $p - 1$, since $f(x)$ divides $x^{p-1} - 1$ but does not divide any $x^k - 1$ if $k < p - 1$.

For example, taking $p = 5$, the sequence $F_n \pmod 5$ is

$$1, 1, 3, 3, 0, 1, 3, 3, 0, 1, 3, 3, 0, \ldots.$$

Indeed periodic of length $p - 1 = 4$ with the pre-period 1 at the beginning.

12.3   The minimal polynomial

If a modulo p sequence satisfies some linear recurrence, it satisfies many others, too. For example, if A_n is a sequence such that[5] $A_{n+k} = A_n$, then it follows that $A_{n+2k} = A_n$. In this case the respective characteristic polynomials are $f_1(x) = x^k - 1$ and $f_2(x) = x^{2k} - 1$. Note that $f_2(x)$ is a multiple of $f_1(x)$.

The most extreme case is of the identically zero sequence: it satisfies *any* homogeneous recurrence relation. In this case the simplest characteristic polynomial is $f(x) = 1$; this polynomial divides *any* polynomial.

We might think that there is, in some sense a *minimal* recurrence that our sequence satisfies. Also, based on the above examples, we might think that the characteristic polynomial belonging to this minimal recurrence *divides* the

---

[5] From now on we leave the $\equiv$ sign and add mod p. Instead, we simply write equality sign, and the modulus is understood to be a prime number.

characteristic polynomials of the other, non-minimal recurrences. We therefore take the following definition.

Definition 12.3.1. *The monic[6] $m(x)$ polynomial of positive degree[7] is the minimal polynomial of the linear recurrence sequence $A_0, A_1, \ldots$ if it divides all the charateristic polynomials of this sequence.*

Here come some facts about minimal polynomials (see [373, Section 8.4]):

1. The minimal polynomial exists for any linear recurrence sequence, and it is unique.

2. The minimal polynomial is the characteristic polynomial belonging to the linear recurrence of the least possible order.

3. The order of the minimal polynomial coincides with the period of the linear recurrence it belongs to.

4. If a characteristic polynomial belonging to a given recurrence of a given sequence is monic irreducible, then it is minimal.

Suppose that we have a sequence A_n satisfying a recurrence of the form (12.1). Then the characteristic polynomial is of degree k, so the minimal polynomial must be of degree $\leq k$. This degree can be bounded from above in another way, too. The size of the maximal non-zero Hankel determinant (if it exists[8]) is a bound, indeed, by what we have said in Section 12.1.

Supposing that this bound is k, to solve Equation (12.2) we needed the first $2k$ terms of the sequence to determine the $a_0, \ldots, a_{k-1}$ coefficients which are the coefficients of the characteristic polynomial. Therefore, ideally, we might think that these first $2k$ terms can be used to find the minimal polynomial, too. This is true, and the construction goes by the Berlekamp–Massey algorithm.

12.3.1 The Berlekamp–Massey algorithm

The Berlekamp–Massey algorithm (see [71, 398] for the original works, and [202, Chapter 9], [373, Section 8.6]) is a recursive algorithm which produces the minimal polynomials in finitely many steps once we know an upper bound for the degree of the minimal polynomial.

Let $A_0, A_1, \ldots$ be a sequence, and

$$G(x) = \sum_{n=0}^{2k-1} A_n x^n$$

---

[6]A polynomial is *monic* if its leading coefficient is one.

[7]This is a necessary condition because the constants always divide any polynomial in a field.

[8]If there are mod p non-zero Hankel determinants of arbitrary size, the sequence is not a linear recurrence, see the Outlook.

be the generating function of the first $2k$ terms.

We recursively define the polynomials $g_j(x)$ and $h_j(x)$, and numbers m_j, b_j $(j = 0, 1, \dots)$ as follows.

$$g_0(x) = 1, \quad h_0(x) = x, \quad m_0 = 0.$$

Then let b_j be the coefficient of x^j in $g_j(x)G(x)$, and set

$$g_{j+1}(x) = g_j(x) - b_j h_j(x),$$
$$h_{j+1}(x) = \begin{cases} \frac{1}{b_j} x g_j(x), & \text{if } b_j \neq 0 \text{ and } m_j \geq 0; \\ x h_j(x), & \text{otherwise,} \end{cases}$$

and

$$m_{j+1} = \begin{cases} -m_j, & \text{if } b_j \neq 0 \text{ and } m_j \geq 0; \\ m_j + 1, & \text{otherwise.} \end{cases}$$

If the degree of the $m(x)$ minimal polynomial is k, then

$$m(x) = x^k g_{2k}\left(\frac{1}{x}\right).$$

If we only know that the degree of $m(x)$ is bounded by k, then set

$$r = \left\lfloor k + \frac{1}{2} - \frac{1}{2} m_{2k} \right\rfloor,$$

and in this case

$$m(x) = x^r g_{2k}\left(\frac{1}{x}\right).$$

In any cases, it is seen that the algorithm uses only the first $2k$ terms of A_n, and thus $G(x)$ indeed suffices, we do not need the generating function in its totality.

The proof that the algorithm indeed yields the minimal polynomial after the indicated number of steps can be found on pp. 441-444. of [373].

Mathematica code for the Berlekamp – Massey algorithm

In the following lines we present the code for the Berlekamp – Massey algorithm written in *Mathematica*:

```
g[x_,0]:=1
h[x_,0]:=x
m[0]:=0

For[
j=0,j<=2k, j++,
```

```
b[j]=SeriesCoefficient[g[x,j]*G[x],{x,0,j}];
g[x_,j+1]=g[x,j]−b[j]h[x,j];
h[x_,j+1]=If[b[j]!=0&&m[j]>=0,1/b[j]*x*g[x,j],x*h[x,j]];
m[j+1]=If[b[j]!=0&&m[j]>=0,−m[j],m[j]+1]
](* End of For *)
```

```
r:=Floor[k+1/2−1/2m[2k]]
PolynomialMod[x^r*g[1/x,2k],p]//Expand
```

If we want to find the minimal polynomial for the Bell polynomials modulo $p = 5$, we recall that

n	0	1	2	3	4	5	6	7	8	9
B_n (mod 5)	1	1	2	0	0	2	3	2	0	2

therefore, we set for our algorithm

```
p:= 5
G[x_]:= 1 + 1x + 2x^2 + 2x^5 + 3x^6 + 2x^7 + 2x^9
```

Running the algorithm, the output (which is printed out be the last line of the above code) will be

```
4 + 4 x + x^5
```

This is the minimal polynomial: $m(x) = x^5 + 4x + 4 \equiv x^5 - x - 1$ (mod 5). In congruence terms

$$B_{n+5} + 4B_{n+1} + 4B_n \equiv 0 \quad (\text{mod } 5).$$

Rearranging,

$$B_{n+5} \equiv -4B_{n+1} - 4B_n \equiv B_{n+1} + B_n \quad (\text{mod } 5).$$

We know more now than Touchard proved: we have that his congruence is *minimal* in the sense of our definition, at least for $p = 5$.

With respect to the period of B_n (mod 5) we can do the following calculation having the minimal polynomial: we determine k (which is the order of $m(x)$ as well as the minimal period B_n (mod 5). By computer check, it can be seen that the remainder of the division

$$\frac{x^{781} - 1}{x^5 - x - 1}$$

is zero, but for any other exponent smaller than 781, the remainder is non-zero modulo 5. Thus, the minimal period for the sequence is $k = 781$ and because $m(0) \neq 0$ there is no pre-period. In formulas,

$$B_{n+781} \equiv B_n \quad (\text{mod } 5) \quad (n \geq 0),$$

and there is no $k < 781$ such that

$$B_{n+k} \equiv B_n \quad (\text{mod } 5) \quad (n \geq 0).$$

12.4 Periodicity with respect to composite moduli

Having information on the periodicity modulo primes, we can deduce periodicity properties with respect to composite moduli.

For example, we know from the previous sections that the Bell numbers are periodic modulo 2, and the period length is three, and we also know that they are periodic modulo 5 with period length 781.

12.4.1 Chinese remainder theorem

It seems to be reasonable to say that once we know how the sequence behaves modulo 2 and 5, we should know how it behaves modulo $2 \cdot 5 = 10$. The certainty is provided by the *Chinese remainder theorem*. This theorem says that if we know the remainders of a number a with respect to some relative prime numbers $n_1, n_2, \ldots n_k$, then we know the remainders of a with respect to $n_1 \cdot n_2 \cdots n_k$. In the language of formulas, we have the congruences

$$a \equiv a_1 \pmod{n_1},$$
$$a \equiv a_2 \pmod{n_2},$$
$$\vdots$$
$$a \equiv a_k \pmod{n_k},$$

and the solution a to the congruences is determined uniquely modulo $n_1 n_2 \cdots n_k$.

We now show the general solution to the above congruence system when $k = 2$, then we apply this knowledge to determine the periodicity of the last digits of the Bell numbers.

Bézout's lemma

We shall need *Bézout's*[9] *lemma*. It says that d is the common divisor of the integers n_1 and n_2 (which are not both zero), then there exist integers x and y such that
$$d = n_1 x + n_2 y.$$
In other words, the greatest common divisor of n_1 and n_2 can always be expressed as an integer linear combination of n_1 and n_2.

In the particular case when n_1 and n_2 are co-primes, then

$$n_1 x + n_2 y = 1 \tag{12.7}$$

---

[9]Étienne Bézout (1730-1783), French mathematician.

for some integers x and y.

The proof of this theorem can be found in any basic number theory text-book (see, for instance, [31, Theorem 1.2], [224, Theorem 1.23], or [305, Section 1.2]).

The solution of a system of two congruences

We state that, having the congruences

$$a \equiv a_1 \quad (\text{mod } n_1),$$
$$a \equiv a_2 \quad (\text{mod } n_2),$$

we consequently also have that

$$a \equiv a_1 n_2 y + a_2 n_1 x \quad (\text{mod } n_1 n_2).$$

Here x and y are the numbers taken from (12.7).

To check the validity of this solution for a, we substitute, again by (12.7), $1 - n_1 x$ in place of $n_2 y$:

$$a \equiv a_1 n_2 y + a_2 n_1 x = a_1(1 - n_1 x) + a_2 n_1 x \equiv a_1 \quad (\text{mod } n_1),$$

as it should be. The modulo n_2 case is similar.

The reader can convince herself/himself, that if, for example, a number a is one modulo two and two modulo five, then its last digit is uniquely determined. In formulas, we fix that a satisfies the congruences

$$a \equiv 1 \quad (\text{mod } 2),$$
$$a \equiv 2 \quad (\text{mod } 5).$$

Owing to the Bézout formula with, say, $x = 3$ and $y = -1$, such that

$$2x + 5y = 6 - 5 = 1 = (2, 5),$$

we get

$$a \equiv 1 \cdot 5 \cdot (-1) + 2 \cdot 2 \cdot 3 \quad (\text{mod } 2 \cdot 5).$$

This is then equivalent to

$$a \equiv 7 \quad (\text{mod } 10).$$

12.4.2   The Bell numbers modulo 10

It is now straightforward to get our periodicity result with respect to the last decimal digit of the Bell number sequence. This sequence is periodic of length three modulo 2, and of length 781 modulo 5. The least common multiple of these two periodicities is 2343. We know that both B_n and $B_{n+2\,343}$ have the same remainder with respect to the moduli 2 and 5; therefore, they must have the same remainder modulo $2 \cdot 5 = 10$, too. Therefore

$$B_{n+2\,343} \equiv B_n \pmod{10},$$

that is, the last digits of the Bell numbers form a periodic sequence with length 2343.

Basically everything is the same for higher moduli, but the solution of the congruence system is more laborious when $k > 2$.

We should note that from prime moduli we can deduce results for the product of these moduli, but not for moduli having prime power divisors. For example, we must know how the Bell number sequence's behavior modulo $2^2 = 4$ in order to get results with respect to $20 = 2^2 \cdot 5$.

12.5   Value distributions modulo p

One can ask the following question: which values can a linear recurrence attain modulo a prime number[10]?

For the Fubini numbers, for instance, the sequence $F_n \pmod{p}$ does not attain all the possible values, since, when $p = 5$ $F_n \pmod 5$ starts as

$$1, 1, 3, 3, 0, 1, 3, 3, 0, \ldots$$

(periodic of length four with pre-period of length one), so 2 and 4 do not occur. For $p = 7$ the avoided value is 4.

Let us take now the Bell numbers, say, modulo 5. We have just seen that this sequence is periodic modulo 5 with a relatively huge period, 781. Still, it is enough to go until B_{17} so that we cover all the possible values modulo 5: the values of $B_1, B_2, \ldots, B_{17}$ are

$$1, 2, 0, 0, 2, 3, 2, 0, 2, 0, 0, 2, 2, 2, 0, 2, 4.$$

Indeed, $0, 1, 2, 3, 4$ all appear among these values. This is not a coincidence, $B_n \pmod p$ always attains all the numbers $0, 1, \ldots, p - 1$. We dedicate the following pages to prove this statement.

12.5.1   The Bell numbers modulo p

Is it always possible to solve the congruence

$$B_n \equiv \lambda \pmod{p}$$

---

[10]This question with respect to some of our combinatorial sequences other than linear recurrences and not modulo p but among the integers is treated in Chapter 13.

for *any* fixed λ, where the unknown is n?

We split this question into two parts: when $\lambda = 0$ and when $\lambda \neq 0$.

Solvability of $B_n \equiv 0 \pmod{p}$

We will prove more than the mere solvability of the above congruence. We show that the sequence $B_n \pmod{p}$ contains $p-1$ consecutive zeros:

$$B_{m_p+k} \equiv 0 \pmod{p} \quad (k = 0, 1, \ldots, p-2),$$

where

$$m_p \equiv 1 - \frac{p^p - p}{(p-1)^2} \pmod{\frac{p^p - 1}{p - 1}}.$$

To prove this statement[11], first recall Exercise 16 of Chapter 11. It states, in particular, that

$$B_{pn} \equiv \sum_{i=0}^{n} \binom{n}{i} B_i \pmod{p}.$$

On the right-hand side we can recognize the basic recurrence (1.1) for the Bell numbers. Therefore, for any prime p,

$$B_{pn} \equiv B_{n+1} \pmod{p} \quad (n = 0, 1, 2, \ldots).$$

We now prove another auxiliary statement. Let M be an integer and p be a prime. Then $B_{M+k} \equiv 0 \pmod{p}$ for $k = 0, 1, \ldots, p-2$ if and only if $B_{M+k} \equiv B_{M+pk} \pmod{p}$ for $k = 1, 2, \ldots, p-1$.

Again, by Exercise 16 of Chapter 11 we have

$$B_{M+kp} \equiv \sum_{i=0}^{k} \binom{k}{i} B_{M+i} \pmod{p} \quad (k = 0, 1, 2, \ldots). \qquad (12.8)$$

If $B_{M+k} \equiv 0$ for $k = 0, 1, \ldots, p-2$ then $B_{M+pk} \equiv 0$ and thus it is also true that $B_{M+k} \equiv B_{M+pk}$. When $k = p-1$ (and still $B_{M+k} \equiv 0$ for $k = 0, 1, \ldots, p-2$) the above congruence reduces to $B_{M+(p-1)p} \equiv B_{M+p-1} \pmod{p}$. Thus, the "if" statement is proved.

We continue by proving the "only if" part. If $B_{M+k} \equiv B_{M+pk} \pmod{p}$ for $k = 1, 2, \ldots, p-1$, then (12.8) is the same as

$$B_{M+k} \equiv B_{M+k} + \sum_{i=0}^{k-1} \binom{k}{i} B_{M+i} \pmod{p} \quad (k = 1, 2, \ldots, p-1).$$

---

[11]The proof is taken from [359], another proof can be found in [312].

From here, after canceling out B_{M+k} we get a system of equations:

$$0 \equiv \binom{1}{0} B_M,$$

$$0 \equiv \binom{2}{0} B_M + \binom{2}{1} B_{M+1},$$

$$\vdots$$

$$0 \equiv \binom{p-1}{0} B_M + \binom{p-1}{1} B_{M+1} + \cdots + \binom{p-1}{p-2} B_{M+p-2}.$$

The coefficient matrix is triangular, and non-singular (its diagonal contains the binomial coefficients $\binom{k}{k-1} \neq 0$ $(k \geq 1)$). We get that the solution is

$$B_{M+i} \equiv 0 \pmod{p} \quad (k = 0, 1, \dots, p-2).$$

This is what we stated.

Solvability of $B_n \equiv \lambda \pmod{p}$ $(\lambda \neq 0)$

The statement is that the congruence in the title is always solvable for any fixed λ, and for the smallest solution we have the upper bound

$$n \leq \binom{2p-1}{p-1}.$$

This is Theorem 2 in [518].

Let us define the sequence W_n as

$$W_n = (B_n - \lambda)^{p-1} - 1.$$

It can easily be seen that the $B_n - \lambda \equiv 0$ congruence is solvable if and only if $W_n \not\equiv 0$. W_n is a periodic sequence, and, considering (12.3) applied to the Bell numbers, we have, by the polynomial theorem, that

$$W_n = \sum_{k_0 + \cdots + k_p = p-1} \frac{(p-1)!}{k_0! \cdots k_p!} \lambda^{k_0} \beta_1^{k_1} \cdots \beta_p^{k_p} \left(\alpha_1^{k_1} \cdots \alpha_p^{k_p} \right)^n$$

in the splitting field of the characteristic polynomial $x^p - x - 1$ of the Bell numbers. (Here $\alpha_1, \dots, \alpha_p$ are the zeros of this characteristic polynomial.)

The sum contains $\binom{2p-1}{p-1}$ terms. For two different sets of indices, say $l_0 + \cdots + l_p = p-1$ and $r_0 + \cdots + r_p = p-1$ the products

$$\alpha_1^{l_1} \cdots \alpha_p^{l_p} \quad \text{and} \quad \alpha_1^{r_1} \cdots \alpha_p^{r_p}$$

are different. This must be so; otherwise, the polynomial $x^p - x - 1$ and the multi-variable polynomial

$$q(x_1, \dots, x_p) = x_1^{l_1} \cdots x_p^{l_p} - x_1^{r_1} \cdots x_p^{r_p}$$

would have common zeros. In fact, the polynomial q is single-variable, because if α_1 is a zero in the splitting field, then the other zeros are $\alpha_1 + k$, where k runs from zero to $p - 1$. Thus, the polynomial q is equal to

$$q(x) = x^{l_1} \cdots (x + p - 1)^{l_p} - x^{r_1} \cdots (x + p - 1)^{r_p}.$$

Now we see that the degree of q is less than p. Because of this, and for the irreducibility of $x^p - x - 1$, we see that these polynomials can have no common zeros in $\mathbb{F}_p$.

It is also true that

$$\frac{(p-1)!}{k_0! \cdots k_p!} \lambda^{k_0} \beta_1^{k_1} \cdots \beta_p^{k_p} \neq 0$$

for any $\lambda \neq 0$. Thus, W_n (as a linear combination of terms of the powers of $\alpha_1^{k_1} \cdots \alpha_p^{k_p}$) is a non-zero linear recurrence sequence of order $\binom{2p-1}{p-1}$. By this non-zero property there is an

$$n \leq \binom{2p - 1}{p - 1}$$

such that $W_n \neq 0$. This, as we said, is equivalent to the congruence $B_n \equiv \lambda$.

Exercises

1. Let $A_0 = A_1 = 1$ and $A_{n+2} = A_{n+1} + A_n$ $(n \geq 0)$ be a linear recurrence in $\mathbb{F}_2$ (note that these are the Fibonacci numbers modulo 2). Prove that A_n can be represented as

$$A_n = \alpha^{n+1} + (1 + \alpha)^{n+1} \quad (n \geq 0).$$

Then show that $A_{n+3} = A_n$ for all $n \geq 0$. (So the Fibonacci and Bell numbers coincide modulo 2.)

2. Based on Exercise 24 of Chapter 11, show that the period of the horizontal L_n sums of Lah numbers is $p - 1$ without pre-period.

3. Construct a sequence A_n such that it is not periodic modulo p.

4. Run the Berlekamp–Massey algorithm to find the minimal polynomial for the Euler numbers modulo $p = 5$. (Note that this sequence has a proper pre-period. The result you should get is $m(x) = x^4 + 4$.)

5. Let $N_p = \frac{p^p - 1}{p - 1}$. Show that any N_p consecutive Bell numbers sum up to zero modulo p. (This result can be proven by elementary methods [313].)

6. Let $W_n = (B_n - \lambda)^{p-1} - 1$, as in Subsection 12.5.1. Supposing that $\lambda \neq 0$, show that the $B_n - \lambda \equiv 0$ congruence is solvable if and only if $W_n \not\equiv 0$.

Outlook

1. The modulo p periodicity of the Bell numbers has been studied by several authors. The first result about this was given by Hall [275]. He showed that the sequence of Bell numbers has a period $N_p = \frac{p^p-1}{p-1}$, which is not necessarily minimal. This result was rediscovered by Williams [599], who also showed that the minimum period is exactly N_p for $p = 2, 3, 5$. Levine and Dalton [372] showed that the period is exactly N_p for $p = 7, 11, 13$ and 17. Afterwards, Radoux [475] conjectured that for any prime p, the number N_p is the minimum period of the Bell sequence. Since then, several authors have shown this conjecture for some particular primes. For example, Wagstaff [577] proved that the period is exactly N_p for all primes $p < 102$ and several larger ps. Montgomery et al. [443] improved Wagstaff's result for most primes $p < 180$.

   Lower bounds for the period of the sequence $B_n \pmod{p}$ are known. Denoting the shortes period by $a(p)$, we have that [384]:

   $$\frac{1}{2}\binom{2p}{p} < a(p),$$

   and even that [575]

   $$2^{2.54p} < a(p).$$

   Here "even" means that $\frac{1}{2}\binom{2p}{p} < 2^{2.54p}$. See also [240], where further results on $a(p)$ are given, including divisibility relations between this sequence and N_p (under the assumption, of course, that accidentally $a(p) \neq N_p$).

   Note that Radoux conjecture is equivalent to saying that the smallest number k for which $x^p - x - 1$ divides $x^k - 1$ is $k = N_p$.

   These considerations can be extended to composite modulus. To be able to write Touchard's congruence in a compact form, we use the $Ea_n = a_{n+1}$ right shift operator. Then, if the prime decomposition of the positive integer m is $p_1^{\alpha_1} \ldots p_k^{\alpha_k}$

   $$\left(\prod_{i=1}^{k}(E^{p_i} - E - 1)\right) B_n \equiv 0 \pmod{m}.$$

   Therefore, the period of B_n modulo m is the least common multiple of the periods modulo p_i. The latter is, as we have just seen is $N_{p_i} = \frac{p_i^{p_i}-1}{p_i-1}$.

2. The modulo p periodicity of the r-Bell numbers was studied in [427].

3. Carlitz [121] proved that the Bell numbers have a period $p^k\frac{p^p-1}{p-1}$ modulo p^k. Note that when $k = 1$ this gives back Hall's result.

4. Linear recurrence sequences may not be periodic at all. It can be seen that a sequence is a linear recurrence sequence if and only if there exists a positive integer n such that $h_n^{(r)} = 0$ for all but finitely many $r \geq 0$. Here $h_n(r)$ is the Hankel determinant of the sequence A_{n+r} of size $n \times n$. (See Theorem 8.74 in [373].)

5. See the Notes after Chapter 8 in [373] for the interesting historical remarks on linear recurrence sequences.

6. The mod p periodicity of the sequence $\left(\left\{ {n+k \atop k} \right\} \right)_{n=0}^{\infty}$ was studied by Nijenhuis and Wilf [455]. Results for composite modulus periodicity of the Stirling numbers of the second kind can be found in [348].

Chapter 13

Diophantic results

In this chapter we will study diophantic-type questions regarding our combinatorial numbers.

Taking a triangle of numbers, like the Pascal, Stirling, and Euler triangle, it is an interesting question whether all the numbers appear in these triangles; and if a given number a appears, how many times does it do so?

Since $\binom{n}{1} = n$, it is obvious that every positive integer appears in the Pascal triangle. How many times they do is another question. We study these problems in this section regarding some of our combinatorial triangles.

13.1 Value distribution in the Pascal triangle

As we have just seen, every positive integer appears in the Pascal triangle. How many times a given integer is present? One appears infinitely often, as it appears in every line twice (and once in the zeroth line). In turn, two appears only once, $\binom{2}{1} = 2$. In addition, by the symmetry of the Pascal triangle, every number (except 2) appears at least twice. What is more, there are infinitely many numbers appearing even four times: let a be the value of the binomial coefficient $\binom{n}{k}$. Then, by symmetry, a can appear once again as $\binom{n}{n-k}$, and twice more as $\binom{a}{1}$ and $\binom{a}{a-1}$. To make easier the reference to these cases, we will say that such values are *trivially related*.

We show now that there are infinitely many numbers which appear six times among the binomial coefficients.

13.1.1 Lind's construction

The above-described numbers, which appear at least six times, interestingly, can be parametrized by the Fibonacci and Lucas sequences. The *Lucas sequence* is defined similarly as the Fibonacci sequence: $L_n = L_{n-1} + L_{n-2}$, except that the initial values are $L_0 = 2$, $L_1 = 1$.

The construction was first made by Lind [375] in 1968, then Singmaster [522] gave another proof in 1975. Tovey [565] arrived at the same discovery independently in 1985. As Tovey's proof is the simplest, we present that one here.

For any positive *even* integer i, let

$$n = \frac{F_i L_i + F_i^2}{2}, \tag{13.1}$$

$$k = \frac{F_i L_i - F_i^2}{2} - 1. \tag{13.2}$$

Then

$$\binom{n}{k} = \binom{n-1}{k+1}. \tag{13.3}$$

This infinite family of numbers appears not only twice but, because of

$$\binom{n}{n-k} = \binom{n}{k} = \binom{\binom{n}{k}}{1} = \binom{\binom{n}{k}}{\binom{n}{k}-1},$$

and

$$\binom{n-1}{k+1} = \binom{n-1}{n-k-2}$$

it occurs altogether six times, as we stated.

The proof is as follows. Writing out the factorials in (13.3), we see that this equation is actually equivalent to the simple

$$n(k+1) = (n-k)(n-k-1).$$

Setting

$$x = n - k - 1, \tag{13.4}$$

we have that (13.3) is equivalent to

$$n(n-x) = x(x+1).$$

Solving this quadratic equation for n, we get that

$$n = \frac{x + \sqrt{5x^2 + 4x}}{2}. \tag{13.5}$$

(As n is positive, the positive sign is taken.) We need to find x such that the above expression is an integer. To reach this aim, $5x^2 + 4x$ must be a perfect square (i.e., of the form m^2 for some integer m). Since $5x^2 + 4x = x(5x+4)$, it is natural to claim that both x and $5x + 4$ are perfect squares (although this is a stronger condition than simply requiring $5x^2 + 4x$ to be a perfect square).

Here is the point where the Fibonacci numbers enter into the picture. The following equality will be used:

$$5F_i^2 + 4(-1)^i = (F_{i-1} + F_{i+1})^2, \tag{13.6}$$

more precisely, its version when i is even:

$$5F_i^2 + 4 = (F_{i-1} + F_{i+1})^2 \quad (i = 2, 4, 6, \dots), \tag{13.7}$$

The proof of (13.6) is based on the equality

$$(F_{i+1} + F_{i-1})^2 - 5F_i^2 = 4(F_{i-1}^2 + F_i F_{i-1} - F_i^2) = -4(F_i^2 + F_{i+1}F_i + F_{i+1}^2),$$

which can be proven by using the basic recurrence for the Fibonacci numbers. Now (13.6) follows by induction.

What (13.7) tells us is that when $x = F_i^2$ is a perfect square (with i even), then

$$5x + 4 = 5F_i^2 = 4 = (F_{i-1} + F_{i+1})^2$$

is a perfect square, too.

The parameters n and k can be recovered from (13.5) and (13.4):

$$n = \frac{F_i^2 + F_i(F_{i-1} + F_{i+1})}{2},$$

$$k = \frac{F_i(F_{i-1} + F_{i+1}) - F_i^2}{2} - 1.$$

The basic identity

$$L_i = F_{i-1} + F_{i+1} \tag{13.8}$$

shows that (13.1)-(13.2) indeed hold.

One more form can be given by using another standard identity:

$$F_i L_i = F_{2i},$$

whence

$$n = \frac{F_{2i} + F_i^2}{2},$$

$$k = \frac{F_{2i} - F_i^2}{2} - 1.$$

Note that in these expressions the nominator is always even, because in (13.5) x and $5x^2 + 4x$ always have the same parity:

$$5x^2 + 4x \equiv 5x^2 \equiv x^2 \equiv x \pmod 2.$$

Tovey also proved that whenever (13.3) holds, n and k must be of the form as given in (13.1)-(13.2).

A related problem proposed by Ira Gessel[1] is a test for a number n being a Fibonacci number. Gessel asked a proof for the following statement: the positive integer n is a Fibonacci number if and only if either $5n^2 + 4$ or $5n^2 - 4$ is a Fibonacci number. For the details, see [251].

---

[1] Ira Martin Gessel (1951-), American mathematician.

13.1.2   The number of occurrences of a positive integer

In the preceding subsections, we constructed numbers which occurred multiple times in the Pascal triangle. In turn, one can easily show that every number greater than one can only finitely many times occur in the Pascal triangle: an upper bound for the row index of the last occurrence is found below. What is more, an estimation can also be given for the number of occurrences.

Every number > 1 can only finitely many times occur in the Pascal triangle

We know from p. 108, every line in the binomial coefficient table forms a log-concave sequence. This, in particular, means that the minimal elements in a given line are the extremal elements (the ones, $\binom{n}{0} = \binom{n}{n}$), and the second smallest elements are the neighboring elements to ones ($\binom{n}{1} = \binom{n}{n-1} = n$). Therefore, a given number $a > 1$ can appear only up to the ath line. By this very reason, each number $a > 1$ can occur only finitely many times in the Pascal triangle.

Let us denote the number of occurrences of a given a by $N(a)$. We know already that $N(1) = \infty$, $N(2) = 1$, and that there are infinitely many a such that $N(a) \in \{2, 4\}$. We are going to give an upper bound for $N(a)$. We saw that there are infinitely many a with $N(a) = 6$.

The estimation $N(a) \leq 2 + 2\log_2(a)$

We will need an elementary fact about the sequences of binomial coefficients of the form $\binom{i+j}{i}$: such sequences are monotonically increasing both in i and in j. Indeed, appealing to the basic recurrence

$$\binom{n}{k} = \binom{n-1}{k} + \binom{n-1}{k-1},$$

we get that

$$\binom{i+j}{i} = \binom{i+j-1}{i} + \binom{i+j-1}{i-1} \geq \binom{i-1+j}{i-1},$$

and similarly with respect to j.

To proceed, we consider the central binomial coefficient sequence $\binom{2b}{b}$. It is strictly increasing if $b \geq 0$; therefore, there is a smallest b such that $\binom{2b}{b} > a$. By the above-described monotonicity property,

$$\binom{b+i+b+j}{b+i} \geq \binom{b+b+j}{b} \geq \binom{2b}{b} > a$$

for all non-negative i and j.

Since every binomial coefficient can be written in the form $\binom{i+j}{j}$, to localize a in the Pascal triangle we need to find those i and j for which

$$\binom{i+j}{j} = a. \tag{13.9}$$

Because of the definition of b, it is not possible that both i and j be greater than b. Thus, there are at most $2b$ solutions of (13.9): $N(a) \leq 2b$. Still, b is a kind of implicit quantity, we would like to express $N(a)$ in terms of a. To this end, note that

$$\binom{2b}{b} \geq 2^b \quad (b \geq 0). \tag{13.10}$$

So,

$$a \geq \binom{2(b-1)}{b-1} \geq 2^{b-1},$$

thus $b \leq 1 + \log_2(a)$. By the above estimation, $N(a) \leq 2b = 2 + 2\log_2(a)$, and this is what we wanted to prove.

The above proof was presented by D. Singmaster.

13.1.3 Singmaster's conjecture

Subsequently, better bounds were found for $N(a)$, see [1, 315, 316]. The best bound up to now is [316]

$$N(a) = O\left(\frac{(\log a)(\log\log\log a)}{(\log\log a)^3}\right).$$

It can even be true that $N(a)$ is finite for all $a > 1$, that is, $N(a) = O(1)$. This is Singmaster's conjecture from 1971 [521].

There are some values occurring non-trivially (in the sense of p. 363) as common values of two or three binomial coefficients [590]:

$$\binom{16}{2} = \binom{10}{3} = 120, \quad \binom{56}{2} = \binom{22}{3} = 1540,$$

$$\binom{153}{2} = \binom{19}{5} = 11\,628, \quad \binom{221}{2} = \binom{17}{8} = 24\,310,$$

$$\binom{21}{2} = \binom{10}{4} = 210, \quad \binom{120}{2} = \binom{36}{3} = 7\,140.$$

These values therefore appear six times in the Pascal triangle. Moreover, because of

$$\binom{78}{2} = \binom{15}{5} = \binom{14}{6} = 3\,003$$

the number $3\,003$ appears eight times. There are no known examples of higher number of occurrences up to n, $\binom{n}{k} \leq 10^{60}$ [81]. This leads to the question whether

$$N(a) \leq 8 \quad (a \geq 2).$$

13.2   Equal values in the Pascal triangle

13.2.1   The history of the equation $\binom{n}{k} = \binom{m}{l}$

We have seen in the previous section that the equation

$$\binom{n}{k} = \binom{m}{l} \tag{13.11}$$

(and especially its non-trivial solutions) leads to results on the number of occurrences of numbers in the Pascal triangle. But Equation (13.11) is interesting in itself. Much effort has been put into the solution of (13.11), but only partial results are known, because as easy the shape of this equation is, as hard is the machinery needed to solve it.

Complete solutions are known only in the cases when

$$(k, l) \in \{(2, 3), (2, 4), (2, 5), (2, 6), (2, 8), (3, 4), (3, 6), (4, 6), (4, 8)\}.$$

The case $(k, l) = (2, 3)$ was solved by Avanesov [40] in 1966, $(k, l) = (2, 4)$ was settled by de Weger [591] and Pintér [463] independently in 1996 and 1995, respectively. de Weger [590] gave the complete solution when $(k, l) = (3, 4)$ in 1997. A result of Mordell[2] [444] easily yields this case but this fact was seemingly not realized by Mordell or by others until de Weger published his paper 30 years after the publication of Mordell's result. Bugeaud and his co-authors solved the case $(k, l) = (2, 5)$ in 2008 [104]. The rest of the cases were solved by Stroeker and de Weger [541] in 1999.

In order to show what type of tools are necessary, we highlight a simple case below which was discovered by the above-mentioned de Weger (and essentially solved by Mordell).

13.2.2   The particular case $\binom{n}{3} = \binom{m}{4}$

Writing out

$$\binom{n}{3} = \binom{m}{4} \tag{13.12}$$

in its simplified form, we face to the equation

$$\frac{n(n-1)(n-2)}{6} = \frac{m(m-1)(m-2)(m-3)}{24}.$$

The symmetry of this equation dictates that we should introduce the new variables

$$x = n - 1, \quad y = \frac{m(m-3)}{2},$$

---

[2]Louis Joel Mordell (1888-1972) American-British number theorist.

which then lead to the indeed simpler diophantic equation

$$x^3 - x = y^2 + y,$$

or

$$(x-1)x(x+1) = y(y+1).$$

That is, when solving (13.12) we are, in fact, looking for numbers which are simultaneously products of two consecutive integers and three consecutive integers. This problem was solved by Mordell in 1963 [444]. His proof uses algebraic number theory (properties of the class group of the unit group of a certain number field). See [590] for details, extensions to rational numbers, and references to other solutions.

13.3   Value distribution in the Stirling triangles

We now turn to the Stirling triangle and will study the similar questions we raised above with respect to the binomial triangle: we study the occurrence of numbers, and we study those repeated values in the Stirling triangles from which only a very few are known.

13.3.1   Stirling triangle of the second kind

We are now going to give a bound for the occurrence numbers $M_1(a)$ and $M_2(a)$ in the first kind, and second kind Stirling number triangle. We elaborate on the calculations in the second kind case. The following are based on the work [51].

The approach is basically the same as Singmaster's, but the estimation of the quantity b needs a bit more work.

First note that the sequence $\left\{ {i+j \atop i} \right\}$ is monotone increasing both in i and in j. Also, the sequence $\left\{ {2n \atop n} \right\}$ of the central Stirling numbers of the second kind is strictly increasing.

The proofs are straightforwardly coming from the recurrence

$$\left\{ {n \atop k} \right\} = \left\{ {n-1 \atop k-1} \right\} + k \left\{ {n-1 \atop k} \right\}.$$

For example,

$$\left\{ {i+j \atop i} \right\} = \left\{ {i+j-1 \atop i-1} \right\} + i \left\{ {i+j-1 \atop i} \right\} \geq \left\{ {i-1+j \atop i-1} \right\},$$

so $\left\{ {i+j \atop i} \right\}$ is indeed monotone increasing in i. The other statements are similar.

We will also need the Lambert W function $W(a)$, and its estimation from the Appendix.

The estimation of $M_2(a)$

We are going to prove that for all $a \geq 2$

$$M_2(a) \leq 2 + 2\frac{\log a}{W\left(\frac{1}{2}\log a\right)}.$$

If a tends to infinity,

$$M_2(a) = O\left(\frac{\log a}{\log\log a - \log\log\log a}\right). \tag{13.13}$$

The proof is as follows. Let us fix an integer $a > 1$. By the second point of the above monotonicity property of the Stirling numbers, there exists a unique smallest integer b such that

$$\left\{{2b \atop b}\right\} > a$$

still holds. If we attempt to solve the equation

$$\left\{{i + j \atop j}\right\} = a, \tag{13.14}$$

we see that it can have at most one solution in i for all fixed j.

Because of the definition of b, it cannot hold true that both i and j are greater than equal to b. That is, we must have that for our potential solutions $i < b$ or $j < b$. It therefore follows that (13.14) has at most $2b$ solutions: $M_2(a) \leq 2b$.

If we can find an upper bound for b in terms of a, we are ready. Here the present proof is a bit more complicated than the proof for the binomial coefficient case. To find a suitable lower bound for the central Stirling numbers, we make use of the expression (10.2) for the Stirling numbers:

$$\left\{{n \atop n-k}\right\} = \sum_{j=0}^{k}\left\{{k+j \atop j}\right\}_{\geq 2}\binom{n}{k+j} \quad (n \geq k+1).$$

In particular

$$\left\{{2b \atop b}\right\} = \sum_{j=0}^{b}\left\{{b+j \atop j}\right\}_{\geq 2}\binom{2b}{b+j} \geq \left\{{2b \atop b}\right\}_{\geq 2}.$$

This is actually an obvious inequality, taking into account the combinatorial definition of these numbers.

By Exercise 6 of Chapter 10 we see that

$$\begin{Bmatrix} 2b \\ b \end{Bmatrix}_{\geq 2} = \frac{1}{2^b} \frac{(2b)!}{b!};$$

whence

$$a \geq \begin{Bmatrix} 2(b-1) \\ b-1 \end{Bmatrix} \geq \begin{Bmatrix} 2(b-1) \\ b-1 \end{Bmatrix}_{\geq 2} \geq \left(\frac{b-1}{2}\right)^{b-1}.$$

The definition of the Lambert W function yields that the equation $a = \left(\frac{b-1}{2}\right)^{b-1}$ is invertible (expressible in terms of W). By monotonicity,

$$b \leq 1 + \frac{\log a}{W\left(\frac{1}{2}\log a\right)};$$

therefore,

$$M_2(a) \leq 2b \leq 2 + 2\frac{\log a}{W\left(\frac{1}{2}\log a\right)}.$$

This is our first statement. The second one, (13.13), follows by recalling the asymptotics of the W function (see the Appendix).

13.3.2 Stirling triangle of the first kind

Everything that we have done in the previous section can be applied to the Stirling numbers of the first kind. The monotonicity of $\begin{bmatrix} i+j \\ i \end{bmatrix}$ and $\begin{bmatrix} 2n \\ n \end{bmatrix}$ can be proven easily. We also need

$$\begin{bmatrix} n \\ n-k \end{bmatrix} = \sum_{j=0}^{k} \begin{bmatrix} k+j \\ j \end{bmatrix}_{\geq 2} \binom{n}{k+j} \quad (n \geq k+1),$$

which was presented in (10.2). In particular,

$$\begin{bmatrix} 2n \\ n \end{bmatrix} \geq \begin{bmatrix} 2n \\ n \end{bmatrix}_{\geq 2}.$$

It is clear that $\begin{bmatrix} 2n \\ n \end{bmatrix}_{\geq 2} = \begin{Bmatrix} 2n \\ n \end{Bmatrix}_{\geq 2}$, so our method to estimate b and $M_1(a)$ can be repeated verbatim. Thus, we have the following statement.

For all $a \geq 2$

$$M_1(a) \leq 2 + 2\frac{\log a}{W\left(\frac{1}{2}\log a\right)}.$$

If a tends to infinity,

$$M_1(a) = O\left(\frac{\log a}{\log\log a - \log\log\log a}\right). \tag{13.15}$$

13.4   Equal values in the Stirling triangles and some related diophantine equations

Although $M_i(a)$ seems to be sublogarithmic (by (13.15) and (13.13)), it can well be true that these are bounded functions. Below we present some calculations supporting these suspects.

By numerical calculations it turns out that

$$M_2(a) \leq 2 \quad (2 \leq a \leq 100\,000).$$

The only numbers in this interval which occur twice are

$$15, \quad 4\,095, \quad 66\,066.$$

They appear as follows:

$$\begin{Bmatrix} 5 \\ 2 \end{Bmatrix} = \begin{Bmatrix} 6 \\ 5 \end{Bmatrix} = 15, \tag{13.16}$$

$$\begin{Bmatrix} 13 \\ 2 \end{Bmatrix} = \begin{Bmatrix} 91 \\ 90 \end{Bmatrix} = 4\,095, \tag{13.17}$$

and

$$\begin{Bmatrix} 14 \\ 11 \end{Bmatrix} = \begin{Bmatrix} 364 \\ 363 \end{Bmatrix} = 66\,066. \tag{13.18}$$

The equal values were studied in [229], but the authors' conjecture does not hold: they proposed that there are no multiple values in the Stirling triangle of the second kind apart of 15 and 4\,095. As we have just seen, there is at least one more: 66\,066.

13.4.1   The Ramanujan-Nagell equation

Note that the Stirling numbers in (13.16) and (13.17) are of the form $\begin{Bmatrix} n \\ 2 \end{Bmatrix}$ or $\begin{Bmatrix} n \\ n-1 \end{Bmatrix}$. these have simple closed-form expressions, as we have shown before:

$$\begin{Bmatrix} n \\ 2 \end{Bmatrix} = 2^{n-1} - 1, \quad \begin{Bmatrix} n \\ n-1 \end{Bmatrix} = \binom{n}{2} = \frac{n(n-1)}{2}.$$

The numbers of the form $2^n - 1$ are often called *Mersenne numbers*, while the numbers of the form $\frac{n(n-1)}{2}$ are named *triangular numbers*.

Therefore 15 and 4\,095 are two remarkable numbers in (13.16) and (13.17): they are both Mersenne and triangular numbers. In other words, they are solutions to the diophantine equation

$$2^n - 1 = \frac{m(m-1)}{2}. \tag{13.19}$$

This equation is well known in the literature, called *Ramanujan-Nagell equation*. Ramanujan[3] conjectured in 1913[4] [483] that that there are only five positive integers that are both of the form $2^n - 1$ and $\frac{m(m-1)}{2}$:

$$0, \quad 1, \quad 3, \quad 15, \quad 4\,095.$$

Nagell[5] proved this conjecture in 1948. See [449] for the original (quite elementary) proof in Norwegian, and [450] for an English version from 1961. Other proofs are found in [280], [445, pp. 205-206], [538, pp. 99-102], and in [304, 433, 570].

In fact, the Ramanujan-Nagell equation was originally written in the form

$$2^n - 7 = x^2,$$

but it is easy to see that it is equivalent to (13.19). The latter equation has only the following solutions:

$$1, \quad 9, \quad 25, \quad 121, \quad 32\,761.$$

The Ramanujan-Nagell equation has been subsequently found useful in coding theory [515] and in differential algebra [401]. Some generalizations of the Ramanujan-Nagell equation can be found in [74, 75].

13.4.2 The diophantine equation $\left\{ {n \atop n-3} \right\} = \left\{ {m \atop m-1} \right\}$

The repeated value $66\,066$ in (13.18) is not a Ramanujan-Nagell number. Instead, it is a solution to the equation

$$\left\{ \begin{matrix} n \\ n-3 \end{matrix} \right\} = \left\{ \begin{matrix} m \\ m-1 \end{matrix} \right\} \tag{13.20}$$

belonging to

$$(n, m) = (14, 364).$$

The $\left\{ {n \atop n-3} \right\}$ special Stirling numbers have a closed form:

$$\left\{ \begin{matrix} n \\ n-3 \end{matrix} \right\} = \binom{n}{4} + 10\binom{n}{5} + 15\binom{n}{6},$$

as it follows from (10.2). Using this expression, we transform (13.20) into the diophantic equation

$$\binom{n}{4} + 10\binom{n}{5} + 15\binom{n}{6} = \binom{m}{2}.$$

This equation was recently studied in [51].

---

[3]Srinivasa Ramanujan (1887-1920), Indian mathematician, whose "insight into formulae was quite amazing, and altogether beyond anything I [G. H. Hardy] have met with in any European mathematician."

[4]Wilhelm Ljunggren proposed the same conjecture in 1943 [376].

[5]Trygve Nagell (1895-1988), Norwegian mathematician, mostly known for his contributions in the theory of diophantine equations.

13.4.3   A diophantic equation involving factorials and triangular numbers

In the first kind Stirling triangle, apart one, only two numbers appear more than once (up to $a \le 100\,000$): 6 and 120, in the form

$$\begin{bmatrix} 4 \\ 1 \end{bmatrix} = \begin{bmatrix} 4 \\ 3 \end{bmatrix} = 6,$$

$$\begin{bmatrix} 6 \\ 1 \end{bmatrix} = \begin{bmatrix} 16 \\ 15 \end{bmatrix} = 120.$$

From these we can write down an interesting diophantic equation which, it seems, has not been studied in the literature. Note that $\begin{bmatrix} n \\ 1 \end{bmatrix} = (n-1)!$, and $\begin{bmatrix} n \\ n-1 \end{bmatrix} = \binom{n}{2}$, thus the above two special values are solutions of

$$n! = \frac{m(m-1)}{2}. \tag{13.21}$$

So 6 and 120 are two factorial-triangular numbers[6]. These belong to the pairs

$$(n, m) = \{(1, 2), (3, 4), (5, 16)\}.$$

Are there any other factorial-triangular numbers?

Equation (13.21) is quadratic in m and equivalent to

$$m = \frac{1 + \sqrt{8n! + 1}}{2}.$$

Note that $8n! + 1$ is odd for all $n \ge 0$, thus $\sqrt{8n! + 1}$ is also odd if it is an integer. Thus, one should consider those n integers for which $8n! + 1$ is a perfect square. The fraction in the expression of m then will automatically be an integer. Therefore, we have that finding the solutions of (13.21) is the same as finding all those ns for which $8n! + 1$ is a perfect square. Computer calculations show that there are no such numbers apart of $1, 3, 5$ up to $1\,000$.

13.4.4   The Klazar – Luca theorem

Still concentrating on equal values, one can ask the question of when it is possible to find equal values in a given *row* in our triangles. For the Pascal triangle, the answer is trivial: by the symmetry $\binom{n}{k} = \binom{n}{n-k}$ we have that any value that is not a peak appears twice (if $n > 2$), and other value $\binom{n}{l}$ cannot be equal with these because of the strict log-concavity property.

The Stirling triangles lack this symmetry, but it still can happen that there are some k_1 and k_2 indices such that

$$\begin{bmatrix} n \\ k_1 \end{bmatrix} = \begin{bmatrix} n \\ k_2 \end{bmatrix}, \quad \text{or} \quad \begin{Bmatrix} n \\ k_1 \end{Bmatrix} = \begin{Bmatrix} n \\ k_2 \end{Bmatrix}.$$

---

[6] Note that 1 is also a triangular-factorial number.

This question was studied with respect to the second kind case by M. Klazar and F. Luca. In the first kind case, such studies are not known.

In fact, Klazar and Luca studied the even more general question: how often can it happen that

$$\left\{ {n \atop k_2} \right\} = c \left\{ {n \atop k_1} \right\}$$

for some integer c and $1 < k_1 < k_2 < n$? It turns out that such an equality can be valid only for *finitely many* n. In other words, only in finitely many rows it can happen that some values are multiples of another.

Now we are going to prove the Klazar – Luca theorem.

The main ingredient of the proof - a result of Corvaja and Zannier

Recall the (2.20) closed-form expression for the Stirling numbers of the second kind:

$$\left\{ {n \atop k} \right\} = \frac{1}{k!} \sum_{l=0}^{k} \binom{k}{l} l^n (-1)^{k-l}.$$

This expression is a special *power sum* of l^n with factors $\frac{1}{k!}\binom{k}{l}(-1)^{k-l}$. In general, we will consider the set $\mathcal{E}$ of power sums with rational number coefficients:

$$\mathcal{E} = \left\{ \sum_{l=0}^{k} c_l a_l^n \,\middle|\, k \geq 1,\, c_l \in \mathbb{Q},\, a_l \in \mathbb{N},\, a_1 > a_2 > \cdots > a_k > 0 \right\}.$$

Here $\mathbb{N}$ is the set of positive integers and $\mathbb{Q}$ is the set of rational numbers.

In these sums k is the *rank*, c_i's are the coefficients, and the a_i's are the *roots* of the given power sum. In particular, by (2.20), $\left\{ {n \atop k} \right\} \in \mathcal{E}$ is a power sum of rank k, and the roots are $1, 2, \ldots, k$.

We now describe a mapping which is important in the study of general power sums. Let $\mathbb{Q}[x_p \mid p \text{ is a prime}]$ be the set of polynomials with rational coefficients and variables (indeterminates) indexed by the primes. For example,

$$\frac{3}{2}x_2^5 - 14x_5^8 + \frac{4}{3}x_{11}^2$$

is an element of the above set. Such polynomials can be added and multiplied in the usual way.

We define $\psi : \mathcal{E} \to \mathbb{Q}[X_p \mid p \text{ is a prime}]$ which maps power sums into the above-defined polynomials such that

$$\psi(c_1 1^n) = c, \quad \psi(cp^n) = cX_p,$$

and we extend ψ multiplicatively and additively to more general sums in $\mathcal{E}$.

For example,

$$\psi\left(2\cdot 1^5 - 3\cdot 5^5 + \frac{1}{2}\cdot 10^5 + 4\cdot 121^5\right) = 2 - 3x_5 + \frac{1}{2}x_2 x_5 + 4x_{11}^2.$$

This map is clearly bijective. Probably the first appearance of this mapping is in [496]. Section 3.2 in [470] contains more information about ψ.

We will need a proposition with respect to the general power sums. Let $\alpha, \beta \in \mathcal{E}$ such that α/β is an integer number for infinitely many exponents n. Then there exists a third power sum $\gamma \in \mathcal{E}$ such that

$$\alpha = \beta\gamma.$$

The above statement, which is due to Corvaja and Zannier, is Theorem 1 in [177].

The proof of the Klazar – Luca theorem

The Klazar–Luca theorem is an application of the Corvaja–Zannier result [333]. (For similar applications, see [383].)

We need, however, one more ingredient: the Bertrand postulate. It is a statement saying that for each $n \geq 2$ there is a prime number p such that $n/2 < p \leq n$. For a proof, see [11].

To prove the Klazar–Luca theorem, we suppose that

$$\left\{ {n \atop k_2} \right\} = c\left\{ {n \atop k_1} \right\} \tag{13.22}$$

holds for infinitely many n $(1 < k_1 < k_2)$. Then by the Corvaja-Zannier result there is a non-zero γ power sum in $\mathcal{E}$ such that

$$\left\{ {n \atop k_2} \right\} = \left\{ {n \atop k_1} \right\}\gamma.$$

The leading roots of the left- and right-hand sides are k_2^n and $k_1^n \cdot a^n$, where a is the leading root of γ, and is a positive integer. It comes that $k_2 = ak_1$. Since $k_1 < k_2$, it also follows that $k_2 \geq 2k_1$. We use Bertrand's postulate to fix a prime p such that $k_1 < p \leq k_2$. Next, we take the above-introduced ψ map:

$$\psi\left(\left\{ {n \atop k_2} \right\}\right) = \psi\left(\left\{ {n \atop k_1} \right\}\right)\psi(\gamma). \tag{13.23}$$

This is an equality among polynomials in $\mathbb{Q}[X_p \mid p \text{ is a prime}]$. Consider the variable x_p (which is the image of p^n by ψ). Since $k_1 < p$, x_p does not appear in $\psi\left(\left\{ {n \atop k_1} \right\}\right)$, but it appears only once in $\psi\left(\left\{ {n \atop k_2} \right\}\right)$ as the image of

$$\psi\left(\frac{1}{k_2!}\binom{k_2}{p}(-1)^{k_2-p}p^n\right).$$

If x_p does not appear in $\psi(\gamma)$, then equality (13.23) is impossible to hold. If, in turn, $\psi(\gamma)$ appears a term with x_p, then (13.23) is still not possible, because then $\psi\left(\left\{{n \atop k_1}\right\}\right)\psi(\gamma)$ would contain a monomial of the form Mx_p with a non-constant monomial coming from $\psi\left(\left\{{n \atop k_1}\right\}\right)$ and containing some x_q. But, as we saw, the left-hand side contains only monomial of the form constant times x_p. This contradiction results in the fact that our original assumption about the existence of infinitely many n's in (13.22) cannot hold.

Exercises

1. Prove inequality (13.10).

2. Show that $N(3) = N(4) = N(5) = 2$, $N(6) = 3$.

3. Based on Tovey's result, prove the following equalities (which are the original ones, found by Lind). Let i be a positive integer, and let

$$n = F_{2i}F_{2i+1},$$
$$k = F_{2i-2}F_{2i+1}.$$

Then

$$\binom{n}{k} = \binom{n-1}{k+1}.$$

4. Prove that (13.8) indeed holds.

5. Give a proof, independent from the one in the text, that $F_i L_i + F_i^2$ (or, equivalently, $F_{2i} + F_i^2$) is always even.

6. Prove that a given number a can occur in the first or second kind Stirling triangle only up to the nth line, where

$$n = \left\lceil \frac{1}{2}(1 + \sqrt{1 + 8a}) \right\rceil.$$

(See [51] for the details.)

7. (Research problem.) Are there any triangular-factorial numbers apart from 1, 6 and 120?

Outlook

1. D. M. Kane [316] studied the number of solutions to $\binom{n}{m} = t$ by using the interesting function f defined implicitly by

$$\binom{f(x)}{x} = t.$$

2. See [81] for the most recent treatise on the problem $\binom{n}{k} = \binom{m}{l}$, and also on the generalization $\binom{n}{k} - \binom{m}{l} = d > 0$. In the first case, the quadruple (n, k, m, l) is called *collision*, in the second case it is a *near collision*.

3. Bertrand's postulate has far-reaching generalizations. For example, in place of the interval n and $2n = n + n$ we can ask whether there is a prime between n and $n + n^\alpha$ where $\alpha < 1$. The most recent result [43] in this direction says that $\alpha = 21/40$ still guarantees the existence of a prime between n and $n + n^\alpha$, at least for large n.

   Another possibility of generalization was observed by Ramanujan and Erdős, independently. For any positive integer k there is an integer $n(k)$ such that there are k primes between n and $2n$ when $n > n(k)$ [220, 484].

4. There is a statement related to but different from Bertrand's postulate: if $n > 1$, then there is a prime between n^2 and $(n+1)^2$. This statement, originally conjectured by Legendre[7], has not been proved yet.

5. The non-integer property of the harmonic numbers can be proven by using Bertrand's postulate. See Exercise 21 of Section 6 in [262] and its solution in the "Answers to Exercises" section.

6. The Klazar-Luca theorem was proved to have much more general power sums than that of the second kind Stirling numbers. Moreover, it is proved also that the diophantine equation

$$\left\{ {n \atop k_1} \right\}^{a_1} \left\{ {n \atop k_2} \right\}^{a_2} \cdots \left\{ {n \atop k_m} \right\}^{a_m} = x^d$$

   has only finitely many solutions in $n > 0$ and $x > 0$ for $d > 1$, and $a_1, \ldots, a_m$ positive integers not all divisible by d, and $1 < k_1 < k_2 \cdots < k_m$. See [333] for the details.

---

[7]Adrien-Marie Legendre (1752-1833), is a French mathematician who worked on a wide range of mathematical problems. He wrote the book *Éléments de Géométrie* on elementary geometry, simplifying and rearranging the statements and proofs of Euclid's *Elements*. This text was the standard book on the subject for a long time.

7. There are some effective finiteness results[8] with respect to the Stirling numbers.

    Pintér [464] proved that the number of solutions of the equation

$$\left\{ {x \atop a} \right\} = \left\{ {y \atop b} \right\}$$

    (with $1 < a < b$ given and $x > a$, $y > b$ unknowns) is finite, and the maximum of x and y is bounded from above by an expression depending only on a and b.

    Another result of Pintér [465] is the following: if $\left\{ {n \atop n-a} \right\}$ is an m-th power ($m \geq 3$) for some $n > a$, then n is bounded from above by a constant which is computable explicitly if one follows the steps of the proof. Such constants are called *effectively computable*. An analogous result was proved in the same paper, stating that if $\left\{ {n \atop n-a} \right\}$ ($n > a$) falls into a set S of positive integers which are divisible only by a fixed finite collection of primes, then n is bounded from above by an effectively computable constant. Remarkably, the bounds in the statements depend only on a and S, but not on the magnitude of the right-hand side.

    Brindza and Pintér [98] established an effective finiteness result with respect to both equations

$$\left[{x \atop x-k} \right] = ay^z,$$

$$\left\{ {x \atop x-k} \right\} = ay^z$$

    with k, a fixed and $x > k$, $|y| > 1$, and $z \geq 2$ unknowns.

8. The Erdős-Moser equation is

$$\sum_{i=1}^{m-1} i^n = m^n.$$

    It was verified [105] up to very large m that there are no solutions in n and m except

$$1^1 + 2^1 = 3^1.$$

    There are a plenty of variations on this theme, not only equations but congruences were also studied. See, for example, [14].

9. Diophantine questions regarding the power sums of Chapter 5 were recently studied in [49, 78, 274, 481, 482].

---

[8]In the theory of diophantine equations "effective finiteness" means that the solutions are finite in number and they can be bounded by an explicit bound which is a function of the fixed parameters. A finiteness result is ineffective if it is known that there are finitely many results but no bound is given.

Appendix

Basic combinatorial notions

Three basic notions in elementary combinatorics are *permutation*, *variation*, and *combination*.

To be able to follow most parts of this book, it is necessary that the reader be familiar with these notions.

Definition 14.4.1. *A given order of n objects is a* permutation *of them.*

The total number of permutations on n objects is denoted by $n!$, and it can be calculated by multiplying the first n positive integers:

$$n! = 1 \cdot 2 \cdots n.$$

It is easy to see why this holds: to put n elements in an order we first determine where the first element goes. We have n positions, so n choices. To choose a position for the second element, we have $n-1$ places, and so on. To put the last element down, we have only one position.

Definition 14.4.2. *If among the n objects there are $l_1, l_2, \ldots, l_k$ not distinguishable[9], then a given order of these is a* permutation with repetition.

Suppose that $l_1 + l_2 + \cdots = l_k = n$. If it is necessary, we can add ones to the sequence. The total number of these is

$$\binom{n}{l_1, l_2, \ldots, l_k} = \frac{n!}{l_1! l_2! \cdots l_k!}.$$

This is so, because in any permutation the order of the non-distinguishable elements is not counted. Therefore, the total number of permutations, $n!$, must be divided by the permutation of these non-distinguishable objects. These are $l_1!, l_2!, \ldots, l_k!$, respectively. The symbol on the left-hand side is the *multinomial coefficient*.

Definition 14.4.3. *If we choose k objects from n different objects such that the order of the k elements matters, then the chosen objects form a* variation.

---

[9]Imagine 3 red balls and 2 blue balls and a green ball. Then $i_1 = 3$, $i_2 = 2$, $i_3 = 1$.

There are

$$n(n-1)(n-2)\cdots(n-(k-1)) = \frac{n!}{(n-k)!} = n^{\underline{k}}$$

variations without repetition for any given n and k. (For the last notation, see the falling factorial on p. 61.)

We can choose the first element from that of n in n ways. The second element is chosen $n-1$ ways, and so on. The kth element can be chosen $n-(k-1)$ ways. This argument justifies the formula for $n^{\underline{k}}$.

Definition 14.4.4. *If we choose k objects from n such that the order counts, and an element can be chosen multiple times*[10]*, then we get a* variation with repetition.

In each step, we choose from n elements and we altogether choose k times, so the number of variations with repetition is

$$n^k.$$

If we do not consider the order of the chosen objects we get combination instead of variation.

Definition 14.4.5. *If we choose k objects from n such that we do not regard the order, then we have a* combination.

For given n and k there are

$$\frac{n(n-1)(n-2)\cdots(n-(k-1))}{k!} = \frac{V_{n,k}}{k!} = \frac{n!}{k!(n-k)!} = \binom{n}{k}$$

combinations without repetition. The symbol $\binom{n}{k}$ is called *binomial coefficient*.

The order of the objects can be "forgotten" by dividing by their $k!$ order in a variation, so the above formula for $\binom{n}{k}$ comes.

If $k=2$ in the multinomial coefficient such that $l_1 + l_2 = n$, we have that

$$\binom{n}{l_1} = \binom{n}{l_2} = \binom{n}{l_1, l_2}.$$

Definition 14.4.6. *If we choose k objects from n such that the order of the elements does not count, and an element can be chosen repeatedly, then we get a* combination with repetition.

The number of repetitions is

$$\binom{n+k-1}{k}.$$

---

[10]We choose an element, register it on a piece of paper, then throw it back to the urn.

The proof of this is slightly more complicated than the above formulas for permutations and variations. Note that

$$\binom{n+k-1}{k} = \frac{n(n+1)\cdots(n+k-1)}{k!}.$$

Label the chosen objects (which can be repeated) and put them in order:

$$0 \leq o_1 \leq o_2 \leq \cdots \leq o_k \leq n.$$

Next, increase the labels like this: $o_1, o_2 + 1, \ldots, o_k + k - 1$. We now have different labels for each object:

$$0 \leq o_1 < o_2 + 1 < \cdots < o_k + k - 1 \leq n + k - 1.$$

Thus, the problem is transformed to choose k labels without repetition while there are $n + k - 1$ labels to choose from. This is a combination, hence the result follows.

The polynomial theorem

The often used binomial theorem

$$(x_1 + x_2)^k = \sum_{j=0}^{k} \binom{k}{j} x_1^k x_2^{n-k}$$

easily generalizes to sums of more terms. This is the *polynomial theorem*:

$$(x_1 + x_2 + \cdots + x_n)^k = \sum_{j_1+j_2+\cdots+j_n=k} \frac{k!}{j_1! j_2! \cdots j_n!} x_1^{j_1} x_2^{j_2} \cdots x_n^{j_n}.$$

Writing out the power as

$$(x_1 + x_2 + \cdots + x_n)(x_1 + x_2 + \cdots + x_n) \cdots (x_1 + x_2 + \cdots + x_n)$$

we see that we must multiply each term with every other term. This means that the expanded power constitutes products of k factors. If x_1 is chosen j_1-times from the k parentheses, x_2 is chosen j_2-times from the $k - j_1$ parentheses, and finally x_n is chosen j_n-times from the $k - j_1 - j_2 - \cdots - j_{n-1}$ parentheses, then the number of possible choices is

$$\binom{k}{j_1}\binom{k - j_1}{j_2}\binom{k - j_1 - j_2}{j_3} \cdots \binom{k - j_1 - j_2 - \cdots - j_{n-1}}{j_n}.$$

It can be seen by the definition of the binomial coefficients (see the combinations above) that this binomial product equals to the much simpler

$$\frac{k!}{j_1! j_2! \cdots j_n!}.$$

Since the sum of powers of the several x_is must be k, it is necessary that $j_1 + \cdots + j_n = k$. Summing over these possibilities, we get the polynomial theorem.

The above long binomial product can be written in the much more compact form, by applying the multinomial coefficient:

$$\binom{k}{j_1}\binom{k - j_1}{j_2} \cdots \binom{k - j_1 - j_2 - \cdots - j_{n-1}}{j_n} = \binom{k}{j_1, j_2, \ldots, j_n}.$$

When $n = 2$, we get the *binomial theorem*:

$$(x_1 + x_2)^k = \sum_{j_1+j_2=k} \frac{k!}{j_1! j_2!} x_1^{j_1} x_2^{j_2}.$$

The sum can be reduced to a simple one index sum:

$$(x_1 + x_2)^k = \sum_{j=0}^{k} \frac{k!}{j!(k-j)!} x_1^j x_2^k{}^j = \sum_{j=0}^{k} \binom{k}{j} x_1^j x_2^{k-j},$$

which, after setting $x_1 = x_2 = 1$ gives the nice identity

$$2^k = \sum_{j=0}^{k} \binom{k}{j} \tag{14.1}$$

An independent combinatorial proof is worth presenting. Take a set of k elements, and choose $0, 1, 2, \ldots, k$ elements from it (this is represented by the right-hand side). In each occasion, we can make a list in which we mark in the different positions from 1 to k which elements were chosen and which were not. There are 2^k such possible list (for each element, we register 1 or 0, depending on whether it was chosen or not).

The Lambert W function

The Lambert W function [176] can be defined implicitly as the solution of the transcendental equation

$$W(z)e^{W(z)} = z. \tag{14.2}$$

Equation (14.2) has infinitely many solutions when $z \neq 0$, and these solutions can be indexed by the integers. In function theory language, we say that the W function has infinitely many branches. These are denoted by $W_k(z)$. Except for W_{-1} and W_0, all the branches are complex-valued and do not attain real values at all. $W_{-1}(z)$ is real when $-\frac{1}{e} \leq z < 0$, and $W_0(z)$ is real when $z \in [-1/e, +\infty[$. This latter branch is called the "principal branch." If we do not write out the branch index, it is understood that we are talking about the principal branch.

It can be seen that

$$W(0) = 0, \quad W(e) = 1, \quad W\left(-\frac{1}{e}\right) = -1, \quad W\left(-\frac{\log(2)}{2}\right) = -\log(2).$$

A notable special constant appears at $z = 1$ [230, Section 6.11]:

$$W(1) = \Omega = 0.567143290409784\ldots.$$

The W function is widely applied in physics and mathematics, see the classical paper [176], and [169, 422] for newer citations.

In combinatorics, the W function often appears in approximations. Therefore, it is useful to know that [176]

$$W(z) \sim \log z - \log \log z \quad (z \to \infty). \tag{14.3}$$

Formulas

Formulas with respect to the Stirling numbers of the first kind

Recursion:

$$\begin{bmatrix} n \\ k \end{bmatrix} = (n-1) \begin{bmatrix} n-1 \\ k \end{bmatrix} + \begin{bmatrix} n-1 \\ k-1 \end{bmatrix}; \quad \begin{bmatrix} 0 \\ 0 \end{bmatrix} = 1; \quad \begin{bmatrix} n \\ 0 \end{bmatrix} = 0 \ (n > 0).$$

Special values:

$$\begin{bmatrix} n \\ 1 \end{bmatrix} = (n-1)!,$$

$$\begin{bmatrix} n \\ 2 \end{bmatrix} = (n-1)! H_{n-1},$$

$$\begin{bmatrix} n \\ 3 \end{bmatrix} = \frac{(n-1)!}{2} \left[H_{n-1}^2 - \left(\frac{1}{1^2} + \frac{1}{2^2} + \cdots + \frac{1}{(n-1)^2} \right) \right],$$

$$\begin{bmatrix} n \\ 4 \end{bmatrix} = \frac{(n-1)!}{6} \left[H_{n-1}^3 - 3 H_n \left(\frac{1}{1^2} + \frac{1}{2^2} + \cdots + \frac{1}{(n-1)^2} \right) + \left(\frac{1}{1^3} + \frac{1}{2^3} + \cdots + \frac{1}{(n-1)^3} \right) \right],$$

$$\begin{bmatrix} n \\ n-3 \end{bmatrix} = \binom{n}{6} + 8 \binom{n+1}{6} + 6 \binom{n+2}{6},$$

$$\begin{bmatrix} n \\ n-2 \end{bmatrix} = 2 \binom{n}{3} + \frac{1}{2} \binom{n}{2} \binom{n-2}{2},$$

$$\begin{bmatrix} n \\ n-1 \end{bmatrix} = \binom{n}{2},$$

$$\begin{bmatrix} n \\ n \end{bmatrix} = 1.$$

Generating functions:

$$\sum_{k=0}^{n} \begin{bmatrix} n \\ k \end{bmatrix} x^k = x(x+1)(x+2)\cdots(x+n-1) = x^{\overline{n}},$$

$$\sum_{n=0}^{\infty} \begin{bmatrix} n \\ k \end{bmatrix} \frac{x^n}{n!} = \frac{1}{k!} \ln\left(\frac{1}{1-x}\right)^k,$$

$$\sum_{n=0}^{\infty} \left(\sum_{k=0}^{n} \begin{bmatrix} n \\ k \end{bmatrix} y^k \right) \frac{x^n}{n!} = \frac{1}{(1-x)^y},$$

$$\sum_{k=0}^{n} \begin{bmatrix} n+1 \\ n+1-k \end{bmatrix} x^k = (1+x)(1+2x)\cdots(1+nx).$$

Orthogonality:

$$\sum_{n=k}^{m} \begin{bmatrix} m \\ n \end{bmatrix} \begin{Bmatrix} n \\ k \end{Bmatrix} = \begin{cases} 1, & \text{if } m = k; \\ 0, & \text{otherwise.} \end{cases}$$

Asymptotics (more on p. 177):

$$\begin{bmatrix} n \\ k \end{bmatrix} \sim \frac{(n-1)!}{(k-1)!} (\ln n + \gamma)^{k-1}.$$

Congruences:

$$\begin{bmatrix} n \\ k \end{bmatrix} \equiv \binom{\lfloor \frac{n}{2} \rfloor}{k - \lceil \frac{n}{2} \rceil} \quad (\text{mod } 2),$$

$$\begin{bmatrix} p \\ k \end{bmatrix} \equiv 0 \quad (\text{mod } p) \quad (1 < k < p),$$

$$\begin{bmatrix} p \\ 1 \end{bmatrix} \equiv -1 \quad (\text{mod } p),$$

$$\begin{bmatrix} p \\ p \end{bmatrix} \equiv 1 \quad (\text{mod } p),$$

$$\begin{bmatrix} p-1 \\ k \end{bmatrix} \equiv 1 \quad (\text{mod } p) \quad (1 \le k \le p),$$

$$\begin{bmatrix} mp \\ k \end{bmatrix} \equiv 0 \quad (\text{mod } p) \quad (m > 1, \ k \ne jp),$$

$$\begin{bmatrix} p+1 \\ k \end{bmatrix} \equiv 0 \quad (\text{mod } p) \quad (2 < k \le p),$$

$$\begin{bmatrix} p+2 \\ k \end{bmatrix} \equiv 0 \quad (\text{mod } p) \quad (3 < k \le p),$$

$$\vdots$$

$$\begin{bmatrix} 2p-2 \\ p \end{bmatrix} \equiv 0 \quad (\text{mod } p).$$

Wolstenholme-congruence:

$$\begin{bmatrix} p \\ 2 \end{bmatrix} \equiv 0 \quad (\text{mod } p^2).$$

Other formulas:

$$\begin{bmatrix} n \\ k \end{bmatrix} = \left\{ \begin{matrix} -k \\ -n \end{matrix} \right\},$$

$$\begin{bmatrix} n+1 \\ k+1 \end{bmatrix} = \sum_{m=0}^{n} \begin{bmatrix} n \\ m \end{bmatrix} \binom{m}{k},$$

$$\begin{bmatrix} n \\ k \end{bmatrix} = \sum_{m=0}^{n} \begin{bmatrix} n+1 \\ m+1 \end{bmatrix} \binom{m}{k} (-1)^{m-k},$$

$$\begin{bmatrix} n+1 \\ k+1 \end{bmatrix} = \sum_{m=0}^{n} \begin{bmatrix} m \\ k \end{bmatrix} n^{\underline{n-m}} = n! \sum_{m=0}^{n} \begin{bmatrix} m \\ k \end{bmatrix} / m!,$$

$$\begin{bmatrix} n+k+1 \\ k \end{bmatrix} = \sum_{m=0}^{n} (n+m) \begin{bmatrix} n+m \\ m \end{bmatrix},$$

$$\begin{bmatrix} n \\ n-k \end{bmatrix} = \sum_{m=0}^{n} \binom{k-n}{k+m} \binom{k+n}{n+m} \left\{ \begin{matrix} k+m \\ m \end{matrix} \right\},$$

$$\begin{bmatrix} n \\ k \end{bmatrix} = \frac{n!}{k!} \sum_{j_1+j_2+\cdots+j_k=n} \frac{1}{j_1 j_2 \cdots j_k},$$

$$\begin{bmatrix} n \\ k \end{bmatrix} = \sum_{\substack{j_1+2j_2+\cdots+nj_n=n \\ j_1+j_2+\cdots+j_n=k}} \frac{n!}{j_1! j_2! \cdots j_n!} \left(\frac{1}{1}\right)^{j_1} \left(\frac{1}{2}\right)^{j_2} \cdots \left(\frac{1}{n}\right)^{j_n},$$

$$\begin{bmatrix} n \\ k \end{bmatrix} = \sum_{j=0}^{k} \begin{bmatrix} n-j \\ k-j \end{bmatrix}_{\geq 2} \binom{n}{j}.$$

Schlömilch (1852) [511], see [83, pp. 113-114.] or [134, p. 290] for proofs:

$$\begin{bmatrix} n \\ k \end{bmatrix} = \sum_{m=0}^{n-k} (-1)^{n+m-k} \binom{n-1+m}{n-k+m} \binom{2n-k}{n-k-m} \left\{ \begin{matrix} n-k+m \\ m \end{matrix} \right\},$$

Determinantal formula:

$$\begin{bmatrix} n+1 \\ k \end{bmatrix} = \frac{1}{1! 2! 3! \cdots (n-1)!} \begin{vmatrix} 1 & 1 & 1 & \cdots & 1 & 1 & \cdots & 1 \\ 1 & 2 & 2^2 & \cdots & 2^{k-1} & 2^{k+1} & \cdots & 2^n \\ 1 & 3 & 3^2 & \cdots & 3^{k-1} & 3^{k+1} & \cdots & 3^n \\ \cdots & \cdots & \cdots & \cdots & \cdots & \cdots & \cdots & \cdots \\ 1 & n & n^2 & \cdots & n^{k-1} & n^{k+1} & \cdots & n^n \end{vmatrix}.$$

Schläfli's formula (1852) [510], see also [474]:

$$\overline{\begin{bmatrix} n \\ n-k \end{bmatrix}} = \binom{n-1}{k} \sum_{j=0}^{k} (-1)^j \binom{k+1}{j+1} \frac{\left\{ \begin{smallmatrix} k+jn \\ jn \end{smallmatrix} \right\}}{\binom{k+jn}{k}}.$$

The dual of Spivey's formula:

$$(n+m)! = \sum_{k=0}^{n} \sum_{j=0}^{m} m^{\overline{n-k}} \begin{bmatrix} m \\ j \end{bmatrix} \binom{n}{k} k!.$$

Formulas with respect to the Stirling numbers of the second kind

Recursion:

$$\left\{{n \atop k}\right\} = \left\{{n-1 \atop k-1}\right\} + k\left\{{n-1 \atop k}\right\}; \quad \left\{{0 \atop 0}\right\} = 1, \quad \left\{{n \atop 0}\right\} = 0 \,(n > 0).$$

Special values:

$$\left\{{n \atop 1}\right\} = 1,$$

$$\left\{{n \atop 2}\right\} = 2^{n-1} - 1,$$

$$\left\{{n \atop 3}\right\} = \frac{1}{2}\left(3^{n-1} - 2^n + 1\right),$$

$$\left\{{n \atop 4}\right\} = \frac{1}{6}4^{n-1} - \frac{1}{2}3^{n-1} + 2^{n-2} - \frac{1}{6},$$

$$\left\{{n \atop n-3}\right\} = \binom{n}{4} + 10\binom{n}{5} + 15\binom{n}{6},$$

$$\left\{{n \atop n-2}\right\} = \binom{n}{3} + 3\binom{n}{4},$$

$$\left\{{n \atop n-1}\right\} = \binom{n}{2},$$

$$\left\{{n \atop n}\right\} = 1.$$

Generating functions:

$$\sum_{k=0}^{n}\left\{{n \atop k}\right\}x(x-1)(x-2)\cdots(x-k+1) = x^n,$$

$$\sum_{n=0}^{\infty}\left\{{n \atop k}\right\}\frac{x^n}{n!} = \frac{1}{k!}(e^x - 1)^k,$$

$$\sum_{n=0}^{\infty}\left\{{n \atop k}\right\}x^n = \frac{x^k}{(1-x)(1-2x)\cdots(1-kx)},$$

$$\sum_{n=0}^{\infty}\left(\sum_{k=0}^{n}\left\{{n \atop k}\right\}y^k\right)\frac{x^n}{n!} = e^{y(e^x - 1)}.$$

Orthogonality:

$$\sum_{n=k}^{m} \overline{\begin{bmatrix} m \\ n \end{bmatrix}} \begin{Bmatrix} n \\ k \end{Bmatrix} = \begin{cases} 1, & \text{if } m = k; \\ 0, & \text{otherwise.} \end{cases}$$

Asymptotics:

$$\begin{Bmatrix} n \\ k \end{Bmatrix} \sim \frac{k^n}{k!}.$$

Congruences:

$$\begin{Bmatrix} n \\ k \end{Bmatrix} \equiv \binom{\lceil \frac{k}{2} \rceil + n - k - 1}{n - k} \pmod 2,$$

$$\begin{Bmatrix} p \\ k \end{Bmatrix} \equiv 0 \pmod p \quad (1 < k < p),$$

$$\begin{Bmatrix} p \\ 1 \end{Bmatrix} \equiv 1 \pmod p,$$

$$\begin{Bmatrix} p \\ p \end{Bmatrix} \equiv 1 \pmod p.$$

$$\begin{Bmatrix} p + 1 \\ k \end{Bmatrix} \equiv 0 \pmod p \quad (2 < k \le p),$$

$$\begin{Bmatrix} p + 2 \\ k \end{Bmatrix} \equiv 0 \pmod p \quad (3 < k \le p),$$

$$\vdots$$

$$\begin{Bmatrix} 2p - 2 \\ p \end{Bmatrix} \equiv 0 \pmod p.$$

Other formulas:

$$\left\{ {n \atop k} \right\} = \left[{-k \atop -n} \right],$$

$$\left\{ {n \atop k} \right\} = \frac{1}{k!} \sum_{l=0}^{k} \binom{k}{l} l^n (-1)^{k-l},$$

$$\left\{ {n \atop k} \right\} = \sum_{l=0}^{n} \binom{n}{l} \sum_{m=0}^{l-1} (-1)^m \left\{ {l-1-m \atop k-1} \right\},$$

$$\left\{ {n+1 \atop k+1} \right\} = \sum_{m=0}^{n} \binom{n}{m} \left\{ {m \atop k} \right\},$$

$$\left\{ {n \atop k} \right\} = \sum_{m=0}^{n} \left\{ {m+1 \atop k+1} \right\} \binom{m}{k} (-1)^{m-k},$$

$$\left\{ {n+1 \atop k+1} \right\} = \sum_{m=0}^{n} \left\{ {m \atop k} \right\} (k+1)^{n-m},$$

$$\left\{ {n+k+1 \atop k} \right\} = \sum_{m=0}^{n} m \left\{ {n+m \atop m} \right\},$$

$$\left\{ {n \atop n-k} \right\} = \sum_{m=0}^{n} \binom{k-n}{k+m} \binom{k+n}{n+m} \left[{k+m \atop m} \right],$$

$$\left\{ {n \atop k+l} \right\} \binom{k+l}{l} = \sum_{m=0}^{n} \left\{ {m \atop l} \right\} \left\{ {n-m \atop k} \right\} \binom{n}{k},$$

$$\left\{ {n \atop k} \right\} = \frac{n!}{k!} \sum_{j_1+j_2+\cdots+j_k=n} \frac{1}{j_1! j_2! \cdots j_k!},$$

$$\left[{n \atop k} \right] = \sum_{\substack{j_1+2j_2+\cdots+nj_n=n \\ j_1+j_2+\cdots+j_n=k}} \frac{n!}{j_1! j_2! \cdots j_n!} \left(\frac{1}{1!} \right)^{j_1} \left(\frac{1}{2!} \right)^{j_2} \cdots \left(\frac{1}{n!} \right)^{j_n},$$

$$\left\{ {n \atop k} \right\} = \sum_{j=0}^{k} \left\{ {n-j \atop k-j} \right\}_{\geq 2} \binom{n}{j},$$

$$\left\{ {n+m \atop k} \right\} = \sum_{i=0}^{k} \sum_{j=i}^{k} \binom{j}{i} \frac{(k-i)!}{(k-j)!} \left\{ {n \atop k-i} \right\} \left\{ {m \atop j} \right\}. \quad [588]$$

Determinantal formula:

$$\left\{{n\atop k}\right\} = \frac{1}{k!}\begin{vmatrix} 1 & 1 & 1 & \cdots & 1 & 1 & 1 \\ 1 & 2 & 2^2 & \cdots & 2^{k-3} & 2^{k-2} & 2^n \\ 1 & 3 & 3^2 & \cdots & 3^{k-3} & 3^{k-2} & 3^n \\ \cdots & \cdots & \cdots & \cdots & \cdots & \cdots & \cdots \\ 1 & k & k^2 & \cdots & k^{k-3} & k^{k-2} & k^n \end{vmatrix}.$$

Relations between the Stirling numbers of the second kind and Eulerian numbers:

$$\left\{{n\atop k}\right\} = \frac{1}{k!}\sum_{m=0}^{n}\left\langle{n\atop m}\right\rangle\binom{m}{n-k},$$

$$\left\langle{n\atop k}\right\rangle = \sum_{m=0}^{n} m!\left\{{n\atop m}\right\}\binom{n-m}{k}(-1)^{n-m-k}.$$

The inverse of Schläfli's formula [474]:

$$\left\{{n+k\atop n}\right\} = \binom{n+k}{k}\sum_{j=1}^{k}(-1)^{j}\binom{k+1}{j+1}\frac{\left[{jn\atop jn-k}\right]}{\binom{jn-1}{k}}.$$

Spivey's formula (2008):

$$B_{n+m} = \sum_{k=0}^{n}\sum_{j=0}^{m} j^{n-k}\left\{{m\atop j}\right\}\binom{n}{k}B_k.$$

Formulas with respect to the r-Stirling numbers of the first kind

Recursion:

$$\left[\begin{matrix}n\\k\end{matrix}\right]_r = 0, \quad \text{if} \quad n < r;$$

$$\left[\begin{matrix}n\\k\end{matrix}\right]_r = 1, \quad \text{if} \quad n = r \text{ and } k = r;$$

$$\left[\begin{matrix}n\\k\end{matrix}\right]_r = 0, \quad \text{if} \quad n = r \text{ and } k \neq r.$$

Special values:

$$\left[\begin{matrix}n\\r\end{matrix}\right]_r = r^{\overline{n-r}},$$

$$\left[\begin{matrix}n\\n\end{matrix}\right]_r = 1 \quad (n \geq r),$$

$$\left[\begin{matrix}n+r\\r+1\end{matrix}\right]_r = n! H_n^r.$$

Generating functions:

$$\sum_{k=0}^{n} \left[\begin{matrix}n\\k\end{matrix}\right]_r x^k = x^r(x+r)(x+r+1)\cdots(x+n-1),$$

$$\sum_{n=0}^{\infty} \left[\begin{matrix}n+r\\k+r\end{matrix}\right]_r \frac{x^n}{n!} = \frac{1}{k!}\left(\frac{1}{1-x}\right)^r \ln^k\left(\frac{1}{1-x}\right),$$

$$\sum_{n=0}^{\infty}\left(\sum_{k=0}^{n}\left[\begin{matrix}n\\k\end{matrix}\right]y^k\right)\frac{x^n}{n!} = \frac{1}{(1-x)^{r+y}}.$$

Orthogonality:

$$\sum_{n=k}^{m}\left[\begin{matrix}m\\n\end{matrix}\right]_r\left\{\begin{matrix}n\\k\end{matrix}\right\}_r = \begin{cases} 1, & \text{if } m = k; \\ 0, & \text{otherwise.} \end{cases}$$

Generalized orthogonality:

$$\sum_{k}\left[\begin{matrix}n+r\\k+r\end{matrix}\right]_r\left\{\begin{matrix}k+p\\m+p\end{matrix}\right\}_p(-1)^{m+k} = \binom{n}{m}(r-p)^{\overline{n-m}}.$$

Other formulas[11]:

$$\left[{n\atop k}\right]_r = \frac{1}{r-1}\left(\left[{n\atop k-1}\right]_{r-1} - \left[{n\atop k-1}\right]_r\right),$$

$$\left[{n\atop n-k}\right]_r = \sum_{r<i_1<\cdots<i_k<n} i_1 i_2\cdots i_k,$$

$$\left[{n\atop k}\right]_r = \sum_{m=0}^{n-r}\binom{n-r}{m}\left\{{n-p-m\atop k-p}\right\}_{r-p} p^{\overline{m}}\quad (r\geq p\geq 0),$$

$$\left[{n\atop k}\right]_r = \sum_{m=0}^{n-r}\binom{n-r}{m}\left\{{n-p-m\atop k-p}\right\}_{r-p} p^{\overline{m}}\quad (r\geq p\geq 0),$$

$$\left[{n+r\atop k+r}\right]_r = \sum_{m}\binom{n}{m}\left[{n-m+p\atop k+p}\right]_p (r-p)^{\overline{m}},$$

$$\left[{n+r\atop k}\right]_r = n\left[{n+r-1\atop k}\right]_r + \left[{n+r-1\atop k-1}\right]_{r-1},$$

$$\left[{n\atop k}\right]_r = \sum_{t=r}^{k}\left[{m\atop t}\right]_r\left[{n\atop k+m-t}\right]_m,$$

$$\left[{n\atop k}\right]_r = \sum_{m}\left[{p\atop p-k}\right]_r\left[{n\atop k+m}\right]_p\quad (r\leq p\leq n),$$

$$\left[{n\atop k}\right]_r = \sum_{m}\left[{n\atop k-r+m}\right]_p\left\{{m-1\atop r-1}\right\}_p (-1)^{k+r}\quad (n\geq r>p\geq 0),$$

$$\binom{k+m}{m}\left[{n+r+s\atop k+m+r+s}\right]_{r+s} = \sum_{l=0}^{n}\binom{n}{l}\left[{l+r\atop k+r}\right]_r\left[{n-l+s\atop m+s}\right]_s,$$

$$\binom{k-r}{l}\left[{n+r\atop k}\right]_r = \sum_{t=k-r-l}^{n\ l}\binom{n}{t}\left[{t+r-m\atop k-m-l}\right]_{r-m}\left[{n+m-t\atop m+l}\right]_m,$$

$$\left[{n\atop r+1}\right]_r = \left[{n\atop m+1}\right]_m\left[{m\atop r}\right]_r + \left[{m\atop r+1}\right]_r\left[{n\atop m}\right]_m\quad (0\geq r\geq m\geq n),$$

$$\left[{n+r\atop k}\right]_r = \left[{n\atop k-r}\right]_1 + n\sum_{t=1}^{r}\left[{n+r-t\atop k-t+1}\right]_{r-t+1},$$

$$\left[{n+r\atop k}\right]_r = \sum_{t=0}^{m}\binom{m}{t}\frac{n!}{(n+t-m)!}\left[{n+r-m\atop k-t}\right]_{r-t}\quad (0\leq m\leq r).$$

---

[11]Those summations which do not contain explicit limits are always running on the widest possible range.

The r-version of the Schlömilch formula [172]:

$$\begin{bmatrix} n+r \\ k+r \end{bmatrix}_r = \sum_{m=k}^{n} \sum_{h=0}^{n-k} \sum_{j=0}^{h} (-1)^{m+k+h+j} \binom{n}{m} \binom{h}{j} \cdot$$

$$\binom{m-1+h}{m-k+h} \binom{2m-k}{m-k-h} \frac{(h-j)^{m-k+h}}{h!} r^{\overline{n-m}}.$$

The r-Stirling numbers of the first kind are expressible in terms of the classical Stirling numbers:

$$\begin{bmatrix} n+r \\ k+r \end{bmatrix}_r = \sum_{j=k}^{n} \binom{n}{j} \begin{bmatrix} j \\ k \end{bmatrix} r^{\overline{n-j}};$$

and vice versa:

$$\begin{bmatrix} n+r \\ k \end{bmatrix} = \sum_{l=0}^{k} \begin{bmatrix} r \\ l \end{bmatrix} \begin{bmatrix} n+r \\ k-l+r \end{bmatrix}_r.$$

Formulas with respect to the r-Stirling numbers of the second kind

Recursion:

$$\left\{ {n \atop k} \right\}_r = 0, \quad \text{if} \quad n < r;$$

$$\left\{ {n \atop k} \right\}_r = 1, \quad \text{if} \quad n = r \text{ and } k = r;$$

$$\left\{ {n \atop k} \right\}_r = 0, \quad \text{if} \quad n = r \text{ and } k \neq r.$$

$$\left\{ {n \atop k} \right\}_r = \left\{ {n-1 \atop k-1} \right\}_r + k \left\{ {n-1 \atop k} \right\}_r$$

Special values:

$$\left\{ {n \atop r} \right\}_r = r^{n-r},$$

$$\left\{ {n \atop r+1} \right\}_r = (r+1)^{n-r} - r^{n-r},$$

$$\left\{ {n \atop r+2} \right\}_r = \frac{1}{2}(r+2)^{n-r} - (r+1)^{n-r} + \frac{1}{2}r^{n-r},$$

$$\left\{ {n \atop n} \right\}_r = 1 \quad (n \geq r).$$

Generating functions:

$$\sum_{k=0}^{n} \left\{ {n+r \atop k+r} \right\}_r x(x-1)\cdots(x-k+1) = (x+r)^n,$$

$$\sum_{n=0}^{\infty} \left\{ {n+r \atop k+r} \right\}_r \frac{x^n}{n!} = \frac{e^{rx}}{k!}(e^x - 1)^k,$$

$$\sum_{n=0}^{\infty} \left\{ {n \atop k} \right\}_r x^n = \frac{x^k}{(1-rx)(1-(r+1)x)\cdots(1-kx)},$$

$$\sum_{n=0}^{\infty} \left(\sum_{k=0}^{n} \left\{ {n+r \atop k+r} \right\}_r y^k \right) \frac{x^n}{n!} = e^{y(e^x - 1)+rx}.$$

Orthogonality:

$$\sum_{n=k}^{m} \left[{m \atop n} \right]_r \left\{ {n \atop k} \right\}_r = \begin{cases} 1, & \text{if } m = k; \\ 0, & \text{otherwise.} \end{cases}$$

Generalized orthogonality:

$$\sum_k \begin{bmatrix} n+r \\ k+r \end{bmatrix}_r \begin{Bmatrix} k+p \\ m+p \end{Bmatrix}_p (-1)^{m+k} = \binom{n}{m} (r-p)^{\overline{n-m}}.$$

Asymptotics:

$$\begin{Bmatrix} n+r \\ k+r \end{Bmatrix}_r \sim \frac{(k+r)^n}{k!}.$$

Other formulas:

$$\left\{ {n+r \atop k+r} \right\}_r = \frac{1}{k!} \sum_{l=0}^{k} \binom{k}{l} (l+r)^n (-1)^{k-l},$$

$$\left\{ {n \atop k} \right\}_r = \left\{ {n \atop k} \right\}_{r-1} - (r-1) \left\{ {n-1 \atop k} \right\}_{r-1} \quad (n \geq r \geq 1),$$

$$\left\{ {n+m \atop n} \right\}_r = \sum_{r \leq i_1 \leq \cdots \leq i_m \leq n} i_1 i_2 \cdots i_m,$$

$$\left\{ {n+r \atop k+r} \right\}_r = \sum_{j=k}^{n} \binom{n}{j} \left\{ {j \atop k} \right\} r^{n-j},$$

$$\left\{ {n \atop k} \right\}_r = \sum_{m=0}^{n-r} \binom{n-r}{m} \left\{ {n-p-m \atop k-p} \right\}_{r-p} p^m \quad (r \geq p \geq 0),$$

$$\left\{ {n+r \atop k+r} \right\}_r = \sum_{m} \binom{n}{m} \left\{ {n-m+p \atop k+p} \right\}_p (r-p)^m,$$

$$\left\{ {n \atop k} \right\}_r = \sum_{m} \left\{ {p+m \atop p} \right\}_r \left\{ {n-m \atop k} \right\}_{p+1} \quad (r \leq p < n),$$

$$\left\{ {n \atop k} \right\}_r = \sum_{m} \left[{r \atop m} \right]_p \left\{ {n-r+m \atop k} \right\}_p (-1)^{k+r} \quad (n \geq r \geq p \geq 0),$$

$$\left\{ {n \atop k} \right\}_r = \sum_{m=0}^{n-r} \binom{n-r}{m} \left\{ {n-p-m \atop k-p} \right\}_{r-p} p^m \quad (r \geq p \geq 0),$$

$$\left\{ {n \atop m} \right\}_r = \sum_{k=2}^{n} \binom{n}{k} \sum_{l=1}^{k-1} (-1)^{l-1} \binom{l \mid r \quad 2}{l-1} \left\{ {k-l \atop m-1} \right\}_{r-1},$$

$$\left\{ {n \atop k} \right\}_r = \sum_{i=0}^{r} \binom{r}{i} (k-i)^{r-i} \left\{ {n-r \atop k-i} \right\},$$

$$\left\{ {n+r \atop k+r} \right\}_r = \sum_{i=0}^{r} k^i \left\{ {n+r-i \atop k+r} \right\}_{r-i}.$$

Properties of the horizontal generating polynomials

Bell polynomials

$$\sum_{n=0}^{\infty} B_n(x)\frac{y^n}{n!} = \exp(x(e^y - 1)),$$

$$\sum_{n=0}^{\infty} B_n(x)y^n = \frac{1}{e^x}\,_1F_1\left(\begin{array}{c}\frac{-1}{z} \\ \frac{z-1}{z}\end{array}\middle| x\right).$$

$$B_n(x) = xB_{n-1}(x) + xB'_{n-1}(x) = e^{-x}x(e^x B_{n-1}(x))',$$

$$B_n(x+y) = \sum_{k=0}^{n}\binom{n}{k}B_k(x)B_{n-k}(y),$$

$$B_{n+1}(x) = x\sum_{k=0}^{n}\binom{n}{k}B_k(x),$$

$$B_n(x) = \frac{1}{e^x}\sum_{k=0}^{\infty}\frac{k^n}{k!}x^k,$$

$$B_{n+m}(x) = \sum_{k=0}^{n}\sum_{j=0}^{m}\binom{n}{k}x^j B_k(x)\left\{\begin{array}{c}m\\j\end{array}\right\}j^{n-k}.$$

The Bell polynomials have only real and negative zeros (except $B_0(x) = 1$), the leftmost zero does not grow faster than $c_1 \cdot n^{3/2}$, see [175]; and does not grow slower than $c_2 n$ [432].

r-Bell polynomials

$$\sum_{n=0}^{\infty} B_{n,r}(x)\frac{y^n}{n!} = e^{x(e^y - 1)+ry},$$

$$\sum_{n=0}^{\infty} B_{n,r}(x)z^n = \frac{-1}{rz - 1}\frac{1}{e^x}\,_1F_1\left(\begin{array}{c}\frac{rz-1}{z} \\ \frac{rz+z-1}{z}\end{array}\middle| x\right).$$

$$B_{n,r}(x) = \sum_{k=0}^{n} \binom{n}{k} B_k(x) r^{n-k},$$

$$B_{n,r}(x) = \sum_{k=0}^{n} \binom{n}{k} B_{k,t}(x) s^{n-k} \quad (s+t=r),$$

$$B_{n,r}(x) = x\left(\frac{d}{dx} B_{n-1,r}(x) + B_{n-1,r}(x)\right) + r B_{n-1,r}(x),$$

$$e^x x^r B_{n,r}(x) = x\frac{d}{dx}\left(e^x x^r B_{n-1,r}(x)\right),$$

$$B_{n,r}(x) = x B_{n-1,r+1}(x) + r B_{n-1,r}(x),$$

$$B_{n,r}(x) = \frac{1}{e^x} \sum_{k=0}^{\infty} \frac{(k+r)^n}{k!} x^k.$$

For the estimations of the leftmost zeros, see [175, 432].

Fubini polynomials

$$\sum_{n=0}^{\infty} F_n(x) \frac{y^n}{n!} = \frac{1}{1 - x(e^y - 1)}.$$

$$F_n(x) = x[F_{n-1}(x) + (1+x)F'_{n-1}(x)] = x((1+x)F_{n-1}(x))',$$

$$F_n(x) = x^n E_n\left(\frac{1+x}{x}\right).$$

The Fubini polynomials have only real and negative zeros, and all the zeros belong to $]-1,0]$.

Eulerian polynomials

$$\sum_{n=0}^{\infty} E_n(x) \frac{y^n}{n!} = \frac{x-1}{x - e^{(x-1)y}}.$$

$$E_n(x) = (1 + (n-1)x)E_{n-1}(x) + (x - x^2)E'_{n-1}(x),$$

$$E_n(x) = (x-1)^n F_n\left(\frac{1}{x-1}\right),$$

$$\frac{x^n}{(1-x)^{n+1}} E_n\left(\frac{1}{x}\right) = \sum_{i=0}^{\infty} i^n x^i.$$

The Euler polynomials have only real and negative zeros.

r-Fubini polynomials

$$\sum_{n=0}^{\infty} F_{n,r}(x)\frac{t^n}{n!} = \frac{r!e^{rt}}{(1-x(e^t-1))^{r+1}} = r!e^{rt}\left(\sum_{n=0}^{\infty} F_n(x)\frac{t^n}{n!}\right)^{r+1}.$$

$$F_{n,r}(x) = x[(r+1)F_{n-1,r}(x) + (1+x)F'_{n-1,r}(x)] + rF_{n-1,r}(x),$$

$$F_{n,r}(x) = x^{1-r}\left[(x^{r+1}+x^r)F_{n-1,r}(x)\right]',$$

$$F_{n,r+1}(x) = (r+1)\sum_{k=0}^{n}\binom{n}{k}\sum_{l=0}^{k}\binom{k}{l}F_l(x)F_{k-l,r}(x),$$

$$F_{n,r}(x) = x^n E_{n,r}\left(\frac{1+x}{x}\right).$$

The r-Fubini polynomials have only real and negative zeros, and all the zeros belong to $]-1,0[$ $(r>0)$.

r-Eulerian polynomials

$$\sum_{n=0}^{\infty} E_{n,r}(x)\frac{t^n}{n!} = r!e^{r(x-1)t}\left(\frac{x-1}{x-e^{(x-1)z}}\right)^{r+1}.$$

$$E_{n,r}(x) = (1+(n+r-1)x)E_{n-1,r}(x) + (x-x^2)E'_{n-1,r}(x),$$

$$E_{n,r}(x) = (x-1)^n F_{n,r}\left(\frac{1}{x-1}\right),$$

$$\frac{x^{n+r}}{(1-x)^{n+r+1}}E_{n,r}\left(\frac{1}{x}\right) = \sum_{i=0}^{\infty} i^n(i)^r_x x^i.$$

The r-Euler polynomials have only real and negative zeros.

Tables

The headlines over the tables contain the Sloane On-Line Encyclopedia of Integer Sequences[12] (OEIS) ID of the given sequence where available. For some sequences, more members are calculated below than in the Sloane Encyclopedia (but for other sequences, the contrary is true, in OEIS you can find sometimes up to hundreds of members).

The Stirling numbers of the first kind (A132393)

$\left[{n\atop k}\right]$	$k=1$	$k=2$	$k=3$	$k=4$	$k=5$	$k=6$	$k=7$	$k=8$	$k=9$
$n=1$	1								
$n=2$	1	1							
$n=3$	2	3	1						
$n=4$	6	11	6	1					
$n=5$	24	50	35	10	1				
$n=6$	120	274	225	85	15	1			
$n=7$	720	1764	1624	735	175	21	1		
$n=8$	5040	13068	13132	6769	1960	322	28	1	
$n=9$	40320	109584	118124	67284	22449	4536	546	36	1

$\left[{0\atop 0}\right]=1.$

The Stirling numbers of the second kind (A008277)

$\left\{{n\atop k}\right\}$	$k=1$	$k=2$	$k=3$	$k=4$	$k=5$	$k=6$	$k=7$	$k=8$	$k=9$
$n=1$	1								
$n=2$	1	1							
$n=3$	1	3	1						
$n=4$	1	7	6	1					
$n=5$	1	15	25	10	1				
$n=6$	1	31	90	65	15	1			
$n=7$	1	63	301	350	140	21	1		
$n=8$	1	127	966	1701	1050	266	28	1	
$n=9$	1	255	3025	7770	6951	2646	462	36	1

$\left\{{0\atop 0}\right\}=1.$

---

[12]http://oeis.org/

The 2-Stirling numbers of the first kind (A143491)

$\begin{bmatrix} n \\ k \end{bmatrix}_2$	$k=2$	$k=3$	$k=4$	$k=5$	$k=6$	$k=7$	$k=8$	$k=9$
$n=2$	1							
$n=3$	2	1						
$n=4$	6	5	1					
$n=5$	24	26	9	1				
$n=6$	120	154	71	14	1			
$n=7$	720	1044	580	155	20	1		
$n=8$	5040	8028	5104	1665	295	27	1	
$n=9$	40320	69264	48860	18424	4025	511	35	1
$n=10$	362880	663696	509004	214676	54649	8624	826	44

$\begin{bmatrix} 10 \\ 10 \end{bmatrix}_2 = 1.$

The 2-Stirling numbers of the second kind (A143494)

$\left\{ \begin{matrix} n \\ k \end{matrix} \right\}_2$	$k=2$	$k=3$	$k=4$	$k=5$	$k=6$	$k=7$	$k=8$	$k=9$	$k=10$
$n=2$	1								
$n=3$	2	1							
$n=4$	4	5	1						
$n=5$	8	19	9	1					
$n=6$	16	65	55	14	1				
$n=7$	32	211	285	125	20	1			
$n=8$	64	665	1351	910	245	27	1		
$n=9$	128	2059	6069	5901	2380	434	35	1	
$n=10$	256	6305	26335	35574	20181	5418	714	44	1

The 3-Stirling numbers of the first kind (A143492)

$\left[{n\atop k}\right]_3$	$k = 3$	$k = 4$	$k = 5$	$k = 6$	$k = 7$	$k = 8$	$k = 9$	$k = 10$
$n = 3$	1							
$n = 4$	3	1						
$n = 5$	12	7	1					
$n = 6$	60	47	12	1				
$n = 7$	360	342	119	18	1			
$n = 8$	2520	2754	1175	245	25	1		
$n = 9$	20160	24552	12154	3135	445	33	1	
$n = 10$	181440	241128	133938	40369	7140	742	42	1
$n = 11$	1814400	2592720	1580508	537628	111769	14560	1162	52

$\left[{11\atop 11}\right]_3 = 1.$

The 3-Stirling numbers of the second kind (A143495)

$\left\{{n\atop k}\right\}_3$	$k = 3$	$k = 4$	$k = 5$	$k = 6$	$k = 7$	$k = 8$	$k = 9$	$k = 10$
$n = 3$	1							
$n = 4$	3	1						
$n = 5$	9	7	1					
$n = 6$	27	37	12	1				
$n = 7$	81	175	97	18	1			
$n = 8$	243	781	660	205	25	1		
$n = 9$	729	3367	4081	1890	380	33	1	
$n = 10$	2187	14197	23772	15421	4550	644	42	1
$n = 11$	6561	58975	133057	116298	47271	9702	1022	52	1

The 2-restricted Stirling numbers of the first and second kind and Bessel numbers (A100861)

$\begin{bmatrix} n \\ k \end{bmatrix}_{\leq 2}$	$k=1$	$k=2$	$k=3$	$k=4$	$k=5$	$k=6$	$k=7$	$k=8$
$n=1$	1							
$n=2$	1	1						
$n=3$		3	1					
$n=4$		3	6	1				
$n=5$			15	10	1			
$n=6$			15	45	15	1		
$n=7$				105	105	21	1	
$n=8$				105	420	210	28	1
$n=9$					945	1260	378	36
$n=10$					945	4725	3150	630
$n=11$						10395	17325	6930
$n=12$						10395	62370	51975
$n=13$							135135	270270
$n=14$							135135	945945
$n=15$								2027025
$n=16$								2027025

$$\begin{bmatrix} n \\ k \end{bmatrix}_{\leq 2} = \left\{ n \atop k \right\}_{\leq 2} = B(n,k).$$

The 3-restricted Stirling numbers of the first kind (|A171996|)

$\left[{n\atop k}\right]_{<3}$	$k=1$	$k=2$	$k=3$	$k=4$	$k=5$	$k=6$
$n=1$	1					
$n=2$	1	1				
$n=3$	2	3	1			
$n=4$		11	6	1		
$n=5$		20	35	10	1	
$n=6$		40	135	85	15	1
$n=7$			490	525	175	21
$n=8$			1120	2905	1540	322
$n=9$			2240	12600	11865	3780
$n=10$				47600	76545	38325
$n=11$				123200	435050	333795
$n=12$				246400	2032800	2582195
$n=13$					8008000	17357340
$n=14$					22422400	102302200
$n=15$					44844800	504504000
$n=16$						2062860800
$n=17$						6098892800
$n=18$						12197785600

The 3-restricted Stirling numbers of the second kind (A144385)

$\left\{{n\atop k}\right\}_{<3}$	$k=1$	$k=2$	$k=3$	$k=4$	$k=5$	$k=6$
$n=1$	1					
$n=2$	1	1				
$n=3$	1	3	1			
$n=4$		7	6	1		
$n=5$		10	25	10	1	
$n-6$		10	75	65	15	1
$n=7$			175	315	140	21
$n=8$			280	1225	980	266
$n=9$			280	3780	5565	2520
$n=10$				9100	26145	19425
$n=11$				15400	102025	125895
$n=12$				15400	323400	695695
$n=13$					800800	3273270
$n=14$					1401400	12962950
$n=15$					1401400	42042000
$n=16$						106506400
$n=17$						190590400
$n=18$						190590400

The 2-associated Stirling numbers of the first kind (A008306)

$\left[\begin{smallmatrix}n\\k\end{smallmatrix}\right]_{>2}$	$k=1$	$k=2$	$k=3$	$k=4$	$k=5$	$k=6$
$n=1$	0					
$n=2$	1					
$n=3$	2					
$n=4$	6	3				
$n=5$	24	20				
$n=6$	120	130	15			
$n=7$	720	924	210			
$n=8$	5040	7308	2380	105		
$n=9$	40320	64224	26432	2520		
$n=10$	362880	623376	303660	44110	945	
$n=11$	3628880	6636960	3678840	705320	34650	
$n=12$	39916800	76998240	47324376	11098780	866250	10395
$n=13$	479001600	967524480	647536032	177331440	18858840	540540

The 3-associated Stirling numbers of the first kind (A050211)

$\left[\begin{smallmatrix}n\\k\end{smallmatrix}\right]_{>3}$	$k=1$	$k=2$	$k=3$	$k=4$	$k=5$
$n=1$	0				
$n=2$	0				
$n=3$	6				
$n=4$	6				
$n=5$	24				
$n=6$	120	40			
$n=7$	720	420			
$n=8$	5040	3948			
$n=9$	40320	38304	2240		
$n=10$	362880	396576	5040		
$n=11$	3628880	4419360	859320		
$n=12$	39916800	53048160	13665960	246400	
$n=13$	479001600	684478080	216339552	9609600	
$n=14$	6227020800	9464307840	3501834336	258978720	
$n=15$	87178291200	139765167360	58680445824	6112906800	44844800
$n=16$	1307674368000	2197067846400	1023947084160	137124907920	2690688000

The 2-associated Stirling numbers of the second kind (A008299)

${n \brace k}_{>2}$	$k=1$	$k=2$	$k=3$	$k=4$	$k=5$	$k=6$	$k=7$
$n=1$	0						
$n=2$	1						
$n=3$	1						
$n=4$	1	3					
$n=5$	1	10					
$n=6$	1	25	15				
$n=7$	1	56	105				
$n=8$	1	119	490	105			
$n=9$	1	246	1918	1260			
$n=10$	1	501	6825	9450	945		
$n=11$	1	1012	22935	56980	17325		
$n=12$	1	2305	74316	302995	190575	10395	
$n=13$	1	4082	235092	1487200	1636635	270270	
$n=14$	1	8177	731731	6914908	12122110	4099095	135135
$n=15$	1	16368	2252341	30950920	81431350	47507460	4729725

The 3-associated Stirling numbers of the second kind (A059022)

${n \brace k}_{>3}$	$k=1$	$k=2$	$k=3$	$k=4$	$k=5$	$k=6$	$k=7$
$n=1$	0						
$n=2$	0						
$n=3$	1						
$n=4$	1						
$n=5$	1						
$n=6$	1	10					
$n=7$	1	35					
$n=8$	1	91					
$n=9$	1	210	280				
$n=10$	1	456	2100				
$n=11$	1	957	10365				
$n=12$	1	1969	42735	15400			
$n=13$	1	4004	158301	200200			
$n=14$	1	8086	549549	1611610			
$n=15$	1	16263	1827826	10335325	1401400		
$n=16$	1	32631	5903898	57962905	28028000		
$n=17$	1	65382	18682014	297797500	333533200		
$n=18$	1	130900	58257810	1439774336	3073270200	190590400	

The Eulerian numbers (A008292)

$\langle{n \atop k}\rangle$	$k=1$	$k=2$	$k=3$	$k=4$	$k=5$	$k=6$	$k=7$	$k=8$	$k=9$
$n=1$	0								
$n=2$	1	0							
$n=3$	4	1	0						
$n=4$	11	11	1	0					
$n=5$	26	66	26	1	0				
$n=6$	57	302	302	57	1	0			
$n=7$	120	1191	2416	1191	120	1	0		
$n=8$	247	4293	15619	15619	4293	247	1	0	
$n=9$	502	14608	88234	156190	88234	14608	502	1	0

$\langle{n \atop 0}\rangle = 1$ for all $n \geq 0$.

The 2-Eulerian numbers[13]

$\langle{n \atop k}\rangle_2$	$k=0$	$k=1$	$k=2$	$k=3$	$k=4$	$k=5$	$k=6$
$n=0$	2						
$n=1$	2	4					
$n=2$	2	14	8				
$n=3$	2	36	66	16			
$n=4$	2	82	342	262	32		
$n=5$	2	176	1436	2416	946	64	
$n=6$	2	366	5364	16844	14394	3222	128
$n=7$	2	748	18654	99560	156190	76908	10562
$n=8$	2	1514	61946	528818	1378310	1242398	33734
$n=9$	2	3048	199464	2610840	10593276	15724248	8882952

$\langle{8 \atop 7}\rangle_2 = 512, \langle{9 \atop 7}\rangle_2 = 1796136, \langle{9 \atop 8}\rangle_2 = 105810, \langle{9 \atop 9}\rangle_2 = 1024.$

The 3-Eulerian numbers

$\langle{n \atop k}\rangle_3$	$k=0$	$k=1$	$k=2$	$k=3$	$k=4$	$k=5$	$k=6$	$k=7$
$n=0$	6							
$n=1$	6	18						
$n=2$	6	60	54					
$n=3$	6	150	402	162				
$n=4$	6	336	1956	2256	486			
$n=5$	6	714	7884	18804	11454	1458		
$n=6$	6	1476	28650	122520	151290	54564	4374	
$n=7$	6	3006	97758	690630	1491570	1083834	248874	13122

---

[13]The r-Eulerian numbers are still not added to the Sloane encyclopedia.

The Bell numbers (A000110)

B_0 = 1
B_1 = 1
B_2 = 2
B_3 = 5
B_4 = 15
B_5 = 52
B_6 = 203
B_7 = 877
B_8 = 4140
B_9 = 21147
B_{10} = 115975
B_{11} = 678570
B_{12} = 4213597
B_{13} = 27644437
B_{14} = 190899322
B_{15} = 1382958545
B_{16} = 10480142147
B_{17} = 82864869804
B_{18} = 682076806159
B_{19} = 5832742205057
B_{20} = 51724158235372
B_{21} = 474869816156751
B_{22} = 4506715738447323
B_{23} = 44152005855084346
B_{24} = 445958869294805289
B_{25} = 4638590332229999353
B_{26} = 49631246523618756274
B_{27} = 545717047936059989389
B_{28} = 6160539404599934652455
B_{29} = 71339801938860275191172
B_{30} = 846749014511809332450147

The Fubini numbers (ordered Bell numbers) (A000670)

$$F_0 = 1$$
$$F_1 = 1$$
$$F_2 = 3$$
$$F_3 = 13$$
$$F_4 = 75$$
$$F_5 = 541$$
$$F_6 = 4683$$
$$F_7 = 47293$$
$$F_8 = 545835$$
$$F_9 = 7087261$$
$$F_{10} = 102247563$$
$$F_{11} = 1622632573$$
$$F_{12} = 28091567595$$
$$F_{13} = 526858348381$$
$$F_{14} = 10641342970443$$
$$F_{15} = 230283190977853$$
$$F_{16} = 5315654681981355$$
$$F_{17} = 130370767029135901$$
$$F_{18} = 3385534663256845323$$
$$F_{19} = 92801587319328411133$$
$$F_{20} = 2677687796244384203115$$
$$F_{21} = 81124824998504073881821$$
$$F_{22} = 2574844419803190384544203$$
$$F_{23} = 85438451336745709294580413$$
$$F_{24} = 2958279121074145472650648875$$
$$F_{25} = 106697365438475775825583498141$$
$$F_{26} = 4002225759844168492486127539083$$
$$F_{27} = 155897763918621623249276226253693$$
$$F_{28} = 6297562064950066033518373935334635$$
$$F_{29} = 263478385263023690020893329044576861$$

The odd indexed Bernoulli numbers are all zero, except $B_1 = -\frac{1}{2}$. The numerator and denominator of these numbers are separately indexed in OEIS under the ID A027641 and A027642, respectively.

The Bernoulli numbers

$$B_0 = 1$$

$$B_1 = -\frac{1}{2}$$

$$B_2 = \frac{1}{6}$$

$$B_4 = -\frac{1}{30}$$

$$B_6 = \frac{1}{42}$$

$$B_8 = -\frac{1}{30}$$

$$B_{10} = \frac{5}{66}$$

$$B_{12} = -\frac{691}{2\,730}$$

$$B_{14} = \frac{7}{6}$$

$$B_{16} = -\frac{3\,617}{510}$$

$$B_{18} = \frac{43\,867}{798}$$

$$B_{20} = -\frac{174\,611}{330}$$

$$B_{22} = \frac{854\,513}{138}$$

$$B_{24} = -\frac{236\,364\,091}{2\,730}$$

$$B_{26} = \frac{8\,553\,103}{6}$$

$$B_{28} = -\frac{23\,749\,461\,029}{870}$$

$$B_{30} = \frac{8\,615\,841\,276\,005}{14\,322}$$

$$B_{32} = -\frac{7\,709\,321\,041\,217}{510}$$

$$B_{34} = \frac{2\,577\,687\,858\,367}{6}$$

$$B_{36} = -\frac{26\,315\,271\,553\,053\,477\,373}{1\,919\,190}$$

The numerator and denominator of the Cauchy numbers of the first kind are separately indexed in OEIS under the ID A006232 and A006233, respectively.

The Cauchy numbers of the first kind

$$c_0 = 1$$

$$c_1 = \frac{1}{2}$$

$$c_2 = -\frac{1}{6}$$

$$c_3 = \frac{1}{4}$$

$$c_4 = -\frac{19}{30}$$

$$c_5 = \frac{9}{4}$$

$$c_6 = -\frac{863}{84}$$

$$c_7 = \frac{1\,375}{24}$$

$$c_8 = -\frac{33\,953}{90}$$

$$c_9 = \frac{57\,281}{20}$$

$$c_{10} = -\frac{3\,250\,433}{132}$$

$$c_{11} = \frac{1\,891\,755}{8}$$

$$c_{12} = -\frac{13\,695\,779\,093}{5\,460}$$

$$c_{13} = \frac{24\,466\,579\,093}{840}$$

$$c_{14} = -\frac{132\,282\,840\,127}{360}$$

$$c_{15} = \frac{240\,208\,245\,823}{48}$$

$$c_{16} = -\frac{111\,956\,703\,448\,001}{1\,530}$$

$$c_{17} = \frac{4\,573\,423\,873\,125}{4}$$

$$c_{18} = -\frac{30\,342\,376\,302\,478\,019}{1\,596}$$

$$c_{19} = \frac{56\,310\,194\,579\,604\,163}{168}$$

The numerator and denominator of the Cauchy numbers of the second kind are separately indexed in OEIS under the ID A002657 and A002790, respectively.

The Cauchy numbers of the second kind

$$C_0 = 1$$

$$C_1 = \frac{1}{2}$$

$$C_2 = \frac{5}{6}$$

$$C_3 = \frac{9}{4}$$

$$C_4 = \frac{251}{30}$$

$$C_5 = \frac{475}{12}$$

$$C_6 = \frac{19\,087}{84}$$

$$C_7 = \frac{36\,799}{24}$$

$$C_8 = \frac{1\,070\,017}{90}$$

$$C_9 = \frac{2\,082\,753}{20}$$

$$C_{10} = \frac{134\,211\,265}{132}$$

$$C_{11} = \frac{262\,747\,265}{24}$$

$$C_{12} = \frac{703\,604\,254\,357}{5\,460}$$

$$C_{13} = \frac{1\,382\,741\,929\,621}{840}$$

$$C_{14} = \frac{8\,164\,168\,737\,599}{360}$$

$$C_{15} = \frac{5\,362\,709\,743\,125}{16}$$

$$C_{16} = \frac{8\,092\,989\,203\,533\,249}{1\,530}$$

$$C_{17} = \frac{15\,980\,174\,332\,775\,873}{180}$$

$$C_{18} = \frac{12\,600\,467\,236\,042\,756\,559}{7\,980}$$

$$C_{19} = \frac{24\,919\,383\,499\,187\,492\,303}{840}$$

The 2-Fubini numbers (A232472)

$$F_{1,2} = 10$$
$$F_{2,2} = 62$$
$$F_{3,2} = 466$$
$$F_{4,2} = 4\,142$$
$$F_{5,2} = 42\,610$$
$$F_{6,2} = 498\,542$$
$$F_{7,2} = 6\,541\,426$$
$$F_{8,2} = 95\,160\,302$$
$$F_{9,2} = 1\,520\,385\,010$$
$$F_{10,2} = 26\,468\,935\,022$$
$$F_{11,2} = 498\,766\,780\,786$$
$$F_{12,2} = 10\,114\,484\,622\,062$$
$$F_{13,2} = 219\,641\,848\,007\,410$$
$$F_{14,2} = 5\,085\,371\,491\,003\,502$$
$$F_{15,2} = 125\,055\,112\,347\,154\,546$$
$$F_{16,2} = 3\,255\,163\,896\,227\,709\,422$$
$$F_{17,2} = 89\,416\,052\,656\,071\,565\,810$$
$$F_{18,2} = 2\,584\,886\,208\,925\,055\,791\,982$$
$$F_{19,2} = 78\,447\,137\,202\,259\,689\,678\,706$$
$$F_{20,2} = 2\,493\,719\,594\,804\,686\,310\,662\,382$$
$$F_{21,2} = 82\,863\,606\,916\,942\,518\,910\,036\,210$$
$$F_{22,2} = 2\,872\,840\,669\,737\,399\,763\,356\,068\,462$$
$$F_{23,2} = 103\,739\,086\,317\,401\,630\,352\,932\,849\,266$$
$$F_{24,2} = 3\,895\,528\,394\,405\,692\,716\,660\,544\,040\,942$$
$$F_{25,2} = 151\,895\,538\,158\,777\,454\,756\,790\,098\,714\,610$$
$$F_{26,2} = 6\,141\,664\,301\,031\,444\,410\,269\,097\,709\,080\,942$$
$$F_{27,2} = 257\,180\,823\,198\,073\,623\,987\,374\,955\,109\,242\,226$$
$$F_{28,2} = 11\,140\,090\,408\,748\,856\,793\,721\,570\,867\,140\,325\,102$$
$$F_{29,2} = 498\,604\,467\,780\,257\,507\,947\,098\,559\,879\,733\,217\,010$$
$$F_{30,2} = 23\,035\,146\,049\,160\,627\,260\,753\,649\,613\,068\,937\,357\,422$$

The 1-Fubini numbers are just re-indexed Fubini numbers: $F_{n,1} = F_{n+1}$.

The 3-Fubini numbers (A232473)

$F_{1,3}$ = 42

$F_{2,3}$ = 342

$F_{3,3}$ = 3 210

$F_{4,3}$ = 34 326

$F_{5,3}$ = 413 322

$F_{6,3}$ = 5 544 342

$F_{7,3}$ = 82 077 450

$F_{8,3}$ = 1 330 064 406

$F_{9,3}$ = 23 428 165 002

$F_{10,3}$ = 445 828 910 742

$F_{11,3}$ = 9 116 951 060 490

$F_{12,3}$ = 199 412 878 763 286

$F_{13,3}$ = 4 646 087 794 988 682

$F_{14,3}$ = 114 884 369 365 147 542

$F_{15,3}$ = 3 005 053 671 533 400 330

$F_{16,3}$ = 82 905 724 863 616 146 966

$F_{17,3}$ = 2 406 054 103 612 912 660 362

$F_{18,3}$ = 73 277 364 784 409 578 094 742

$F_{19,3}$ = 2 336 825 320 400 166 931 304 970

$F_{20,3}$ = 77 876 167 727 333 146 288 711 446

$F_{21,3}$ = 2 707 113 455 903 514 725 535 996 042

$F_{22,3}$ = 97 993 404 977 926 830 826 220 712 342

$F_{23,3}$ = 3 688 050 221 770 889 455 954 678 342 410

$F_{24,3}$ = 144 104 481 369 966 069 323 469 010 632 726

$F_{25,3}$ = 5 837 873 224 713 889 500 755 517 511 651 722

$F_{26,3}$ = 244 897 494 596 010 735 166 836 759 691 080 342

$F_{27,3}$ = 10 625 728 762 352 709 545 746 820 956 921 840 650

$F_{28,3}$ = 476 324 286 962 759 794 359 655 418 145 452 566 806

$F_{29,3}$ = 22 037 937 113 600 112 244 859 452 493 309 470 923 402

$F_{30,3}$ = 1 051 344 296 019 000 117 096 802 419 429 724 152 398 742

The factorials (A000142)

$$
\begin{aligned}
0! &= 1 \\
1! &= 1 \\
2! &= 2 \\
3! &= 6 \\
4! &= 24 \\
5! &= 120 \\
6! &= 720 \\
7! &= 5\,040 \\
8! &= 40\,320 \\
9! &= 362\,880 \\
10! &= 3\,628\,800 \\
11! &= 39\,916\,800 \\
12! &= 479\,001\,600 \\
13! &= 6\,227\,020\,800 \\
14! &= 87\,178\,291\,200 \\
15! &= 1\,307\,674\,368\,000 \\
16! &= 20\,922\,789\,888\,000 \\
17! &= 355\,687\,428\,096\,000 \\
18! &= 6\,402\,373\,705\,728\,000 \\
19! &= 121\,645\,100\,408\,832\,000 \\
20! &= 2\,432\,902\,008\,176\,640\,000 \\
21! &= 51\,090\,942\,171\,709\,440\,000 \\
22! &= 1\,124\,000\,727\,777\,607\,680\,000 \\
23! &= 25\,852\,016\,738\,884\,976\,640\,000 \\
24! &= 620\,448\,401\,733\,239\,439\,360\,000 \\
25! &= 15\,511\,210\,043\,330\,985\,984\,000\,000 \\
26! &= 403\,291\,461\,126\,605\,635\,584\,000\,000 \\
27! &= 10\,888\,869\,450\,418\,352\,160\,768\,000\,000 \\
28! &= 304\,888\,344\,611\,713\,860\,501\,504\,000\,000 \\
29! &= 8\,841\,761\,993\,739\,701\,954\,543\,616\,000\,000 \\
30! &= 265\,252\,859\,812\,191\,058\,636\,308\,480\,000\,000
\end{aligned}
$$

The harmonic and hyperharmonic numbers

$H_0 = 0$	$H_0^2 = 0$	$H_0^3 = 0$	$H_0^4 = 0$
$H_1 = 1$	$H_1^2 = 1$	$H_1^3 = 1$	$H_1^4 = 1$
$H_2 = \dfrac{3}{2}$	$H_2^2 = \dfrac{5}{2}$	$H_2^3 = \dfrac{7}{2}$	$H_2^4 = \dfrac{9}{2}$
$H_3 = \dfrac{11}{6}$	$H_3^2 = \dfrac{13}{3}$	$H_3^3 = \dfrac{47}{6}$	$H_3^4 = \dfrac{37}{3}$
$H_4 = \dfrac{25}{12}$	$H_4^2 = \dfrac{77}{12}$	$H_4^3 = \dfrac{57}{4}$	$H_4^4 = \dfrac{319}{12}$
$H_5 = \dfrac{137}{60}1$	$H_5^2 = \dfrac{87}{10}$	$H_5^3 = \dfrac{459}{20}$	$H_5^4 = \dfrac{743}{15}$
$H_6 = \dfrac{49}{20}$	$H_6^2 = \dfrac{223}{20}$	$H_6^3 = \dfrac{341}{10}$	$H_6^4 = \dfrac{2509}{30}$
$H_7 = \dfrac{363}{140}$	$H_7^2 = \dfrac{481}{35}$	$H_7^3 = \dfrac{3349}{70}$	$H_7^4 = \dfrac{2761}{21}$
$H_8 = \dfrac{761}{280}$	$H_8^2 = \dfrac{4609}{280}$	$H_8^3 = \dfrac{3601}{56}$	$H_8^4 = \dfrac{32891}{168}$
$H_9 = \dfrac{7129}{2520}$	$H_9^2 = \dfrac{4861}{252}$	$H_9^3 = \dfrac{42131}{504}$	$H_9^4 = \dfrac{35201}{126}$
$H_{10} = \dfrac{7381}{2520}$	$H_{10}^2 = \dfrac{55991}{2520}$	$H_{10}^3 = \dfrac{44441}{420}$	$H_{10}^4 = \dfrac{485333}{1260}$
$H_{11} = \dfrac{83711}{27720}$	$H_{11}^2 = \dfrac{58301}{2310}$	$H_{11}^3 = \dfrac{605453}{4620}$	$H_{11}^4 = \dfrac{511073}{990}$
$H_{12} = \dfrac{86021}{27720}$	$H_{12}^2 = \dfrac{785633}{27720}$	$H_{12}^3 = \dfrac{631193}{3960}$	$H_{12}^4 = \dfrac{535097}{792}$
$H_{13} = \dfrac{1145993}{360360}1$	$H_{13}^2 = \dfrac{811373}{25740}$	$H_{13}^3 = \dfrac{655217}{3432}$	$H_{13}^4 = \dfrac{1115239}{1287}$
$H_{14} = \dfrac{1171733}{360360}$	$H_{14}^2 = \dfrac{835397}{24024}$	$H_{14}^3 = \dfrac{1355479}{6006}$	$H_{14}^4 = \dfrac{19679783}{18018}$
$H_{15} = \dfrac{1195757}{360360}$	$H_{15}^2 = \dfrac{1715839}{45045}$	$H_{15}^3 = \dfrac{23763863}{90090}$	$H_{15}^4 = \dfrac{6786821}{5005}$
$H_{16} = \dfrac{2436559}{720720}$	$H_{16}^2 = \dfrac{29889983}{720720}$	$H_{16}^3 = \dfrac{24444543}{80080}$	$H_{16}^4 = \dfrac{133033679}{80080}$
$H_{17} = \dfrac{42142223}{12252240}$	$H_{17}^2 = \dfrac{30570663}{680680}$	$H_{17}^3 = \dfrac{476698557}{1361360}$	$H_{17}^4 = \dfrac{136913555}{68068}$
$H_{18} = \dfrac{14274301}{4084080}$	$H_{18}^2 = \dfrac{197698279}{4084080}$	$H_{18}^3 = \dfrac{162779395}{408408}$	$H_{18}^4 = \dfrac{140608675}{58344}$
$H_{19} = \dfrac{275295799}{77597520}$	$H_{19}^2 = \dfrac{201578155}{3879876}$	$H_{19}^3 = \dfrac{166474515}{369512}$	$H_{19}^4 = \dfrac{144135835}{50388}$
$H_{20} = \dfrac{55835135}{15519504}$	$H_{20}^2 = \dfrac{41054655}{739024}$	$H_{20}^3 = \dfrac{34000335}{67184}$	$H_{20}^4 = \dfrac{678544345}{201552}$

The idempotent numbers (A000248)

$$1_0 \;=\; 1$$
$$1_1 \;=\; 1$$
$$1_2 \;=\; 3$$
$$1_3 \;=\; 10$$
$$1_4 \;=\; 41$$
$$1_5 \;=\; 196$$
$$1_6 \;=\; 1\,057$$
$$1_7 \;=\; 6\,322$$
$$1_8 \;=\; 41\,393$$
$$1_9 \;=\; 293\,608$$
$$1_{10} \;=\; 2\,237\,921$$
$$1_{11} \;=\; 18\,210\,094$$
$$1_{12} \;=\; 157\,329\,097$$
$$1_{13} \;=\; 1\,436\,630\,092$$
$$1_{14} \;=\; 13\,810\,863\,809$$
$$1_{15} \;=\; 139\,305\,550\,066$$
$$1_{16} \;=\; 1\,469\,959\,371\,233$$
$$1_{17} \;=\; 16\,184\,586\,405\,328$$
$$1_{18} \;=\; 185\,504\,221\,191\,745$$
$$1_{19} \;=\; 2\,208\,841\,954\,063\,318$$
$$1_{20} \;=\; 27\,272\,621\,155\,678\,841$$
$$1_{21} \;=\; 348\,586\,218\,389\,733\,556$$
$$1_{22} \;=\; 4\,605\,223\,387\,997\,411\,873$$
$$1_{23} \;=\; 62\,797\,451\,641\,106\,266\,330$$
$$1_{24} \;=\; 882\,730\,631\,284\,319\,415\,505$$
$$1_{25} \;=\; 12\,776\,077\,318\,891\,628\,112\,376$$
$$1_{26} \;=\; 190\,185\,523\,485\,851\,040\,093\,857$$
$$1_{27} \;=\; 2\,908\,909\,247\,751\,545\,392\,493\,182$$
$$1_{28} \;=\; 45\,671\,882\,246\,215\,264\,120\,864\,553$$
$$1_{29} \;=\; 735\,452\,644\,411\,097\,903\,203\,941\,148$$
$$1_{30} \;=\; 12\,136\,505\,435\,201\,514\,536\,093\,218\,561$$

Involutions, the 2-restricted Bell numbers and factorials (A000085)

$$
\begin{aligned}
I_0 &= 1 \\
I_1 &= 1 \\
I_2 &= 2 \\
I_3 &= 4 \\
I_4 &= 10 \\
I_5 &= 26 \\
I_6 &= 76 \\
I_7 &= 232 \\
I_8 &= 764 \\
I_9 &= 2\,620 \\
I_{10} &= 9\,496 \\
I_{11} &= 35\,696 \\
I_{12} &= 140\,152 \\
I_{13} &= 568\,504 \\
I_{14} &= 2\,390\,480 \\
I_{15} &= 10\,349\,536 \\
I_{16} &= 46\,206\,736 \\
I_{17} &= 211\,799\,312 \\
I_{18} &= 997\,313\,824 \\
I_{19} &= 4\,809\,701\,440 \\
I_{20} &= 23\,758\,664\,096 \\
I_{21} &= 119\,952\,692\,896 \\
I_{22} &= 618\,884\,638\,912 \\
I_{23} &= 3\,257\,843\,882\,624 \\
I_{24} &= 17\,492\,190\,577\,600 \\
I_{25} &= 95\,680\,443\,760\,576 \\
I_{26} &= 532\,985\,208\,200\,576 \\
I_{27} &= 3\,020\,676\,745\,975\,552 \\
I_{28} &= 17\,411\,277\,367\,391\,104 \\
I_{29} &= 101\,990\,226\,254\,706\,560
\end{aligned}
$$

$$I_n = B_{n,\leq 2} = A_{n,\leq 2}.$$

Tables

The 3-restricted Bell numbers (A006505)

$$B_{0,\leq 3} = 1$$
$$B_{1,\leq 3} = 1$$
$$B_{2,\leq 3} = 2$$
$$B_{3,\leq 3} = 5$$
$$B_{4,\leq 3} = 14$$
$$B_{5,\leq 3} = 46$$
$$B_{6,\leq 3} = 166$$
$$B_{7,\leq 3} = 652$$
$$B_{8,\leq 3} = 2\,780$$
$$B_{9,\leq 3} = 12\,644$$
$$B_{10,\leq 3} = 61\,136$$
$$B_{11,\leq 3} = 312\,676$$
$$B_{12,\leq 3} = 1\,680\,592$$
$$B_{13,\leq 3} = 9\,467\,680$$
$$B_{14,\leq 3} = 55\,704\,104$$
$$B_{15,\leq 3} = 341\,185\,496$$
$$B_{16,\leq 3} = 2\,170\,853\,456$$
$$B_{17,\leq 3} = 14\,314\,313\,872$$
$$B_{18,\leq 3} = 97\,620\,050\,080$$
$$B_{19,\leq 3} = 687\,418\,278\,544$$
$$B_{20,\leq 3} = 4\,989\,946\,902\,176$$
$$B_{21,\leq 3} = 37\,286\,121\,988\,256$$
$$B_{22,\leq 3} = 286\,432\,845\,428\,192$$
$$B_{23,\leq 3} = 2\,259\,405\,263\,572\,480$$
$$B_{24,\leq 3} = 18\,280\,749\,571\,449\,664$$
$$B_{25,\leq 3} = 151\,561\,941\,235\,370\,176$$
$$B_{26,\leq 3} = 1\,286\,402\,259\,593\,355\,776$$
$$B_{27,\leq 3} = 11\,168\,256\,342\,434\,121\,152$$
$$B_{28,\leq 3} = 99\,099\,358\,725\,069\,658\,880$$
$$B_{29,\leq 3} = 898\,070\,590\,439\,513\,534\,464$$
$$B_{30,\leq 3} = 8\,306\,264\,068\,494\,786\,829\,696$$

The 3-restricted factorials (A057693)

$$A_{0,\leq 3} = 1$$
$$A_{1,\leq 3} = 1$$
$$A_{2,\leq 3} = 2$$
$$A_{3,\leq 3} = 6$$
$$A_{4,\leq 3} = 18$$
$$A_{5,\leq 3} = 66$$
$$A_{6,\leq 3} = 276$$
$$A_{7,\leq 3} = 1\,212$$
$$A_{8,\leq 3} = 5\,916$$
$$A_{9,\leq 3} = 31\,068$$
$$A_{10,\leq 3} = 171\,576$$
$$A_{11,\leq 3} = 1\,014\,696$$
$$A_{12,\leq 3} = 6\,319\,512$$
$$A_{13,\leq 3} = 41\,143\,896$$
$$A_{14,\leq 3} = 281\,590\,128$$
$$A_{15,\leq 3} = 2\,007\,755\,856$$
$$A_{16,\leq 3} = 14\,871\,825\,936$$
$$A_{17,\leq 3} = 114\,577\,550\,352$$
$$A_{18,\leq 3} = 913\,508\,184\,096$$
$$A_{19,\leq 3} = 7\,526\,682\,826\,848$$
$$A_{20,\leq 3} = 64\,068\,860\,545\,056$$
$$A_{21,\leq 3} = 561\,735\,627\,038\,496$$
$$A_{22,\leq 3} = 5\,068\,388\,485\,760\,832$$
$$A_{23,\leq 3} = 47\,026\,385\,852\,423\,616$$
$$A_{24,\leq 3} = 447\,837\,548\,306\,401\,728$$
$$A_{25,\leq 3} = 4\,374\,221\,252\,904\,547\,776$$
$$A_{26,\leq 3} = 43\,785\,991\,472\,018\,760\,576$$
$$A_{27,\leq 3} = 448\,610\,150\,446\,698\,125\,952$$
$$A_{28,\leq 3} = 4\,701\,535\,239\,730\,197\,200\,256$$
$$A_{29,\leq 3} = 50\,364\,829\,005\,083\,927\,722\,368$$
$$A_{30,\leq 3} = 550\,980\,793\,119\,978\,524\,802\,816$$

The 2-associated Bell numbers (A000296)

$$B_{0,\geq 2} = 1$$
$$B_{1,\geq 2} = 0$$
$$B_{2,\geq 2} = 1$$
$$B_{3,\geq 2} = 1$$
$$B_{4,\geq 2} = 4$$
$$B_{5,\geq 2} = 11$$
$$B_{6,\geq 2} = 41$$
$$B_{7,\geq 2} = 162$$
$$B_{8,\geq 2} = 715$$
$$B_{9,\geq 2} = 3\,425$$
$$B_{10,\geq 2} = 17\,722$$
$$B_{11,\geq 2} = 98\,253$$
$$B_{12,\geq 2} = 580\,317$$
$$B_{13,\geq 2} = 3\,633\,280$$
$$B_{14,\geq 2} = 24\,011\,157$$
$$B_{15,\geq 2} = 166\,888\,165$$
$$B_{16,\geq 2} = 1\,216\,070\,380$$
$$B_{17,\geq 2} = 9\,264\,071\,767$$
$$B_{18,\geq 2} = 73\,600\,798\,037$$
$$B_{19,\geq 2} = 608\,476\,008\,122$$
$$B_{20,\geq 2} = 5\,224\,266\,196\,935$$
$$B_{21,\geq 2} = 46\,499\,892\,038\,437$$
$$B_{22,\geq 2} = 428\,369\,924\,118\,314$$
$$B_{23,\geq 2} = 4\,078\,345\,814\,329\,009$$
$$B_{24,\geq 2} = 40\,073\,660\,040\,755\,337$$
$$B_{25,\geq 2} = 405\,885\,209\,254\,049\,952$$
$$B_{26,\geq 2} = 4\,232\,705\,122\,975\,949\,401$$
$$B_{27,\geq 2} = 45\,398\,541\,400\,642\,806\,873$$
$$B_{28,\geq 2} = 500\,318\,506\,535\,417\,182\,516$$
$$B_{29,\geq 2} = 5\,660\,220\,898\,064\,517\,469\,939$$
$$B_{30,\geq 2} = 65\,679\,581\,040\,795\,757\,721\,233$$

The 3-associated Bell numbers (A000296)

$$B_{0,\geq 3} = 1$$
$$B_{1,\geq 3} = 0$$
$$B_{2,\geq 3} = 0$$
$$B_{3,\geq 3} = 1$$
$$B_{4,\geq 3} = 1$$
$$B_{5,\geq 3} = 1$$
$$B_{6,\geq 3} = 11$$
$$B_{7,\geq 3} = 36$$
$$B_{8,\geq 3} = 92$$
$$B_{9,\geq 3} = 491$$
$$B_{10,\geq 3} = 2\,557$$
$$B_{11,\geq 3} = 11\,353$$
$$B_{12,\geq 3} = 60\,105$$
$$B_{13,\geq 3} = 362\,506$$
$$B_{14,\geq 3} = 2\,169\,246$$
$$B_{15,\geq 3} = 13\,580\,815$$
$$B_{16,\geq 3} = 91\,927\,435$$
$$B_{17,\geq 3} = 650\,078\,097$$
$$B_{18,\geq 3} = 4\,762\,023\,647$$
$$B_{19,\geq 3} = 36\,508\,923\,530$$
$$B_{20,\geq 3} = 292\,117\,087\,090$$
$$B_{21,\geq 3} = 2\,424\,048\,335\,917$$
$$B_{22,\geq 3} = 20\,847\,410\,586\,719$$
$$B_{23,\geq 3} = 185\,754\,044\,235\,873$$
$$B_{24,\geq 3} = 1\,711\,253\,808\,769\,653$$
$$B_{25,\geq 3} = 16\,272\,637\,428\,430\,152$$
$$B_{26,\geq 3} = 159\,561\,718\,111\,166\,776$$
$$B_{27,\geq 3} = 1\,611\,599\,794\,949\,346\,621$$
$$B_{28,\geq 3} = 16\,747\,401\,536\,644\,152\,613$$
$$B_{29,\geq 3} = 178\,881\,496\,831\,139\,695\,357$$
$$B_{30,\geq 3} = 1\,962\,101\,672\,879\,398\,037\,863$$

The 2-associated factorials, or derangements (A000166)

$$A_{0,\geq 2} = 1$$
$$A_{1,\geq 2} = 0$$
$$A_{2,\geq 2} = 1$$
$$A_{3,\geq 2} = 2$$
$$A_{4,\geq 2} = 9$$
$$A_{5,\geq 2} = 44$$
$$A_{6,\geq 2} = 265$$
$$A_{7,\geq 2} = 1\,854$$
$$A_{8,\geq 2} = 14\,833$$
$$A_{9,\geq 2} = 133\,496$$
$$A_{10,\geq 2} = 1\,334\,961$$
$$A_{11,\geq 2} = 14\,684\,570$$
$$A_{12,\geq 2} = 176\,214\,841$$
$$A_{13,\geq 2} = 2\,290\,792\,932$$
$$A_{14,\geq 2} = 32\,071\,101\,049$$
$$A_{15,\geq 2} = 481\,066\,515\,734$$
$$A_{16,\geq 2} = 7\,697\,064\,251\,745$$
$$A_{17,\geq 2} = 130\,850\,092\,279\,664$$
$$A_{18,\geq 2} = 2\,355\,301\,661\,033\,953$$
$$A_{19,\geq 2} = 44\,750\,731\,559\,645\,106$$
$$A_{20,\geq 2} = 895\,014\,631\,192\,902\,121$$
$$A_{21,\geq 2} = 18\,795\,307\,255\,050\,944\,540$$
$$A_{22,\geq 2} = 413\,496\,759\,611\,120\,779\,881$$
$$A_{23,\geq 2} = 9\,510\,425\,471\,055\,777\,937\,262$$
$$A_{24,\geq 2} = 228\,250\,211\,305\,338\,670\,494\,289$$
$$A_{25,\geq 2} = 5\,706\,255\,282\,633\,466\,762\,357\,224$$
$$A_{26,\geq 2} = 148\,362\,637\,348\,470\,135\,821\,287\,825$$
$$A_{27,\geq 2} = 4\,005\,791\,208\,408\,693\,667\,174\,771\,274$$
$$A_{28,\geq 2} = 112\,162\,153\,835\,443\,422\,680\,893\,595\,673$$
$$A_{29,\geq 2} = 3\,252\,702\,461\,227\,859\,257\,745\,914\,274\,516$$
$$A_{30,\geq 2} = 97\,581\,073\,836\,835\,777\,732\,377\,428\,235\,481$$

The 3-associated factorials (A038205)

$$A_{0,\geq 3} = 1$$
$$A_{1,\geq 3} = 0$$
$$A_{2,\geq 3} = 0$$
$$A_{3,\geq 3} = 2$$
$$A_{4,\geq 3} = 6$$
$$A_{5,\geq 3} = 24$$
$$A_{6,\geq 3} = 160$$
$$A_{7,\geq 3} = 1\,140$$
$$A_{8,\geq 3} = 8\,988$$
$$A_{9,\geq 3} = 80\,864$$
$$A_{10,\geq 3} = 809\,856$$
$$A_{11,\geq 3} = 8\,907\,480$$
$$A_{12,\geq 3} = 106\,877\,320$$
$$A_{13,\geq 3} = 1\,389\,428\,832$$
$$A_{14,\geq 3} = 19\,452\,141\,696$$
$$A_{15,\geq 3} = 291\,781\,655\,984$$
$$A_{16,\geq 3} = 4\,668\,504\,894\,480$$
$$A_{17,\geq 3} = 79\,364\,592\,318\,720$$
$$A_{18,\geq 3} = 1\,428\,562\,679\,845\,888$$
$$A_{19,\geq 3} = 27\,142\,690\,734\,936\,864$$
$$A_{20,\geq 3} = 542\,853\,814\,536\,802\,656$$
$$A_{21,\geq 3} = 11\,399\,930\,109\,077\,490\,560$$
$$A_{22,\geq 3} = 250\,798\,462\,399\,300\,784\,640$$
$$A_{23,\geq 3} = 5\,768\,364\,635\,100\,620\,089\,152$$
$$A_{24,\geq 3} = 138\,440\,751\,242\,507\,472\,273\,856$$
$$A_{25,\geq 3} = 3\,461\,018\,781\,064\,593\,367\,693\,824$$
$$A_{26,\geq 3} = 89\,986\,488\,307\,675\,206\,245\,836\,800$$
$$A_{27,\geq 3} = 2\,429\,635\,184\,307\,185\,219\,369\,763\,200$$
$$A_{28,\geq 3} = 68\,029\,785\,160\,601\,345\,467\,104\,670\,848$$
$$A_{29,\geq 3} = 1\,972\,863\,769\,657\,440\,129\,000\,783\,404\,544$$
$$A_{30,\geq 3} = 59\,185\,913\,089\,723\,198\,139\,150\,966\,450\,176$$

The Lah numbers (A105278)

$\lfloor \frac{n}{k} \rfloor$	$k=1$	$k=2$	$k=3$	$k=4$	$k=5$	$k=6$	$k=7$	$k=8$
$n=1$	1							
$n=2$	2	1						
$n=3$	6	6	1					
$n=4$	24	36	12	1				
$n=5$	120	240	120	20	1			
$n=6$	720	1800	1200	300	30	1		
$n=7$	5040	15120	12600	4200	630	42	1	
$n=8$	40320	141120	141120	58800	11760	1176	56	1
$n=9$	362880	1451520	1693440	846720	211680	28224	2016	72

$\lfloor \frac{9}{9} \rfloor = 1.$

The Superfactorials[14] (A000178)

$$sf(0) = 1$$
$$sf(1) = 1$$
$$sf(2) = 2$$
$$sf(3) = 12$$
$$sf(4) = 288$$
$$sf(5) = 34\,560$$
$$sf(6) = 24\,883\,200$$
$$sf(7) = 125\,411\,328\,000$$
$$sf(8) = 5\,056\,584\,744\,960\,000$$
$$sf(9) = 1\,834\,933\,472\,251\,084\,800\,000$$
$$sf(10) = 6\,658\,606\,584\,104\,736\,522\,240\,000\,000$$
$$sf(11) = 265\,790\,267\,296\,391\,946\,810\,949\,632\,000\,000\,000$$
$$sf(12) = 127\,313\,963\,299\,399\,416\,749\,559\,771\,247\,411\,200\,000\,000\,000$$
$$sf(13) = 792\,786\,697\,595\,796\,795\,607\,377\,086\,400\,871\,488\,552\,960\,000\,000\,000\,000\,000$$
$$sf(14) = 69\,113\,789\,582\,492\,712\,943\,486\,800\,506\,462\,734\,562\,847\,413\,501\,952 \cdot 10^{15}$$

---

[14]The notation $n\$$ is also in use.

The horizontal sum of the Lah numbers (A000262)

$$L_0 = 1$$
$$L_1 = 1$$
$$L_2 = 3$$
$$L_3 = 13$$
$$L_4 = 73$$
$$L_5 = 501$$
$$L_6 = 4\,051$$
$$L_7 = 37\,633$$
$$L_8 = 394\,353$$
$$L_9 = 4\,596\,553$$
$$L_{10} = 58\,941\,091$$
$$L_{11} = 824\,073\,141$$
$$L_{12} = 12\,470\,162\,233$$
$$L_{13} = 202\,976\,401\,213$$
$$L_{14} = 3\,535\,017\,524\,403$$
$$L_{15} = 65\,573\,803\,186\,921$$
$$L_{16} = 1\,290\,434\,218\,669\,921$$
$$L_{17} = 26\,846\,616\,451\,246\,353$$
$$L_{18} = 588\,633\,468\,315\,403\,843$$
$$L_{19} = 13\,564\,373\,693\,588\,558\,173$$
$$L_{20} = 327\,697\,927\,886\,085\,654\,441$$
$$L_{21} = 8\,281\,153\,039\,765\,859\,726\,341$$
$$L_{22} = 218\,456\,450\,997\,775\,993\,367\,443$$
$$L_{23} = 6\,004\,647\,590\,528\,092\,507\,965\,393$$
$$L_{24} = 171\,679\,472\,549\,945\,695\,230\,447\,313$$
$$L_{25} = 5\,097\,728\,684\,975\,832\,001\,895\,021\,401$$
$$L_{26} = 156\,976\,479\,403\,800\,014\,958\,377\,703\,651$$
$$L_{27} = 5\,006\,229\,763\,167\,109\,991\,562\,254\,382\,853$$
$$L_{28} = 165\,145\,148\,432\,723\,439\,035\,142\,843\,093\,913$$
$$L_{29} = 5\,628\,563\,759\,710\,900\,871\,382\,077\,742\,916\,173$$
$$L_{30} = 197\,987\,401\,295\,571\,718\,915\,006\,598\,239\,796\,851$$

The Euler numbers (A000111)

$$E_0 = 1$$
$$E_1 = 1$$
$$E_2 = 1$$
$$E_3 = 2$$
$$E_4 = 5$$
$$E_5 = 16$$
$$E_6 = 61$$
$$E_7 = 272$$
$$E_8 = 1\,385$$
$$E_9 = 7\,936$$
$$E_{10} = 50\,521$$
$$E_{11} = 353\,792$$
$$E_{12} = 2\,702\,765$$
$$E_{13} = 22\,368\,256$$
$$E_{14} = 199\,360\,981$$
$$E_{15} = 1\,903\,757\,312$$
$$E_{16} = 19\,391\,512\,145$$
$$E_{17} = 209\,865\,342\,976$$
$$E_{18} = 2\,404\,879\,675\,441$$
$$E_{19} = 29\,088\,885\,112\,832$$
$$E_{20} = 370\,371\,188\,237\,525$$
$$E_{21} = 4\,951\,498\,053\,124\,096$$
$$E_{22} = 69\,348\,874\,393\,137\,901$$
$$E_{23} = 1\,015\,423\,886\,506\,852\,352$$
$$E_{24} = 15\,514\,534\,163\,557\,086\,905$$
$$E_{25} = 246\,921\,480\,190\,207\,983\,616$$
$$E_{26} = 4\,087\,072\,509\,293\,123\,892\,361$$
$$E_{27} = 70\,251\,601\,603\,943\,959\,887\,872$$
$$E_{28} = 1\,252\,259\,641\,403\,629\,865\,468\,285$$
$$E_{29} = 23\,119\,184\,187\,809\,597\,841\,473\,536$$
$$E_{30} = 441\,543\,893\,249\,023\,104\,553\,682\,821$$

Bibliography

[1] H. L. Abbott, P. Erdős, D. Hanson, On the number of times an integer occurs as a binomial coefficient, 81(3) (1974), 256-261.

[2] V. S. Adamchik, *On the Hurwitz function for rational arguments*, Appl. Math. Comput. 187(1) (2007), 3-12.

[3] V. S. Adamchik, *On Stirling numbers and Euler sums*, J. Comput. Appl. Math. 79 (1997), 119-130.

[4] A. Adelberg, *Congruences of p-adic integer order Bernoulli numbers*, J. Number Theory 59 (1996), 374-388.

[5] A. Adelberg, *The p-adic analysis of Stirling numbers via higher order Bernoulli numbers*, arXiv:1805.00995.

[6] J. A. Adell, A. Lekuona, *Closed form expressions for the Stirling numbers of the first kind*, Integers 17 (2017), #A26.

[7] T. Agoh, K. Dilcher, *Representations of Stirling numbers of the first kind by multiple integrals*, Integers 15 (2015), #A8.

[8] T. Agoh, K. Dilcher, L. Skula, *Wilson quotients for composite moduli*, Math. Comput. 67(222) (1998), 843-861.

[9] T. Agoh, L. Skula, *The fourth power of the Fermat quotient*, J. Number Theory 128 (2008), 2865-2873.

[10] M. Aigner, *Combinatorial Theory*, Springer-Verlag, Berlin, Heidelberg, 1997.

[11] M. Aigner, G. Ziegler, *Proofs from the Book*, Springer, 2001.

[12] M. Aigner, *A characterization of the Bell numbers*, Discrete Math. 205 (1999), 207-210.

[13] A. Akbary, S. Siavashi, *The largest known Wieferich numbers*, Integers 18 (2018), #A3.

[14] M. A. Alekseyev, J. M. Grau, A. M. Oller-Marcén, *Computing solutions to the congruence* $1^n + 2^n + \cdots + n^n \equiv p \pmod{n}$, Discrete Appl. Math. 200 (2019), 427-440.

[15] E. Alkan, *Variations on Wolstenholme's theorem,* Amer. Math. Monthly 101 (1994), 1001-1004.

[16] E. Alkan, H. Gōral, D. C. Sertbaş, *Hyperharmonic numbers can rarely be integers,* Integers 18 (2018), #A43.

[17] T. Amdeberhan, V. H. Moll, *Involutions and their progenies,* J. Comb. 6(4) (2015), 483-508.

[18] T. Amdeberhan, V. De Angelis, V. Moll, *Complementary Bell numbers: arithmetical properties and Wilf's conjecture,* Advances in Combinatorics Waterloo Workshop in Computer Algebra, W80, Springer, 2013, 23-56.

[19] T. Amdeberhan, D. Manna, V. Moll, *The 2-adic valuation of Stirling numbers,* Exp. Math. 17 (2009), 69-82.

[20] R. A. Amrane, H. Belbachir, *Non-integerness of class of hyperharmonic numbers,* Ann. Math. Inform. 37 (2010), 7-11.

[21] R. A. Amrane, H. Belbachir, *Are the hyperharmonics integral? A partial answer via the small intervals containing primes,* C. R. Math. Acad. Sci. Paris 349(3-4) (2011), 115-117.

[22] R. A. Amrane, H. Belbachir, *Non-integerness of hyperharmonic numbers by using intervals containing primes,* Int. Conference on Disc. Math. and Comp. Science (Proceedings), 2011, 9.

[23] T. Andreescu, Z. Feng, *A Path to Combinatorics for Undergraduates,* Birkhäuser, 2004.

[24] V. Andrejić, M. Tatarevic, *Searching for a counterexample to Kurepa's conjecture,* Math. Comp. 85 (2016), 3061-3068.

[25] G.E. Andrews, *The Theory of Partitions,* Addison-Wesley, 1976.

[26] D. André, *Développement de* sec x *and* tg x, C. R. Math. Acad. Sci. Paris 88 (1879), 965-979.

[27] D. André, *Sur les permutations alternées,* Journal de Mathématiques Pures et Appliquées, 3e série, 7 (1881) 167-184.

[28] D. André, *Mémoire sur les permutations quasi-alternées,* Journal de Mathématiques Pures et Appliquées, 5e série, 1 (1895), 315-350.

[29] V. De Angelis, D. Marcello, *Wilf's conjecture,* Amer. Math. Monthly 123(6) (2016), 557-573.

[30] M. Apagodu, D. Applegate, N. J. A. Sloane, D. Zeilberger, *Analysis of the gift exchange problem,* Electronic J. Comb. 24(3) (2017), #P3.9.

[31] T. M. Apostol, *An Introduction to Analytic Number Theory*, Springer, 1976.

[32] P. Appell, *Développement en série entière de* $(1 + ax)^{1/x}$, Grunert Archiv 65 (1880), 171-175.

[33] D. Applegate, N. J. A. Sloane, *The gift exchange problem*, arXiv:0907.0513v1

[34] T. Arakawa, T. Ibukiyama, M. Kaneko, *Bernoulli Numbers and Zeta Functions*, Springer, 2014.

[35] E. Artin, *The Gamma Function*, Dover, 2015 (reprint).

[36] N. Asai, I. Kubo, H.-H. Kuo *Bell numbers, log-concavity, and log-convexity*, Acta Appl. Math. 63 (2000), 79-87.

[37] W. Asakly, T. Mansour, M. Schork, *Representing elements of the Weyl algebra by labeled trees*, J. Math. Phys. 54 (2013), 023514.

[38] R. Askey, *Orthogonal Polynomials and Special Functions*, Regional Conference Series in Applied Mathematics, 21 (1975).

[39] R. Askey, M. E. H. Ismail, *Permutation problems and special functions*, Canad. J. Math. XXVIII (4) (1976), 853-875.

[40] È. T. Avanesov, *Solution of a problem on figurate numbers*, (in Russian) Acta Arith. 12 (1966/67), 409-420.

[41] C. Babbage, *Demonstration of a theorem relating to prime numbers*, Edinburgh Philosophical J. 1 (1819), 46-49.

[42] G. Bach, *Über eine Verallgemeinerung der Differenzengleichung der Stirlingschen Zahlen 2. Art und Einige damit zusammenhängende Fragen*, J. Reine Angew. Math. 233 (1968), 213-220.

[43] R. C. Baker, G. Harman, J. Pintz, *The difference between consecutive primes, II*, Proc. London Math. Soc. 83(3) (2001), 532-562.

[44] D. Barsky, B. Benzaghou, *Nombres de Bell et somme de factorielles*, J. Th. Nombres Bordeaux 16(1) (2004), 1-17.

[45] D. Barsky, B. Benzaghou, *Erratum à l'article Nombres de Bell et somme de factorielles*, J. Th. Nombres Bordeaux 23(2) (2011), 527.

[46] A. Barghi, *Stirling numbers of the first kind for graphs*, Australas. J. Combin. 70(2) (2018), 253-268.

[47] P. Barry, *Riordan Arrays: A Primer*, Lulu.com, 2017.

[48] P. Barry, *A note on three families of orthogonal polynomials defined by circular functions, and their moment sequences*, J. Integer Seq. 15 (2012), Article 12.7.2.

[49] A. Bazsó, D. Kreso, F. Luca, Á. Pintér, *On equal values of power sums of arithmetic progressions*, Glas. Mat. Ser. III 47 (2012), 253-263.

[50] A. Bazsó, I. Mező, *On the coefficients of power sums of arithmetic progressions*, J. Number Theory 153 (2015), 117-123.

[51] A. Bazsó, I. Mező, *On the distribution of integers in the Stirling number triangles* (manuscript).

[52] A. Bazsó, Á. Pintér, H. M. Srivastava, *A refinement of Faulhaber's theorem concerning sums of powers of natural numbers*, Appl. Math. Lett. 25 (2012), 486-489.

[53] H. W. Becker, J. Riordan, *The arithmetic of Bell and Stirling numbers*, Amer. J. Math. 70 (1948), 385-394.

[54] H. Belbachir, A. Belkhir, *Cross recurrence relations for r-Lah numbers*, Ars Combin. 110 (2013), 199-203.

[55] H. Belbachir, I. E. Bousbaa, *Associated Lah numbers and r-Stirling numbers*, arXiv:1404.5573v2.

[56] H. Belbachir, I. E. Bousbaa, *Translated Whitney and r-Whitney numbers: a combinatorial approach*, J. Integer Seq. 16 (2013), Article 13.8.6.

[57] H. Belbachir, A. Khelladi, *On a sum involving powers of reciprocals of an arithmetic progression*, Ann. Math. Inform. 34 (2007), 29-31.

[58] H. Belbachir, M. Mihoubi, *A generalized recurrence for Bell polynomials: An alternate approach to Spivey and Gould-Quaintance formulas*, Eur. J. Combin. 30(5) (2009), 1254-1256.

[59] E. A. Bender, E. R. Canfield, *Log-concavity and related properties of the cycle index polynomials*, J. Combin. Theory Ser. A. 74(1) (1996), 57-70.

[60] A. Benjamin, L. Ericksen, P. Jayawant, M. Shattuck, *Combinatorial trigonometry with Chebyshev polynomials*, J. Statist. Plann. Inference 140 (2010), 2157-2160.

[61] A. T. Benjamin, D. Gaebler, R. Gaebler, *A combinatorial approach to hyperharmonic numbers*, Integers, 3 (2003), 1-9.

[62] A. T. Benjamin, J. Ornstein *A bijective proof of a derangement recurrence*, Proceedings of the 17th International Conference on Fibonacci Numbers and Their Applications (2017), 28-29.

[63] A. T. Benjamin, G. O. Preston, J. J. Quinn, *A Stirling encounter with harmonic numbers,* Math. Mag. 75(2) (2002), 95-103.

[64] M. Benoumhani, *On Whitney numbers of Dowling lattices,* Discrete Math. 159 (1996) 13-33.

[65] M. Benoumhani, *On some numbers related to Whitney numbers of Dowling lattices,* Adv. Appl. Math. 19 (1997) 106-116.

[66] M. Benoumhani, *Log-concavity of Whitney numbers of Dowling lattices,* Adv. Appl. Math. 22 (1999), 186-189.

[67] B. Bényi, *Restricted lonesum matrices,* ArXiv:1711.10178v1.

[68] B. Bényi, M. Méndez, J. L. Ramírez, T. Wakhare, *Restricted r-Stirling numbers and their combinatorial applications,* Appl. Math. Comput. 348 (2019), 186-205.

[69] B. Bényi, J. L. Ramírez, *Some applications of S-restricted set partitions,* Period. Math. Hungar. 78(1) (2019), 110-127.

[70] D. Berend, T. Tassa, *Improved bounds on Bell numbers and on moments of sums of random variables,* Probab. Math. Statist. 30(2) (2010), 185-205.

[71] E. R. Berlekamp, *Algebraic Coding Theory,* McGraw Hill, New York, 1968.

[72] M. Bernstein, N. J. A. Sloane, *Some canonical sequences of integers,* Linear Alg. Appl. 226-228 (1995), 57-72.

[73] A. Berrizbeitia, L. A. Medina, A. C. Moll, V. H. Moll, L. Noble, *The p-adic valuation of Stirling numbers,* J. Alg. Num. Th. Acad. 1 (2010), 1-30.

[74] F. Beukers, *On the generalized Ramanujan-Nagell equation, I,* Acta Arith. 38 (1980/81), 389-410.

[75] F. Beukers, *On the generalized Ramanujan-Nagell equation, II,* Acta Arith. 39 (1981), 113-123.

[76] G. Bhatnagar, *In praise of an elementary identity of Euler,* Electron. J. Combin. 18(2) (2011), #P13.

[77] G. Bianchi, R. Sorrentino, *Electronic filter simulation & design,* McGraw-Hill, 2007.

[78] Y. F. Bilu, B. Brindza, P. Kirschenhofer, Á. Pintér, R. F. Tichy, *Diophantine equations and Bernoulli polynomials,* (with an Appendix by A. Schinzel) Compos. Math. 131 (2002), 173-188.

[79] I. V. Blagouchine, *Three notes on Ser's and Hasse's representations for the zeta-functions*, Integers 18A (2018), #A3.

[80] P. Blasiak, P. Flajolet, *Combinatorial models of creation-annihilation*, Sém. Lothar. Combin. 65 (2011), Article B65c.

[81] A. Blokhuis, A. Brouwer, B. de Weger, *Binomial collisions and near collisions*, Integers 17 (2017), #A64.

[82] M. Bóna, *A Walk Through Combinatorics* (third edition), World Scientific Publishing Co., 2011.

[83] M. Bóna, *Combinatorics of Permutations*, Chapman & Hall / CRC, 2004.

[84] M. Bóna (editor), *Handbook of Enumerative Combinatorics*, Chapman & Hall / CRC, 2015.

[85] M. Bóna, *On a balanced property of derangements*, Electron. J. Combin. 13 (2006), R102.

[86] M. Bóna, R. Ehrenborg, *A combinatorial proof of the log-concavity of the numbers of permutations with k runs*, J. Combin. Theory Ser. A 90 (2000), 293-303.

[87] M. Bóna, I. Mező, *Real zeros and partitions without singleton blocks*, Eur. J. Combin. 51 (2016), 500-510.

[88] B. A. Bondarenko, V. V. Karachik, R. B. Tulyaganov, *Distribution of Eulerian and Stirling numbers mod p in arithmetical triangles* (in Russian), Voprosy Vychisl. i Prikl. Mat. 102 (1996), 133-140, and 146.

[89] J. Bonnar, *The Gamma Function*, CreateSpace Independent Publishing Platform, 2014.

[90] G. Boros, V. Moll, *Irresistible Integrals*, Cambridge University Press, 2004.

[91] J. Bourgain, K. Ford, S. V. Konyagin, I. E. Shparlinski, *On the divisibility of Fermat quotients*, Michigan Math. J. 59 (2010), 313-328.

[92] S. Bouroubi, *Bell numbers and Engel's conjecture*, Rostock. Math. Kolloq. 62 (2007), 61-70.

[93] K. Boyadzhiev, *Close encounters with the Stirling numbers of the second kind*, Math. Mag. 85(4) (2012), 252-266.

[94] D. W. Boyd, *A p-adic study of the partial sums of the harmonic series*, Experiment. Math. 3 (1994), 287-302.

[95] F. Brenti, *Permutation enumeration, symmetric functions, and uni-modality,* Pacific J. Math. 157(1) (1993), 1-28.

[96] F. Brenti, *Unimodal, Log-Concave and Pólya Frequency Sequences in Combinatorics,* Amer. Math. Soc. 1989.

[97] R. Breusch, *Solution to Problem 5040,* Amer. Math. Monthly 70 (1963), 769.

[98] B. Brindza, Á. Pintér, *On the power values of Stirling numbers,* Acta Arith. 60 (1991), 169-175.

[99] J. R. Britnell, M. Wildon, *Bell numbers, partition moves and the eigenvalues of the random-to-top shuffle in Dynkin Types A, B and D,* J. Combin. Th. Ser. A 148 (2017), 116-144.

[100] A. Z. Broder, *The r-Stirling numbers,* Disc. Math. 49 (1984), 241-259.

[101] B. H. Brown, B. Rosenbaum, *E46,* Amer. Math. Monthly 41 (1934), 48-49.

[102] D. H. Browne, H. W. Becker, *Problem E461,* Amer. Math. Monthly 48 (1941), proposal: p. 210, solution: p. 701-703.

[103] R. A. Brualdi, *Introductory Combinatorics,* 2nd edition, North-Holland, 1992.

[104] Y. Bugeaud, M. Mignotte, S. Siksek, M. Stoll, Sz. Tengely, *Integral points on hyperelliptic curves,* Alg. Number Th. 2(8) (2008), 859-885.

[105] W. Butske, L. M. Jaje, D. R. Mayernik, *On the equation $\sum_{p|N} \frac{1}{p} + \frac{1}{N} = 1$, pseudoperfect numbers, and perfectly weighted graphs,* Math. Comp. 69 (2000), 407-420.

[106] J. B. Caicedo, V. H. Moll, J. L. Ramírez, D. Villamizar, *Extensions of set partitions and permutations,* Electron. J. Combin. 26(2) (2019), P2.20.

[107] A. R. Calderbank, P. Hanlon, R. W. Robinson, *Partitions into even and odd block size and some unusual characters of the symmetric groups,* Proc. London Math. Soc. s3-53(2) (1986), 288-320.

[108] C. K. Caldwell, G. L. Honaker Jr., *Prime Curios!: The Dictionary of Prime Number Trivia,* CreateSpace Independent Publishing Platform, 2009.

[109] D. Callan, *Sets, lists and non-crossing partitions,* J. Integer Seq. 11 (2008), Article 08.1.3.

[110] D. Callan, *Notes on Stirling cycle numbers,* online note, http://www.stat.wisc.edu/~callan/papersother/

[111] D. Callan, *Cesàro's integral formula for the Bell numbers (corrected)*, online note, http://www.stat.wisc.edu/~callan/papersother/

[112] D. Callan, *A note on downup permutations and increasing 0-1-2 trees*, online note, http://www.stat.wisc.edu/~callan/papersother/

[113] D. Callan, $\sum_{i=1}^{n} i^k = \sum_{j=1}^{k+1} \left\{{k+1 \atop j}\right\} \binom{n}{j}(j-1)!$, online note, http://www.stat.wisc.edu/~callan/papersother/

[114] E. R. Canfield, *Asymptotic Normality in Binomial Type Enumeration* (Doctoral Dissertation), University of California San Diego, 1975.

[115] E. R. Canfield, *Location of the maximum Stirling number(s) of the second kind*, Stud. Appl. Math. 59 (1978), 83-93.

[116] E. R. Canfield, *Engel's inequality for Bell numbers*, J. Combin. Theory Ser. A 72(1) (1995), 184-187.

[117] E. R. Canfield, *The size of the largest antichain in the partition lattice*, J. Combin. Theory Ser. A. 83 (1998), 188-201.

[118] E. R. Canfield, L. H. Harper, *Large antichains in the partition lattice*, Random Structures Algorithms 6 (1995), 89-104.

[119] E. R. Canfield, C. Pomerance, *On the problem of uniqueness for the maximum Stirling number(s) of the second kind*, Integers 2 (2002), paper A01. Corrigendum: 5 (2005), #A09.

[120] L. Carlitz, *On some polynomials of Tricomi*, Boll. Un. Mat. Ital. 13(3) (1958), 58-64.

[121] L. Carlitz, *Congruences for generalized Bell and Stirling numbers*, Duke Math. J. 22 (1955), 193-205.

[122] L. Carlitz, *Single variable Bell polynomials*, Collect. Math. 14 (1962), 13-25.

[123] L. Carlitz, *A problem in partitions related to the Stirling numbers*, Bull. Amer. Math. Soc. 70 (1964), 275-278.

[124] L. Carlitz, *Generating functions and partition problems*, Proc. Symp. Pure Math. vol. VIII, American Mathematical Society (1965), 144-169.

[125] L. Carlitz, *Some partition problems related to the Stirling numbers of the second kind*, Acta Arith. (1965), 409-422.

[126] L. Carlitz, *Weighted Stirling numbers of the first and second kind – I*, Fibonacci Quart. 18 (1980), 147-162.

[127] L. Carlitz, *Weighted Stirling numbers of the first and second kind – II*, Fibonacci Quart. 18 (1980), 242-257.

[128] L. Carlitz, *On a class of finite sums*, Amer. Math. Monthly 37 (1930), 472-479.

[129] L. Carlitz, *On arrays of numbers*, Amer. J. Math. 54 (1932), 739-752.

[130] A. L. Cauchy, *Exercices de mathematique*, Oeuvres 2(9) (1829), 122.

[131] M. E. Cesàro, *Sur une équation aux différences mêlées*, Nouvelles Annales de Math. 4(3) (1885), 36-40.

[132] O-Y. Chan, D. Manna, *Congruences for Stirling numbers of the second kind*, Gems in Experimental Mathematics (Volume 517 of Contemporary Mathematics), Amer. Math. Soc. (2010), 97-111.

[133] R. Chapman, *Evaluating $\sum_{n=1}^{N}(a+nd)^p$ again*, Math. Gaz. 92 (2008) 92-94.

[134] Ch. A. Charalambides, *Enumerative Combinatorics*, Chapman & Hall / CRC, 2002.

[135] Ch. A. Charalambides, *Combinatorial Methods in Discrete Distributions*, John Wiley & Sons, Hoboken, New Jersey, 2005.

[136] R. X. F. Chen, *A note on the generating function for the Stirling numbers of the first kind*, J. Integer Seq. 18 (2015), Article 15.3.8.

[137] W. Y. C. Chen, *The skew, relative, and classical derangements*, Discrete Math. 160 (1996), 235-239.

[138] W. Y. C. Chen, J. C. Y. Zhang, *The skew and relative derangements of type B*, Electron. J. Combin. 14 (2007), #N24.

[139] Y.-G. Chen, M. Tang, *On the elementary symmetric functions of $1, 1/2, \ldots, 1/n$*, Amer. Math. Monthly 119 (2012), 862-867.

[140] W. Y. C. Chen, L. X. W. Wang, A. L. B. Yang, *Recurrence relations for strongly q-log-convex polynomials*, Canad. Math. Bull. 54 (2011), 217-229.

[141] G.-S. Cheon, M. E. A. El-Mikkawy, H.-G. Seol, *New identities for Stirling numbers via Riordan arrays*, J. Korea Soc. Math. Educ. Ser. B: Pure Appl. Math. 13(4) (2006), 311-318.

[142] G.-S. Cheon, J.-H. Jung, *r-Whitney numbers of Dowling lattices*, Discrete Math. 312(15) (2012), 2337-2348.

[143] G.-S. Cheon, Ji-H. Jung, L. W. Shapiro, *Generalized Bessel numbers and some combinatorial settings*, Discrete Math. 313(20) (2013), 2127-2138.

[144] G.-S. Cheon, J.-S. Kim, *Stirling matrix via Pascal matrix*, Linear Algebra Appl. 329 (2001), 49-59.

[145] N. Chheda, M. K. Gupta, *RNA as a permutation*, arXiv:1403.5477.

[146] J. Y. Choi, J. D. H. Smith, *On the combinatorics of multi-restricted numbers*, Ars Combin. 75 (2005), 45-63.

[147] J. Y. Choi, J. D. H. Smith, *On the unimodality and combinatorics of Bessel numbers*, Discrete Math. 264 (2003), 45-53.

[148] J. Y. Choi, J. D. H. Smith, *Recurrences for tri-restricted numbers*, J. Combin. Math. Combin. Comput. 58 (2006), 3-11.

[149] J. Y. Choi, L. Long, S.-H. Ng, J. D. H. Smith, *Reciprocity for multirestricted Stirling numbers*, J. Combin. Theory Ser. A 113 (2006), 1050-1060.

[150] C.-O. Chow, *On derangement polynomials of type B*, Sém. Lothar. Combin. 55 (2006), Article B55b.

[151] S. Chowla, I. N. Herstein, W. K. Moore, *On recursions connected with symmetric groups I*, Canad. J. Math. 3 (1951), 328-334.

[152] F. Chung, R. Graham, D. E. Knuth, *A symmetrical Eulerian identity*, J. Comb. 17(1) (2010), 29-38.

[153] F. Clarke, *Hensel's lemma and the divisibility by primes of Stirling-like numbers*, J. Number Theory 52(1) (1995), 69-84.

[154] P. Codara, O. M. D'Antona, P. Hell, *A simple combinatorial interpretation of certain generalized Bell and Stirling numbers*, Discrete Math. 318 (2014), 53-57.

[155] H. Cohen, *Number Theory, Vol I*. Springer, 2007.

[156] H. Cohen, *Number Theory, Vol II*. Springer, 2007.

[157] A. Colucci, *General maggiorazione dei polinomi e delle derivate e una sua conseguenza*, Boll. Un. Mat. Ital. 8(3) (1953), 258-260.

[158] L. Comtet, *Advanced Combinatorics*, D. Reidel Publishing Company, Dordrecht, Holland, 1974.

[159] L. Comtet, *Une formule explicite pour les puissances successives de l'opérateur de dérivation de Lie*, C. R. Acad. Sci. Paris Sér. A-B 276 (1973), A165-168.

[160] K. Conrad, *The p-adic growth of harmonic sums*, available online at http://www.math.uconn.edu/~kconrad/blurbs/gradnumthy/padicharmonicsum.pdf

[161] F. Constantinescu, *Relations entre les coefficients de deux polynomes dont les racines se séparent,* Časopis Pěst. Mat. 89 (1964), 1-4.

[162] J. H. Conway, R. K. Guy, *The Book of Numbers,* Springer-Verlag, New York, 1996.

[163] C. B. Corcino, *An asymptotic formula for the r-Bell numbers,* Matimyás Mat. 24(1) (2001), 9-18.

[164] C. B. Corcino, R. B. Corcino, *An asymptotic formula for r-Bell numbers with real arguments,* ISRN Discrete Math. 2013 (2013), Article ID 274697.

[165] C. B. Corcino, R. B. Corcino, *Asymptotic estimates for second kind generalized Stirling numbers,* J. Appl. Math. 2013 (2013), Article ID 918513.

[166] C. B. Corcino, R. B. Corcino, N. Acala, *Asymptotic estimates for r-Whitney numbers of the second kind,* J. Appl. Math. 2014 (2014), Article ID 354053.

[167] R. B. Corcino, C. B. Corcino, R. Aldema, *Asymptotic normality of the (r, β)-Stirling numbers,* Ars Combin. 81 (2006), 81-96.

[168] C. B. Corcino, R. B. Corcino, R. J. Gasparin, *Equivalent asymptotic formulas of second kind r-Whitney numbers,* Integral Transforms Spec. Funct. 26(3) (2015), 192-202.

[169] C. B. Corcino, R. B. Corcino, I. Mező, *Integrals and derivatives connected to the r-Lambert function,* Integral Transforms Spec. Funct. 28(11) (2017), 838-845.

[170] C. B. Corcino, R. B. Corcino, I. Mező, J. L. Ramírez, *Some polynomials associated with the r-Whitney numbers,* Proc. Math. Sci. 128:27 (2018).

[171] R. B. Corcino, R. O. Celeste, K. J. M. Gonzales, *Rook theoretic proofs of some identities related to Spivey's Bell number formula,* Ars Combin. 132 (2017), 11-26.

[172] R. B. Corcino, M. B. Montero, S. L. Ballenas, *Schlömilch-type formula for r-Whitney numbers of the first kind,* Matimyás Mat. 37(1-2) (2014), 1-10.

[173] R. B. Corcino, M. B. Montero, C. B. Corcino, *On Generalized Bell Numbers for Complex Argument,* Util. Math. 88 (2012), 267-279.

[174] C. B. Corcino, L. C. Hsu, E. L. Tan, *Asymptotic approximations of r-Stirling numbers,* Approx. Theor. Appl. 15(3) (1999), 13-25.

[175] R. B. Corcino, I. Mező, *The estimation of the zeros of the Bell and r-Bell polynomials*, Appl. Math. Comput. 250 (2015), 727-732.

[176] R. M. Corless, G. H. Gonnet, D. E. G. Hare, D. J. Jeffrey, D. E. Knuth, *On the Lambert W function*, Adv. Comput. Math. 5 (1996), 329-359.

[177] P. Corvaja, U. Zannier, *Diophantine equations with power sums and universal Hilbert sets*, Indag. Math. 9(3) (1998), 317-332.

[178] R. E. Crandall, K. Dilcher, C. Pomerance, *A search for Wieferich and Wilson primes*, Math. Comput. 66 (1997), 433-449.

[179] J. Culver, A. Weingartner, *Set partitions without blocks of certain sizes*, arXiv:1806.02316

[180] T. W. Cusick, Y. Li, *kth order symmetric SAC boolean functions and bisecting binomial coefficients*, Discrete Appl. Math. 149 (2005), 73-86.

[181] D. Cvijović, *The Dattoli-Srivastava conjectures concerning generating functions involving the harmonic numbers*, Appl. Math. Comput. 215(11) (2010), 4040-4043.

[182] D. Cvijović, *Derivative polynomials and closed-form higher derivative formulae*, Appl. Math. Comput. 215(8) (2009), 3002-3006.

[183] S. Daboul, J. Mangaldan, M. Z. Spivey, P. J. Taylor, *The Lah numbers and the nth derivative of $e^{1/x}$*, Math. Mag. 86(1) (2013), 39-47.

[184] K. Dale, I. Skau, *The (generalized) secretary's packet problem and the Bell numbers*, Discrete Math. 137 (1995), 357-360.

[185] G. Dattoli, H. M. Srivastava, *A note on harmonic numbers, umbral calculus and generating functions*, Appl. Math. Lett. 21(7) (2008), 686-693.

[186] H. Davenport, G. Pólya, *On the product of two power series*, Canad. J. Math. 1 (1949), 1-5.

[187] M. Davis, *Quadrant marked mesh patterns and the r-Stirling numbers*, J. Integer Seq. 18 (2015), Article 15.10.1.

[188] K. S. Davis, W. A. Webb, *Lucas' theorem for prime powers*, Eur. J. Combin. 11 (1990), 229-233.

[189] K. S. Davis, W. A. Webb, *A binomial coefficient congruence modulo prime powers*, J. Number Theory 43 (1993), 20-23.

[190] N. G. de Bruijn, *Asymptotic Methods in Analysis*, Dover, 1981.

[191] E. Deutsch, S. Elizalde, *The largest and the smallest fixed points of permutations*, Eur. J. Combin. 31(5) (2010), 1404-1409.

[192] P. Diaconis, M. Shahshahani, *On the eigenvalues of random matrices,* J. Appl. Prob. 31 (1994), 49-62.

[193] T. Diagana, H. Maïga, *Some new identities and congruences for Fubini numbers,* J. Number Theory 173 (2017), 547-569.

[194] R. M. Dickau, *Bell number diagrams,* see the webpage `http:// mathforum.org/advanced/robertd/bell.html`

[195] A. Dil, I. Mező, M. Cenkci, *Evaluation of Euler-like sums via Hurwitz zeta values,* Turkish J. Math. 41 (2017), 1640-1655.

[196] K. Dilcher, L. Skula, *A new criterion for the first case of Fermat's last theorem,* Math. Comp. 64 (1995), 363-392.

[197] K. Dilcher, L. Skula, *Linear relations between certain sums of reciprocals modulo p,* Ann. Sci. Math. Québec 35 (2011), 17-29.

[198] K. Dilcher, L. Skula, *The cube of the Fermat quotient,* Integers 6 (2006), #A24.

[199] J. D. Dixon, *Polynomials with real roots,* Canad. Math. Bull. 5 (1962), 259-263.

[200] A. J. Dobson, *A note on Stirling numbers of the second kind,* J. Combin. Theory 5(2) (1968), 212-214.

[201] J. B. Dobson, *On Lerch's formula for the Fermat quotient,* Arxiv:1103.3907

[202] L. L. Dornhoff, F. E. Hohn, *Applied Modern Algebra,* Macmillan, New York, 1978.

[203] T. A. Dowling, *A class of geometric lattices based on finite groups,* J. Combin. Theory Ser. B 14 (1973), 61-86. Erratum: J. Combin. Theory Ser. B 15 (1973), 211.

[204] B. S. El-Desouky, F. A. Shiha, E. M. Shokr, *The generalized r-Whitney numbers,* Appl. Math. 8 (2017), 117-132.

[205] G. Dobiński, *Summierung der Reihe $\sum n^m/m!$ für $m = 1, 2, 3, 4, 5, \ldots,$* Grunert Archiv (Arch. Math. Phys.) 61 (1877), 333-336.

[206] `https://archive.org/stream/archivdermathem88unkngoog#page/ n349`. Accessed on 11 June 2019.

[207] R. Donaghey, *Alternating permutations and binary increasing trees,* J. Combin. Th. Ser. A 18 (1975), 141-148.

[208] B. Dragovich, A. Yu. Khrennikov, S. V. Kozyrev, I. V. Volovich, E. I. Zelenov, *p-adic mathematical physics: the first 50 years,* arXiv:1705.04758

[209] B. Duncan, R. Peele, *Bell and Stirling numbers for graphs*, J. Integer Seq. 12 (2009), Article 09.7.1.

[210] A. Edrei, *Zeros of successive derivatives of entire functions of the form* $h(z)\exp(-e^z)$, Trans. Amer. Math. Soc. 259(1) (1980), 207-226.

[211] R. Ehrenborg, *The Hankel determinant of exponential polynomials*, Amer. Math. Monthly 107(6) (2000), 557-560.

[212] C. Elbert, *Strong asymptotics for the generating polynomials of the Stirling numbers of the second kind*, J. Approx. Theory 109 (2001), 198-217.

[213] C. Elbert, *Weak asymptotics for the generating polynomials of the Stirling numbers of the second kind*, J. Approx. Theory 109 (2001), 218-228.

[214] J. Engbers, D. Galvin, J. Hilyard, *Combinatorially interpreting generalized Stirling numbers*, Eur. J. Combin. 43 (2015), 32-54.

[215] J. Engbers, D. Galvin, C. Smyth, *Restricted Stirling and Lah number matrices and their inverses*, J. Combin. Th. 161 (2019), 271-298.

[216] K. Engel, *On the average rank of an element in a filter of the partition lattice*, J. Combin. Theory Ser. A 65(1) (1994), 67-78.

[217] E. A. Enneking, J. C. Ahuja *Generalized Bell numbers*, Fibonacci Quart. 14 (1976), 67-73.

[218] P. Erdős, *On a conjecture of Hammersley*, J. London Math. Soc. 28 (1953), 232-236.

[219] P. Erdős, *Verallgemeinerung eines elementar-zahlenteoretischen Satzes von Kürschák*. Mat. Fiz. Lapok 39 (1932), 17-24.

[220] P. Erdős, *A Theorem of Sylvester and Schur*, J. London Math. Soc. 9(4) (1934), 282-288.

[221] R. Ernvall, T. Metsänkylä, *On the p-divisibility of Fermat quotients*, Math. Comp. 66 (1997), 1353-1365.

[222] A. Eswarathasan, E. Levine, *p-integral harmonic sums*, Discrete Math. 91 (1991), 249-257.

[223] A. F. von Ettingshausen, *Die combinatorische Analysis als Vorbereitungslehre zum Studium der theoretischen höhern Mathematik*, J. B. Wallishauffer (1826).

[224] G. Everest, T. Ward, *An Introduction to Number Theory*, Springer, 2005.

[225] N.-E. Fahssi, *Polynomial triangles revisited*, arXiv:1202.0228v7.

[226] J. Favard, *Sur les polynomes de Tchebicheff (in French)*, C. R. Acad. Sci. Paris 200 (1935), 2052-2053.

[227] A. E. Fekete, *Apropos Two notes on notation*, Amer. Math. Monthly 101(8) (1994), 771-778.

[228] L. Fekih-Ahmed, *On the power series expansion of the reciprocal gamma function*, HAL Archives, 01029331v1, 2014.

[229] J. Ferenczik, Á. Pintér, B. Porvázsnyik, *On equal values of Stirling numbers of the second kind*, Appl. Math. Comput. 218 (2011), 980-984.

[230] S. R. Finch, *Mathematical Constants*, Cambridge University Press, 2003.

[231] N. J. Fine, *Binomial coefficients modulo a prime*, Amer. Math. Monthly 54 (1947), 589-592.

[232] S. Fisk, *The secretary's packet problem*, Math. Mag. 61(2) (1988), 103-105.

[233] P. Flajolet, *On congruences and continued fractions for some classical combinatorial quantities*, Discrete Math. 41 (1982), 145-153.

[234] P. Flajolet, *Combinatorial aspects of continued fractions*, Discrete Math. 32 (1980), 125-161.

[235] P. Flajolet, R. Sedgewick, *Analytic Combinatorics*, Cambridge University Press, 2009.

[236] D. Foata, G.-N. Han, *Secant tree calculus*, Cent. Eur. J. Math. 12(12) (2014), 1852-1870.

[237] M. Fujiwara, *Über die obere Schranke des absoluten Betrages der Wurzeln einer algebraischen Gleichung*, Tôhoku Math. J. 10 (1916) 167-171.

[238] W. Fulton, *Young Tableaux: With Applications to Representation Theory and Geometry*, Cambridge University Press, 1996.

[239] H. G. Funkhouser, *A short account of the history of symmetric functions of roots of equations*, Amer. Math. Monthly 37(7) (1930), 357-365.

[240] L. H. Gallardo, *A property of the period of a Bell number modulo a prime number*, Appl. Math. E-Notes 16 (2016), 72-79.

[241] D. Galvin, Do T. Thanh, *Stirling numbers of forests and cycles,* Electron. J. Combin. 20 (2013), #P73.

[242] A. M. Garsia, J. B. Remmel *Q-counting rook configurations and a formula of Frobenius,* J. Combin. Theory Ser. A 41 (1986), 246-275.

[243] M. Gardner, *Fractal Music, Hypercards and More: Mathematical Recreations from Scientific American,* W. H. Freeman and Company, 1992.

[244] V. Gasharov, *On the Neggers-Stanley conjecture and the Eulerian polynomials,* J. Combin. Theory Ser. A 82 (1998), 134-146.

[245] B. J. Gassner, *Sorting by replacement selecting,* Commun. ACM 10 (1967), 89-93.

[246] J. von zur Gathen, J. Roche, *Polynomials with two values,* Combinatorica 17 (1997), 345-362.

[247] A. Gertsch, *Nombres harmoniques généralisés,* C. R. Acad. Sci. Paris (Série 1) 324 (1997), 7-10.

[248] A. Gertsch, *Congruences pour quelques suites classiques de nombres, sommes de factorielles et calcul ombral,* Thesis, Univ. Neuchâtel (1999).

[249] A. Gertsch, A. M. Robert, *Some congruences concerning the Bell numbers,* Bull. Belg. Math. Soc. Simon Stevin 3(4) (1996), 467-475.

[250] I. M. Gessel, *Congruences for Bell and tangent numbers,* Fibonacci Quart. 19 (1981) 137-144.

[251] I. M. Gessel, *Fibonacci is a square,* Fibonacci Quart. 10 (1972) 417-419.

[252] I. M. Gessel, *Combinatorial proofs of congruences,* Joint Mathematics Meeting, 2010. Available online at http://people.brandeis.edu/~gessel/homepage/slides/

[253] I. Gessel, T. Lengyel, *On the order of Stirling numbers and alternating binomial coefficient sums,* Fibonacci Quart. 39 (2001), 444-454.

[254] J. W. L. Glaisher, *On the Residues of r^{p-1} to modulus p^2, p^3, etc.,* Quart. J. Pure Appl. Math. 32 (1901), 1-27.

[255] J. W. L. Glaisher, *On the residues of the sums of products of the first $p-1$ numbers and their powers to modulus p^2 or p^3,* Quart. J. Math. Oxford 31 (1900), 321-353.

[256] J. Goldman, J. Joichi, D. White, *Rook theory III. Rook polynomials and the Chromatic structure of graphs*, J. Combin. Th. Ser. B 25 (1978), 135-142.

[257] H. Gōral, D. C. Sertbaş, *Almost all hyperharmonic numbers are not integers*, J. Number Theory 171 (2017), 495-526.

[258] H. Gōral, D. C. Sertbaş, *Divisibility properties of hyperharmonic numbers*, Acta Math. Hungar. 154(1) (2018), 147-186.

[259] H. W. Gould, *Explicit formulas for Bernoulli numbers*, Amer. Math. Monthly 79 (1972), 44-51.

[260] H. W. Gould, J. Quaintance, *Implications of Spivey's Bell number formula*, J. Integer Seq. 11 (2008), Article 08.3.7.

[261] F. Q. Gouvêa, *p-adic Numbers – An Introduction*, second edition, Springer, 1997.

[262] R. L. Graham, D. E. Knuth, O. Patashnik, *Concrete Mathematics*, Addison-Wesley, 1990.

[263] A. Granville, *Binomial coefficients modulo prime powers*, Canadian Mathematical Society Conference Proceedings 20 (1997), 253-275.

[264] A. Granville, *The square of the Fermat quotient*, Integers 4 (2004), #A22.

[265] G. Grätzer, *Lattice Theory: Foundation*, Birkhäuser, 2011.

[266] J. L. Gross, *Combinatorial Methods with Computer Applications*, Chapman and Hall/CRC, 2007.

[267] O. A. Gross, *Preferential arrangements*, Amer. Math. Monthly 69(1) (1962), 4-8.

[268] E. Grosswald, *Bessel Polynomials*, Springer, 1978.

[269] B.-Ni Guo, I. Mező, F. Qi, *An explicit formula for Bernoulli polynomials in terms of r-Stirling numbers of the second kind*, Rocky Mountain J. Math. 46(6) (2016), 1919-1923.

[270] B.-Ni Guo, F. Qi, *Some identities and an explicit formula for Bernoulli and Stirling numbers*, J. Comput. Appl. Math. 255 (2014), 568-579.

[271] B.-Ni Guo, F. Qi, *Explicit formulae for computing Euler polynomials in terms of Stirling numbers of the second kind*, J. Comput. Appl. Math. 272 (2014), 251-257.

[272] R. Gy *Generalized Lerch primes*, Integers 18 (2018), #A10.

[273] E. Gyimesi, G. Nyul, *New combinatorial interpretations of r-Whitney and r-Whitney-Lah numbers*, Discrete Appl. Math. 255 (2019), 222-233.

[274] K. Győry, Á. Pintér, *On the equation* $1^k + 2^k + \cdots + x^k = y^n$, Publ. Math. Debrecen 62 (2003), 403-414.

[275] M. Hall, *Arithmetic properties of a partition function*, Bull. Amer. Math. Society 40(5) (1934), 387.

[276] H. Han, S. Seo, *Combinatorial proofs of inverse relations and log-concavity for Bessel numbers*, Eur. J. Combin. 29 (2008), 1544-1554.

[277] H. Harborth, *Über das Maximum bei Stirlingschen Zahlen 2. Art*, J. Reine Angew. Math. 230 (1968), 213-214.

[278] H. Harborth, *Number of odd binomial coefficients*, Proc. Amer. Math. Soc. 62(1) (1977), 19-22.

[279] L. H. Harper, *Stirling behavior is asymptotically normal*, Ann. Math. Stat. 38 (1967), 410-414.

[280] H. Hasse, *Uber eine diophantische Gleichung von Ramanujan-Nagell und ihre verallgemeinerung*, Nagoya Math. J. 27 (1966), 77-102.

[281] J. Havil, *Gamma: Exploring Euler's Constant*, Princeton University Press, 2003.

[282] F. Heneghan, T. K. Petersen, *Power series for up-down min-max permutations*, College Math. J. 45(2) (2014), 83-91.

[283] M. D. Hirschhorn, *Evaluating* $\sum_{n=1}^{N}(a + nd)^p$, Math. Gaz. 90 (2006) 114-116.

[284] M. E. Hoffman, *Quasi-symmetric functions and mod p multiple harmonic sums*, Kyushu J. Math. 69 (2015), 345-366.

[285] J. M. Holte, *Carries, combinatorics, and an amazing matrix*, Amer. Math. Monthly 104(2) (1997), 138-149.

[286] Sh.-F. Hong, J-R. Zhao, W. Zhao, *The 2-adic valuations of Stirling numbers of the second kind*, Int. J. Number Theory 8(4) (2012), 1057-1066.

[287] F. T. Howard, *Associated Stirling numbers*, Fibonacci Quart. 18 (1980), 303-315.

[288] F. T. Howard, *Congruences for the Stirling numbers and associated Stirling numbers*, Acta Arith. 55 (1990), 29-41.

[289] F. T. Howard, *Extensions of congruences of Glaisher and Nielsen concerning Stirling numbers*, Fibonacci Quart. 28(4) (1990), 355-362.

[290] F. T. Howard, *Sums of powers of integers via generating functions*, Fibonacci Quart. 34 (1996) 244-256.

[291] L. C. Hsu, *Note on an asymptotic expansion of the n-th difference of zero*, Ann. Math. Statist. 19 (1948), 273-277.

[292] L. C. Hsu, P.J.-S. Shiue, *A unified approach to generalized Stirling numbers*, Adv. Appl. Math. 20 (1998), 366-384.

[293] G. Hurst, A. Schultz, *An elementary (number theory) proof of Touchard's congruence*, arXiv:0906.0696v2

[294] E. J. Ionaşcu, *A variation on bisecting the binomial coefficients*, Discrete Appl. Math. 250 (2018), 276-284.

[295] E. J. Ionaşcu, T. Martinsen, P. Stănică, *Bisecting binomial coefficients*, Discrete Appl. Math. 227 (2017) 70-83.

[296] M. E. H. Ismail, *Determinants with orthogonal polynomial entries*, J. Comput. Appl. Math. 178(1-2) (2005), 255-266.

[297] A. Ivić, Ž. Mijajlović, *On Kurepa's problems in number theory*, Publ. Inst. Math., Nouv. 57(71) (1995), 19-28. Available also on arXiv:math/0312202

[298] S. Jakubec, *Connection between Fermat quotients and Euler numbers*, Math. Slovaca 58(1) (2008), 19-30.

[299] G. James, A. Kerber, *The Representation Theory of the Symmetric Group*, Addison-Wesley, 1981.

[300] S. Janson, *Euler-Frobenius numbers and rounding*, Online J. Anal. Comb. 8 (2013), Article 5.

[301] D. J. Jeffrey, N. Murdoch, *Stirling numbers, Lambert W and the gamma function*, In: J. Blömer, I. Kotsireas, T. Kutsia, D. Simos (eds.) Mathematical Aspects of Computer and Information Sciences. MACIS 2017. Lecture Notes in Computer Science, vol 10693. Springer.

[302] Sh. T. Jensen, *The Laguerre-Samuelson Inequality with Extensions and Applications in Statistics and Matrix Theory* (MSc thesis) Department of Mathematics and Statistics, McGill University, 1999.

[303] W. A. Johnson, *Exponential Hilbert series and the Stirling numbers of the second kind,* Discrete Math. 341 (2018), 1237-1243.

[304] W. Johnson, *The diophantine equation $x^2 + 7 = 2^n$*, Amer. Math. Monthly 94 (1987), 59-62.

[305] G. A. Jones, J. M. Jones, *Elementary Number Theory,* Springer, 1998.

[306] S. József, B. Crstici, *Handbook of Number Theory II,* Kluwer, 2004.

[307] Ji-H. Jung, I. Mező, J. L. Ramírez, *The r-Bessel and restricted r-Bell numbers,* Australas. J. Comb. 70(2) (2017), 202-220.

[308] A. Junod, *Congruences pour les polynômes et nombres de Bell,* Bull. Belg. Math. Soc. 9 (2002) 503-509.

[309] A. Junod, *Congruences par l'analyse p-adique et le calcul symbolique,* Thesis, Univ. de Neuchâtel (2003).

[310] A. Junod, *A generalized trace formula for Bell numbers,* Expo. Math. 23 (2005), 71-79.

[311] A. Junod, *Hankel determinants and orthogonal polynomials,* Expo. Math. 21 (2003), 63-74.

[312] N. Kahale, *New modular properties of Bell numbers,* J. Combin. Theory Ser. A 58 (1991) 147-152.

[313] L. T. Kai, *Bell numbers and Bell numbers modulo a prime number,* Mathematical Medley 42(2) (1997), 55-58.

[314] G. A. Kalugin, D. J. Jeffrey, *Convergence in $\mathbb{C}$ of series for the Lambert W function,* arXiv:1208.0754v1.

[315] D. M. Kane, On the number of representations of t as a binomial coefficient, Integers 4 (2004), #A07.

[316] D. M. Kane, Improved bounds on the number of ways of expressing t as a binomial coefficient, Integers 7 (2007), #A53.

[317] H.-J. Kanold, *Über Stirlingschen Zahlen 2. Art,* J. Reine Angew. Math. 229 (1968), 188-193.

[318] H.-J. Kanold, *Über eine asymptotische Abschätzung bei Stirlingschen Zahlen 2. Art,* J. Reine Angew. Math. 230 (1968), 211-212.

[319] H.-J. Kanold, *Einige neuere Abschätzungen bei Stirlingschen Zahlen 2. Art,* J. Reine Angew. Math. 239 (1969), 148-160.

[320] S. Karlin, *Total Positivity,* Vol. 1., Stanford University Press, 1968.

[321] S. Karlin, G. Szegő, *On certain determinants whose elements are orthogonal polynomials,* J. Anal. Math. 8(1) (1960), 1-157.

[322] A. Kasraoui, *d-regular set partitions and rook placements,* Sém. Lothar. Combin. 62 (2009), Article B62a.

[323] J. Katriel, *On a generalized recurrence for Bell numbers*, J. Integer Seq. 11 (2008), Article 08.3.8.

[324] J. Katriel, *A multitude of expressions for the Stirling numbers of the first kind*, Integers 10 (2010), 273-297.

[325] U. N. Katugampola, *Mellin transforms of generalized fractional integrals and derivatives*, Appl. Math. Comput. 257(15) (2015), 566-580.

[326] M. Kauers, *Summation algorithms for Stirling number identities*, J. Symbolic Comput. 42 (2007), 948-970.

[327] M. Kauers, C. Schneider, *Automated proofs for some Stirling number identities*, Electron. J. Combin. 15 (2008), #R2.

[328] D. H. Kauffman, *Note on preferential arrangements*, Amer. Math. Monthly 70(1) (1963), 62.

[329] B. C. Kellner, J. Sondow, *Power-sum denominators*, Amer. Math. Monthly 124(8) (2017), 695-709.

[330] G. Kemkes, D. Merlini, B. Richmond, *Maximum Stirling numbers of the second kind*, Integers Electron. J. Combin. Num. Theory, 8 (2008), #A27.

[331] Zs. Kereskényi-Balogh, G. Nyul, *Stirling numbers of the second kind and Bell numbers for graphs*, Australas. J. Combin. 58(2) (2014), 264-274.

[332] T. Kim, D. S. Kim, G.-W. Jang, J. Kwon, *A note on some identities of derangement polynomials*, J. Inequal. Appl. 40 (2018), 17 pages.

[333] M. Klazar, F. Luca, *On some arithmetic properties of polynomial expressions involving Stirling numbers of the second kind*, Acta Arith. 107 (2003), 357-372.

[334] A. Knopfmacher, N. Robbins *Some arithmetical properties of Eulerian numbers*, J. Combin. Math. Combin. Comput. 36 (2001), 31-42.

[335] D. E. Knuth, *The Art of Computer Programming*, vol. 3: Sorting and Searching, Addison-Wesley, 1973.

[336] D. E. Knuth, *Two notes on notation*, Amer. Math. Monthly 99(5) (1992), 403-422.

[337] N. Koblitz, *p-adic Numbers, p-adic Analysis, and Zeta Functions*, second edition, Springer, 1984.

[338] R. Koekoek, R. F. Swarttouw, *The Askey-scheme of hypergeometric orthogonal polynomials and its q-analogue,* Delft University of Technology, Faculty of Technical Mathematics and Informatics, Report no. 94-05 (1994). Available in PDF format at http://aw.twi.tudelft.nl/~koekoek/documents/as98.pdf or in HTML here: http://homepage.tudelft.nl/11r49/askey.html

[339] T. Kojima, *On the limits of the zeros of an algebraic equation,* Tôhoku Math. J. 11 (1917) 119-127.

[340] T. Komatsu, K. Liptai, I. Mező, *Incomplete poly-Bernoulli numbers associated with incomplete Stirling numbers,* Publ. Math. Debrecen 88(3-4) (2016), 357-368.

[341] T. Komatsu, I. Mező, L. Szalay *Incomplete Cauchy numbers,* Acta Math. Hungar. 149(2) (2016), 306-323.

[342] T. Komatsu, J. L. Ramírez, *Some determinants involving incomplete Fubini numbers,* An. Şt. Univ. Ovidius Constanţa, 26(3) (2018), 143-170.

[343] T. Komatsu, P. T. Young, *Exact p-adic valuations of Stirling numbers of the first kind,* J. Number Theory 177 (2017), 20-27.

[344] H. L. Krall, O. Frink, *A new class of orthogonal polynomials: the Bessel Polynomials,* Trans. Amer. Math. Soc. 65 (1949), 100-115.

[345] C. Krattenthaler, *Advanced determinant calculus,* Sém. Lothar. Combin. 42 (1999), Article B42q.

[346] C. Krattenthaler, *Advanced determinant calculus: a complement,* Linear Algebra Appl. 411 (2005), 68-166.

[347] J. Kürschák, *A harmonikus sorról (Hungarian),* Mat. Fiz. Lapok 27 (1918), 299-300.

[348] Y. H. Kwong, *Minimum periods of $S(n,k)$ modulo m,* Fibonacci Quart. 27 (1989), 217-221.

[349] E. Laguerre, *Mémoire pour obtenir par approximation les racines d'une équation algébrique qui a toutes les racines réelles,* Nouv. Ann. Math. 2nd series 19 (1880), 161-172, and 193-202.

[350] I. Lah, *A new kind of numbers and its application in the actuarial mathematics,* Bol. Inst. Actuár. Port. 9 (1954), 7-15.

[351] I. Lah, *Eine neue Art von Zahlen, ihre Eigenschaften und Anwendung in der mathematischen Statistik,* Mitteilungsbl. Math. Statist. 7 (1955), 203-212.

[352] N. J. H. Lai, *p-adic properties of recurrences involving Stirling numbers*, manuscript, available online at `https://www.math.ubc.ca/~njhlai/paper/p-adic-properties-of-recurrences-involving-Stirling-numbers.pdf`

[353] D. Laissaui, M. Rahmani, *An explicit formula for sums of powers of integers in terms of Stirling numbers*, J. Integer Seq. 20 (2017), Article 17.4.8

[354] L. D. Landau, E. M. Lifsic, *Quantum Mechanics – Non-relativistic Theory*, Pergamon Press, 1977.

[355] W. Lang, *On generalizations of the Stirling number triangles*, J. Integer Seq. 3 (2000), Article 00.2.4

[356] W. Lang, *Combinatorial interpretation of generalized Stirling numbers*, J. Integer Seq. 12 (2009), Article 09.3.3

[357] G. Larcher, *On the number of odd binomial coefficients*, Acta Math. Hungar. 71(3) (1996), 183-203.

[358] J. W. Layman, *The Hankel transform and some of its properties*, J. Integer Seq. Vol. 4 (2001), Article 01.1.5.

[359] J. W. Layman, *Maximum zero strings of Bell numbers modulo primes*, J. Comb. Theory Ser. A 40 (1985), 161-168.

[360] J. W. Layman, C. L. Prather, *Generalized Bell numbers and zeros of successive derivatives of an entire function*, J. Anal. Math. 96 (1983), 42-51.

[361] E. Lehmer, *On congruences involving Bernoulli numbers and the quotients of Fermat and Wilson*, Ann. of Math. 39(2) (1938), 350-360.

[362] T. Lengyel, *The order of the Fibonacci and Lucas numbers*, Fibonacci Quart. 33(3) (1995) 234-239.

[363] T. Lengyel, *On the divisibility by 2 of the Stirling numbers of the second kind*, Fibonacci Quart. 32 (1994), 194-201.

[364] T. Lengyel, *Alternative proofs on the 2-adic order of Stirling numbers of the second kind*, Integers 10 (2010), #A38.

[365] T. Lengyel, *On a recurrence involving Stirling numbers*, Eur. J. Combin. 5 (1984), 313-321.

[366] T. Lengyel, *On p-adic properties of the Stirling numbers of the first kind*, J. Number Theory 148 (2015), 73-94.

[367] T. Lengyel, *On a recurrence involving Stirling numbers*, Eur. J. Combin. 5(4) (1984), 313-321.

[368] T. Lengyel, *Stirling numbers of the second kind, associated Stirling numbers of the second kind and Newton-Euler sequences*, J. Comb. Number Theory 6(1) (2014), 51-61.

[369] P. Leonetti, C. Sanna, *On the p-dic valuation of Stirling numbers of the first kind*, Acta Math. Hungar. 151(1) (2017), 217-231.

[370] M. Lerch, *Zur Theorie des Fermatschen Quotienten $\frac{a^{p-1}-1}{p} = q(a)$*, Math. Ann. 60 (1905), 471-490.

[371] C. Leudesdorf, *Some results in the elementary theory of numbers*, Proc. London Math. Soc. 20 (1888), 199-212.

[372] J. Levine, R. E. Dalton, *Minimum periods, modulo p, of first-order Bell exponential numbers*, Math. Comp. 16 (1962), 416-423.

[373] R. Lidl, H. Niederreiter, *Finite Fields*, Cambridge University Press, 1997.

[374] E. H. Lieb, *Concavity properties and a generating function for Stirling numbers*, J. Combin. Theory 5(2) (1968), 203-206.

[375] D. A. Lind, *The quadratic field $\mathbb{Q}(\sqrt{5})$ and a certain diophantine equation*, Fibonacci Quart. 6 (1968), 86-93.

[376] W. Ljunggren, *Oppgave nr 2*, Norsk Mat. Tidsskr. 25 (1943), 29.

[377] G.-D. Liu, *An identity involving Lucas numbers and Stirling numbers*, Fibonacci Quart. 46-47(2) (2008/09), 136-139.

[378] L. L. Liu, Ya-N. Li, *Recurrence relations for linear transformations preserving the strong q-log-convexity*, Electron. J. Combin. 23(3) (2016), #P3.44

[379] L. L. Liu, D. Ma, *Some polynomials related to Dowling lattices and x-Stieltjes moment sequences*, Linear Alg. Appl. 533 (2017), 195-209.

[380] L. Li Liu, Yi Wang, *On the log-convexity of combinatorial sequences*, Adv. Appl. Math. 39(4) (2007), 453-476.

[381] G. Louchard, *Number of singletons in involutions of large size: a central range and a large deviation analysis*, manuscript, available online at: http://www.ulb.ac.be/di/mcs/louchard/louchard.papers/sing.pdf

[382] L. Lovász, *Combinatorial Problems and Exercises* North Holland, 1993.

[383] F. Luca, P. G. Walsh, *The product of like-indexed terms in binary recurrences*, J. Number Theory 96(1) (2002), 152-173.

[384] W. F. Lunnon, P. A. B. Pleasants, N. M. Stephens, *Arithmetic properties of Bell numbers to a composite modulus I*, Acta Arith. 35 (1979) 1-16.

[385] M. S. Maamra, M. Mihoubi, *The $(r_1, \ldots, r_p)$-Bell polynomials*, Integers 14 (2004), #A34.

[386] I. G. Macdonald, *Symmetric Functions and Hall Polynomials*, American Mathematical Society, 1988.

[387] B. J. Malešević, *Some considerations in connection with alternating Kurepa's function*, Integral Transforms Spec. Funct. 19(10) (2008), 747-756.

[388] M. M. Mangontarum, *Some theorems and applications of the $(q; r)$-Whitney numbers*, J. Integer Seq. 20 (2017), Article 17.2.5

[389] M. M. Mangontarum, O. I. Cauntongan, A. P. Macodi-Ringia, *The noncentral version of the Whitney numbers: a comprehensive study*, Int. J. Math. Math. Sci. 2016 (2016), Article ID 6206207.

[390] M. M. Mangontarum, A. M. Dibagulun, *On the translated Whitney numbers and their combinatorial properties*, Br. J. Appl. Sci. Tech. 11(5) (2015), 1-15.

[391] M. M. Mangontarum, J. Katriel, *On q-boson operators and q-analogues of the r-Whitney and r-Dowling numbers*, J. Integer Seq. 18 (2015), Article 15.9.8

[392] T. Mansour, *Combinatorics of Set Partitions*, CRC Press, 2013.

[393] T. Mansour, J. L. Ramírez, M. Shattuck, *A generalization of the r-Whitney numbers of the second kind*, J. Comb. 8(1) (2017), 29-55.

[394] T. Mansour, M. Schork *Commutation Relations, Normal Ordering, and Stirling Numbers*, CRC Press, 2016.

[395] T. Mansour, M. Schork, M. Shattuck, *On a new family of generalized Stirling and Bell numbers*, Electronic J. Combin. 18 (2011), #P77.

[396] I. Martinjak, *Two infinite families of terminating binomial sums*, Period. Math. Hung. 75 (2017), 244-254.

[397] I. Martinjak, D. Stanić, *A short combinatorial proof of derangement identity*, Elem. Math. 73 (2018), 29-33.

[398] J. L. Massey, *Shift-register synthesis and BCH decoding*, IEEE Trans. Information Theory IT-15 (1969), 122-127.

[399] R. J. McIntosh, *On the converse of Wolstenholme's Theorem*, Acta Arith. 71 (1995), 381-389.

[400] R. J. McIntosh, E. L. Roettger, *A search for Fibonacci-Wieferich and Wolstenholme primes*, Math. Comput. 76 (2007), 2087-2094.

[401] D. G. Mead, *The equation of Ramanujan-Nagell and $[y^2]$*, Proc. Amer. Math. Soc. 41 (1973), 333-341.

[402] M. Mendez, P. Blasiak, A. Penson, *Combinatorial approach to generalized Bell and Stirling numbers and boson normal ordering problem*, J. Math. Phys. 46 (2005), 083511.

[403] V. V. Menon, *On the maximum of Stirling numbers of the second kind*, J. Combin. Theory 15(1) (1973), 11-24.

[404] M. Merca, *A note on the r-Whitney numbers of Dowling lattices*, C. R. Acad. Sci. Paris Ser I 351 (2013), 649-655.

[405] M. Merca, *Some experiments with complete and elementary symmetric functions*, Period. Math. Hung. 69 (2014), 182-189.

[406] D. Merlini, R. Sprugnoli, M. C. Verri *The Cauchy numbers*, Discrete Math. 306 (2006), 1906-1920.

[407] R. E. Merrifield, H. E. Simmons, *Enumeration of structure-sensitive graphical subsets: theory*, Proc. Natd Acad. Sci. USA 78(2) (1981), 692-695.

[408] R. Merris, *The p-Stirling numbers*, Turkish J. Math. 24 (2000), 379-399.

[409] R. Meštrović, *Lucas' theorem: its generalizations, extensions and applications (1878–2014)*, arXiv:1409.3820.

[410] R. Meštrović, *Wolstenholmes' theorem: its generalizations and extensions in the last hundred and fifty years (1862-2012)*, arXiv:1111.3057

[411] R. Meštrović, *Congruences involving the Fermat quotient*, Czechoslovak Math. J. 63(4) (2013), 949-968.

[412] I. Mező, *The r-Bell numbers*, J. Integer Seq. 14(1) (2011), Article 11.1.1.

[413] I. Mező, *The dual of Spivey's Bell number formula*, J. Integer Seq. 15(2) (2012), Article 12.2.4.

[414] I. Mező, *On the maximum of r-Stirling numbers*, Adv. Appl. Math. 41 (2008), 293-306.

[415] I. Mező, *About the non-integer property of hyperharmonic numbers*, Annales Univ. Sci. Budapest 50 (2007), 1-8.

[416] I. Mező, *Periodicity of the last digits of some combinatorial sequences,* J. Integer Seq. 17 (2014), Article 14.1.1.

[417] I. Mező, *A kind of Eulerian numbers connected to Whitney numbers of Dowling lattices,* Discrete Math. 328 (2014), 88-95.

[418] I. Mező, *A new formula for the Bernoulli polynomials,* Results. Math. 58 (2010), 329-335.

[419] I. Mező, *Kombinatorikus számok általánosításairól,* PhD thesis, University of Debrecen, 2009. Available online (in Hungarian) at https://dea.lib.unideb.hu/dea/bitstream/handle/2437/

94478/Thesis_MezoI_titkositott.pdf_%27?sequence=1!

[420] I. Mező, *Recent developments in the theory of Stirling numbers,* RIMS Kôkyûroku No. 2013, 68-80.

[421] I. Mező, *Exponential generating function of hyperharmonic numbers indexed by arithmetic progressions,* Cent. Eur. J. Math. 11(5) (2013), 931-939.

[422] I. Mező, *On the structure of the solution set of a generalized Euler-Lambert equation,* J. Math. Anal. Appl. 455(1) (2017), 538-553.

[423] I. Mező, *On the Mahler measure of the Bell polynomials,* manuscript. Available at the author's webpage: https://sites.google.com/site/istvanmezo81/others.

[424] I. Mező, *Asymptotics of the modes of the ordered Stirling numbers,* arXiv:1504.06970.

[425] I. Mező, A. Dil, *Euler-Seidel method for certain combinatorial numbers and a new characterization of Fibonacci sequence,* Cent. Eur. J. Math. 7(2) (2009), 310-321.

[426] I. Mező, A. Dil, *Hyperharmonic series involving Hurwitz zeta function,* J. Number Theory 130(2) (2010), 360-369.

[427] I. Mező, J. L. Ramírez, *Divisibility properties of the r-Bell numbers and polynomials,* J. Number Theory 177 (2017), 136-152.

[428] I. Mező, J. L. Ramírez, *The linear algebra of the r-Whitney matrices,* Integral Transforms Spec. Funct. 26(3) (2015), 213-225.

[429] I. Mező, J. L. Ramírez, *Some identities of the r-Whitney numbers,* Aequationes Math. 90(2) (2016), 393-406.

[430] I. Mező, J. L. Ramírez, *On the enumeration of sets of lists,* Appl. Anal. Discrete Math. (submitted)

[431] I. Mező, A. Bazsó, *On the coefficients of polynomials related to alternating power sums of arithmetic progressions*, J. Integer Seq. 21 (2018), Article 18.7.8.

[432] I. Mező, Ch.-Y. Wang, H.-Y. Guan, *The estimation of the zeros of some counting polynomials* (manuscript).

[433] M. Mignotte, *Une nouvelle résolution de l'équation $x^2 + 7 = 2^n$*, Rend. Sem. Fac. Sci. Univ. Cagliari 54(2) (1984), 41-43.

[434] M. Mignotte, D. Stefanescu, *Estimates for polynomial zeros*, Appl. Algebra Eng. Commun. Comput. 12(6) (2001) 437-453.

[435] M. Mihoubi, M. Rahmani *The partial r-Bell polynomials*, arXiv:1308.0863.

[436] M. Mihoubi, M. Tiachachat, *Some applications of the r-Whitney numbers*, C. R. Math. Acad. Sci. Paris 352(12) (2014), 965-969.

[437] H. Miki, *A relation between Bernoulli numbers*, J. Number Theory 10 (1978), 297-302.

[438] F. L. Miksa, L. Moser, M. Wyman, *Restricted partitions of finite sets*, Canad. Math. Bull. 1(2) (1958), 87-96.

[439] P. Miska, *A note on p-adic analytic functions with application to behavior of the p-adic valuations of Stirling numbers*, arXiv:1803.04533.

[440] D. S. Mitrinović, *Analytic Inequalities*, Springer-Verlag, 1970.

[441] V. H. Moll, *Numbers and Functions: From a Classical-Experimental Mathematician's Point of View*, American Mathematical Society, 2012.

[442] V. H. Moll, J. L. Ramírez, D. Villamizar, *Combinatorial and arithmetical properties of the restricted and associated Bell and factorial numbers*, J. Comb. 9(4) (2018), 693-720.

[443] P. L. Montgomery, S. Nahm, S. S. Wagstaff Jr, *The period of the Bell numbers modulo a prime*, Math. Comp. 79 (2010), 1793-1800.

[444] L. J. Mordell, *On the integer solutions of $y(y + 1) = x(x + 1)(x + 2)$*, Pacific J. Math. 13 (1963), 1347-1351.

[445] L. J. Mordell, *Diophantine Equations*, Academic Press, 1969.

[446] L. Moser, J. R. Pounder, *Problem 53*, Canad. Math. Bull. 5 (1962), 70.

[447] L. Moser, M. Wyman, *On solutions of $x^d = 1$ in symmetric groups*, Canad. J. Math. 7 (1955), 159-168.

[448] T. Nagell, *Eine Eigenschaft gewisser Summen*, Skr. Norske Vid. Akad. Kristiania, I(13) (1923), 10-15.

[449] T. Nagell, *Løsning till oppgave nr 2*, Norsk Mat. Tidsskr. 30 (1948), 62-64.

[450] T. Nagell, *The diophantine equation $x^2 + 7 = 2^n$*, Ark. Mat. 4(2-3) (1961), 185-187.

[451] A. Navon, *Combinatorics and fermion algebra*, Nuovo Cimento 16 (1973), 324-330.

[452] D. J. Newman, *Problem 5040*, Amer. Math. Monthly 69 (1962), 670.

[453] C. P. Niculescu, *A new look at Newton's inequalities*, J. Ineq. Pure Appl. Math. 1(2) (2000), Article 17.

[454] N. Nielsen, *Handbuch der Teorie des Gammafunktion*, Leipzig, 1906. (Reprinted by Chelsea in 1966.)

[455] A. Nijenhuis, H. Wilf, *Periodicities of partition functions and Stirling numbers modulo p*, J. Number Theory 25 (1987), 308-312.

[456] G. Nyul, G. Rácz, *The r-Lah numbers*, Discrete Math. 338 (2015), 1660-1666.

[457] R. Peele, A. J. Radcliffe, H. S. Wilf, *Congruence problems involving Stirling numbers of the first kind*, Fibonacci Quart. 31 (1993) 73-80.

[458] S. V. Pemmaraju, S. Skiena, *Computational Discrete Mathematics: Combinatorics and Graph Theory with Mathematica*, Cambridge University Press, 2003.

[459] T. K. Petersen, *Eulerian Numbers*, Birkhäuser, 2015.

[460] A. Petojević, *A note about the Pochhammer symbols*, Mathematica Moravica 12(1) (2008), 37-42.

[461] R. Petuchovas, *Asymptotic Analysis of the cyclic structure of permutations*, Doctoral dissertation, Vilnius University, 2016. arXiv:1611.02934.

[462] K. H. Pilehrood, T. H. Pilehrood, R. Tauraso, *Multiple harmonic sums and multiple harmonic star sums are (nearly) never integers*, Integers 17 (2017), #A10.

[463] Á. Pintér, *A note on the Diophantine equation $\binom{x}{4} = \binom{y}{2}$*, Publ. Math. Debrecen 47 (1995), 411-415.

[464] Á. Pintér, *On a diophantine problem concerning Stirling numbers*, Acts Math. Hungar. 65(4) (1994), 361-364.

462   Bibliography

[465] Á. Pintér, *On some arithmetical properties of Stirling numbers*, Publ.
       Math. Debrecen 40(1-2) (1992), 91-95.

[466] É. Pité, *Problem 11957,* Amer. Math. Monthly 124(2) (2017), 179.

[467] J. Pitman, *Some probabilistic aspects of set partitions*, Amer. Math.
       Monthly 104(3) (1997), 201-209.

[468] G. Pólya, *On the zeros of the derivatives of a function and its analytic
       character*, Bull. Amer. Math. Soc. 49 (1943), 178-191.

[469] B. Poonen, *Periodicity of a combinatorial sequence*, Fibonacci Quart.
       26(1) (1988), 70-76.

[470] A. J. van der Poorten, *Some facts that should be better known,
       especially about rational functions*, In: R. A. Mollin (ed.), Number
       Theory and Applications, Kluwer, 1989.

[471] V. V. Prasolov, *Polynomials*, Springer, 2004.

[472] F. Qi, *An explicit formula for the Bell numbers in terms of the Lah
       and Stirling numbers*, Mediterr. J. Math. 13 (2016), 2795-2800.

[473] F. Qi, *Integral representations and properties of Stirling numbers of
       the first kind*, J. Number Theory 133(7) (2013), 2307-2319.

[474] J. Quaintance, H. Gould, *Combinatorial Identities for Stirling Num-
       bers – The Unpublished Notes of H. W. Gould,* World Scientific, 2016.

[475] C. Radoux, *Nombres de Bell modulo p premier et extensions de degré
       p de F_p*, C. R. Math. Acad. Sci. Paris Ser A, 281 (1975), 879-882.

[476] C. Radoux, *Une congruence pour les polynômes $P_n(x)$ de fonction
       génératrice $e^{x(e^z-1)}$* C. R. Math. Acad. Sci. Paris Ser A, 284 (1977),
       637-639.

[477] C. Radoux, *Calcul effectif de certains déterminants de Hankel*, Bull.
       Soc. Math. Belg. Sér. B 31 (1979) 49-55.

[478] C. Radoux, *Déterminant de Hankel construit sur des polynomes liés
       aux nombres de dérangements*, Eur. J. Combin. 12 (1991) 327-329.

[479] C. Radoux, *Déterminants de Hankel et théorème de Sylvester*, in
       Sém. Lothar. Combin. (Saint-Nabor, 1992), Publ. Inst. Rech. Math.
       Av., 498, Univ. Louis Pasteur, Strasbourg, 1992, 115-122.

[480] M. Rahmani, *Some results on Whitney numbers of Dowling lattices*,
       Arab J. Math. Sci. 20(1) (2014) 11-27.

[481] Cs. Rakaczki, *On the Diophantine equation $S_m(x) = g(y)$*, Publ.
       Math. Debrecen 65 (2004), 439-460.

[482] Cs. Rakaczki, *On some generalizations of the Diophantine equation* $s(1^k + 2^k + \cdots + x^k) + r = dy^n$, Acta Arith. 151 (2012), 201-216.

[483] S. Ramanujan, *Question 464*, J. Indian Math. Soc. 5 (1913), 130.

[484] S. Ramanujan, *A proof of Bertrand's postulate*, J. Indian Math. Soc. 11 (1919), 181-182.

[485] J. L. Ramírez, I. Mező, *The r-alternating permutations*, Aequationes Math. (accepted).

[486] J. L. Ramírez, M. Shattuck, *Generalized r-Whitney numbers of the first kind*, Ann. Math. Inform. 46 (2016), 175-193.

[487] J. L. Ramírez, M. Shattuck, *A (p; q)-analogue of the r-Whitney-Lah numbers*, J. Integer Seq. 19 (2016), Article 16.5.6.

[488] J. L. Ramírez, S. N. Villamarín, D. Villamizar, *Eulerian numbers associated with arithmetical progressions*, Electron J. Combin. 25(1) (2018), #P1.48

[489] J. B. Remmel, *A note on a recursion for the number of derangements*, Eur. J. Combin. 4 (1983), 371-374.

[490] J. B. Remmel, M. L. Wachs, *Rook theory, generalized Stirling numbers and (p, q)-analogues*, Electron. J. Combin. 11 (2004), #R84.

[491] B. C. Rennie, J. Dobson, *On Stirling numbers of the second kind*, J. Combin. Theory 7(2) (1969), 116-121.

[492] B. Reznick, *Regularity properties of the Stern enumeration of the rationals*, J. Integer Seq. 11 (2008), Article 08.4.1

[493] B. Reznick, *Some binary partition functions*, Analytic Number Theory, Proceedings of a conference in honor of Paul T. Bateman (1990), 451-477.

[494] P. Ribenboim, *My Numbers, My Friends: Popular Lectures on Number Theory*, Springer-Verlag, 2000.

[495] J. Riordan, *An Introduction to Combinatorial Analysis*, Wiley, 1958.

[496] J. F. Ritt, *A factorization theory for functions $\sum_{i=1}^{n} a_i e^{\alpha_i z}$*, Trans. Amer. Math. Soc. 29 (1927), 584-596.

[497] A. M. Robert, *A Course in p-adic Analysis*, Springer, 2000.

[498] G.-C. Rota *The number of partitions of a set*, Amer. Math. Monthly 71(5) (1964), 498-504.

[499] G. C. Rota, B. E. Sagan, *Congruences derived from group action*, Eur. J. Combin. 1 (1980), 67-76.

[500] A. Ruciński, B. Voigt, *A local limit theorem for generalized Stirling numbers,* Rev. Roumaine Math. Pures Appl. 35(2) (1990), 161-172.

[501] C. de J. P. Ruiz, *Weighted sum of squares via generalized Eulerian polynomials,* Fibonacci Quart. 55(5) (2017), 149-165.

[502] B. E. Sagan, *Congruences via Abelian groups,* J. Number Theory 20 (1985), 210-237.

[503] B. E. Sagan, *Shellability of exponential structures,* Order 3(1) (1986), 47-54.

[504] B. E. Sagan, *Inductive proofs of q-log-concavity,* Discrete Math. 99 (1992), 289-306.

[505] B. E. Sagan, *Inductive and injective proofs of log-concavity results,* Discrete Math. 68 (1988), 281-292.

[506] P. Samuelson, *How deviant can you be?,* J. Am. Stat. Assoc. 63(324) (1968) 1522-1525.

[507] C. Sanna, *On the p-adic valuation of harmonic numbers,* J. Number Theory 166 (2016), 41-46.

[508] R. Sánchez-Peregrino, *The Lucas congruence for Stirling numbers of the second kind,* Acta Arith. 94 (2000), 41-52.

[509] H. Scherk, *De evolvenda functione* $(yd \cdot yd \cdot yd \cdots ydx)/dx^n$ *disquisitiones nonnullae analyticae,* Ph.D. Thesis, University of Berlin, 1823.

[510] L. Schläfli, *Sur les coefficients du développement du produit* $1(1 + x)(1 + 2x) \cdots (1 + (n - 1)x)$ *suivant les puissance ascendantes de x,* Crelle's J. Reine Angew. Math. 43 (1852), 1-22.

[511] O. Schlömilch, *Recherches sur les coefficients des facultés analytiques,* Crelle 44 (1852), 344-355.

[512] M. Schork, *On the combinatorics of normal ordering bosonic operators and deformations of it,* J. Phys. A: Math. Gen. 36 (2003), 4651-4665.

[513] I. J. Schwatt, *An Introduction to the Operations with Series* (second edition) Chelsea Pub. Co., 1962.

[514] L. W. Shapiro, S. Getu, W. Woan, L. Woodson, The Riordan group, Discrete Appl. Math. 34 (1991), 229-239.

[515] H. S. Shapiro, D. L. Slotnik, *On the mathematical theory of error-correcting codes,* IBM J. Res. Develop. 3 (1959), 25-34.

[516] M. Shattuck, *Generalized r-Lah numbers,* Proc. Math. Sci. 126(4) (2016), 461-478.

[517] S. Shirai, K.-I. Sato, *Some Identities Involving Bernoulli and Stirling Numbers*, J. Number Theory 90 (2001), 130-142.

[518] I. E. Shparlinski, *On the distribution of values of recurring sequences and the Bell numbers in finite fields*, Eur. J. Comb. 12(1) (1991), 81-87.

[519] I. E. Shparlinski, *On the value set of Fermat quotients*, Proc. Amer. Math. Soc. 140(4) (2011), 1199-1206.

[520] A. Silberger, D. Silberger, S. Silberger, *Finite segments of the harmonic series*, Workshop "Algebra Across the Borders", 2011. Available online at the conference's webpage http://www.mini.pw.edu.pl/~aab/.

[521] D. Singmaster, *How often does an integer occur as a binomial coefficient?*, Amer. Math. Monthly 78(4) (1971), 385-386.

[522] D. Singmaster, *Repeated binomial coefficients and Fibonacci numbers*, Fibonacci Quart. 13(4) (1975), 295-298.

[523] D. Singmaster, *Divisibility of binomial and multinomial coefficients by primes and prime powers*, 18th Anniversary Volume of the Fibonacci Association (1980), 98-113.

[524] L. Skula, *Fermat and Wilson quotients for p-adic integers*, Acta Mathematica et Informatica Universitatis Ostraviensi 6(1) (1998), 167-181. Available online at http://dml.cz/dmlcz/120531

[525] A. B. Soble, *Majorants of polynomial derivatives*, Amer. Math. Monthly 64 (1957), 639-643.

[526] A. Solomon, G. Duchamp, P. Blasiak, A. Horzela, K. Penson, *Normal order: combinatorial graphs*, Proc. 3rd Int. Symp. on Quantum Theory and Symmetries, World Scientific (2003), 527-536.

[527] J. Sondow, *Lerch quotients, Lerch primes, Fermat-Wilson quotients, and the Wieferich-non-Wilson primes 2, 3, 14771*, Combinatorial and Additive Number Theory (CANT) 2011-2012, Springer (2014), 243-255. Available also on Arxiv:1110.3113.

[528] J. Sondow, K. MacMillan, *Reducing the Erdős-Moser equation $1^n + 2^n + \cdots + k^n = (k+1)^n$ modulo k and k^2*, Integers 11 (2011), Article A34. An expanded version available at Arxiv:1011.2154.

[529] M. R. Spiegel, *The summation of series involving roots of transcendental equations and related applications*, J. Appl. Phys. 24(9) (1953), 1103-1106.

[530] L. Spiegelhofer, M. Wallner, *Divisibility of binomial coefficients by powers of two*, J. Number Theory 192 (2018), 221-239.

[531] M. Z. Spivey, *A generalized recurrence for Bell numbers*, J. Integer Seq. 11 (2008), Article 08.2.5.

[532] M. Z. Spivey, *Log-convexity of the Bell numbers*, Blog post, https://mikespivey.wordpress.com/2014/10/02/log-convexity-of-the-bell-numbers/

[533] R. Sprugnoli, *Riordan arrays and combinatorial sums*, Discrete Math. 132 (1994), 267-290.

[534] R. P. Stanley, *Enumerative Combinatorics*, Vol. 1., Cambridge University Press, 1997.

[535] R. P. Stanley, *Log-concave and unimodal sequences in algebra, combinatorics, and geometry*, Ann. New York Acad. Sci. 576 (1989), 500-534.

[536] R. P. Stanley, *A survey of alternating permutations*, Contemp. Math. 531 (2010), 165-196. Available online at arXiv:0912.4240v1.

[537] R. P. Stanley, *Exponential structures*, Stud. Appl. Math. 59 (1978), 73-82.

[538] I. N. Stewart, D. O. Tall, *Algebraic Number Theory*, Chapman & Hall, 1979.

[539] J. Stirling, *Methodus Differentialis: sive Tractatus de Summatione et Interpolatione Serierum Infinitarum*, 1730.

[540] K. B. Stolarsky, *Power and exponential sums of digital sums related to binomial coefficient parity*, SIAM J. Appl. Math. 32 (1977), 717-730.

[541] R. J. Stroeker, B. M. M. de Weger, *Elliptic binomial Diophantine equations*, Math. Comp. 68 (1999), 1257-1281.

[542] X.-T. Su, Yi Wang, *On unimodality problems in Pascal?s triangle*, Electron. J. Combin. 15 (2008), R113.

[543] X.-T. Su, Yi Wang, Y.-N. Yeh, *Unimodality problems of multinomial coefficients and symmetric functions*, Electron. J. Combin. 18(1) (2011), P73.

[544] Zhi-H. Sun, *Congruences concerning Bernoulli numbers and Bernoulli polynomials*, Discrete Appl. Math. 105 (2000), 193-223.

[545] Zhi-H. Sun, *On the properties of Newton-Euler pairs*, J. Number Theory 114 (2005), 88-123.

[546] Y. J. Sun, J.G. Hsieh, *A note on circular bound of polynomial zeros*, IEEE Trans. Circuits Syst. I. Regul. Pap. 43 (1996) 476-478.

[547] Y.-D. Sun, X. Wu, *The largest singletons of set partitions*, Eur. J. Combin. 32 (2011), 369-382.

[548] Y.-D. Sun, X. Wu, J. Zhuang, *Congruences on the Bell polynomials and the derangement polynomials*, J. Number Theory 133 (2013) 1564-1571.

[549] Y.-D. Sun, Y. Xu, *The largest singletons in weighted set partitions and its applications*, Discrete Math. Theor. Comput. Sci. 13(3) (2011) 75-86.

[550] Zhi-W. Sun, *Binomial coefficients and quadratic fields*, Proc. Amer. Math. Soc. 134(8) (2006), 2213-2222.

[551] Zhi-W. Sun, *Arithmetic theory of harmonic numbers*, Proc. Amer. Math. Soc. 140(2) (2011), 415-428.

[552] Zhi-W. Sun, D. Zagier, *On a curious property of Bell numbers*, Bull. Aust. Math. Soc. 84 (2011), 153-158.

[553] B. Sury, *A curious polynomial identity*, Nieuw Arch. Wisk. 11 (1993), 93-96.

[554] J. J. Sylvester, *Note relative aux communications faites dans les séances du 28 Janvier et 4 Février 1861.* C. R. Acad. Sci. Paris 52 (1861), 307-308. Reprinted in Math. Papers Vol. 2, 234-235. See also Corrigenda, 241, Cambridge University Press, 1908.

[555] G. S. Sylvester, *Continuous-spin Ising ferromagnets*, PhD thesis, MIT (1976).

[556] G. Szegő, *Orthogonal Polynomials*, fourth edition, American Mathematical Society, 1975.

[557] M. Tan, T. Wang, *Lah matrix and its algebraic properties*, Ars Combin. 70 (2004), 97-108.

[558] S. M. Tanny *On some numbers related to the Bell numbers*, Canad. Math. Bull. 17(5) (1975), 733-738.

[559] A. F. Tebtoub, *Combinatorials aspects linked to the monotony of classical sequences*, PhD thesis, Mathematics Faculty, University of Sciences and Technology Houari Boumediene, 2016.

[560] N. M. Temme, *A class of polynomials related to those of Laguerre*, in Polynômes Orthogonaux et Applications, Proceedings of the Laguerre Symposium held at Bar-le-Duc, 1984.

[561] L. Theisinger, *Bemerkung über die harmonische Reihe*, Monatsch. Math. Phys. 26 (1915), 132-134.

[562] I. Tomescu, *Méthodes combinatoires dans la théorie des automates finis*, PhD thesis, Bucharest, 1971.

[563] J. Touchard, *Propriétés arithmétiques de certains nombres recurrents*, Ann. Soc. Sci. Bruxelles 53A (1933), 21-31.

[564] J. Touchard, *Nombres exponentiels et nombres de Bernoulli*, Canad. J. Math. 8 (1956), 305-320.

[565] C. A. Tovey, *Multiple occurrences of binomial coefficients*, Fibonacci Quart. 23 (1985), 356-358.

[566] D. Treeby, *Further thoughts on a paradoxical tower*, Amer. Math. Monthly 125(1) (2018), 44-60.

[567] F. G. Tricomi, *A class of non-orthogonal polynomials related to those of Laguerre*, J. Analyse Math. 1 (1951), 209-231.

[568] H. Tsumura, *On some congruences for the Bell numbers and for the Stirling numbers*, J. Number Theory 38(2) (1991), 206-211.

[569] L. R. Turner, *Inverse of the Vandermonde matrix with applications*, NASA Technical note, D-3547, 1966.

[570] G. Turnwald, *A note on the Ramanujan-Nagell equation*, in Number-theoretic Analysis, lecture notes in mathematics, volume 1452, Springer, 1988-89.

[571] D. J. Velleman, G. S. Call, *Permutations and combination locks*, Math. Mag. 68(4) (1995), 243-253.

[572] M. V. Subbarao, A. Verma, *Some remarks on a product expansion. An unexplored partition function*, in Symbolic Computations, Number Theory, Special Functions, Physics and Combinatorics, F. G. Garvan and M. E. H. Ismail, eds., Kluwer, Dordrecht, 2001, 267-283.

[573] G. Viennot, *Interprétations combinatoires des nombres d'Euler et de Genocchi*, Seminar on Number Theory, 1981/1982, No. 11, Univ. Bordeaux I, Talence, 1982.

[574] E. B. Vinberg, *A Course in Algebra*, American Mathematical Society, 2003.

[575] J. F. Voloch, *On some subgroups of the multiplicative group of finite rings*, J. Théor. Nombres Bordeaux 16 (2004), 233-239.

[576] M. L. Wachs, Whitney homology of semipure shellable posets, J. Alg. Comb. 9 (1999), 173-207.

[577] S. S. Wagstaff Jr, *Aurifeuillian factorizations and the period of the Bell numbers modulo a prime*, Math. Comp. 65 (1996), 383-391.

[578] T. Wakhare, *Refinements of the Bell and Stirling numbers,* Trans. Comb. 7(4) (2018), 25-42.

[579] H. S. Wall, *Analytic Theory of Continued Fractions,* Chelsea Publishing Company, 1948.

[580] D. Walsh, *The number of permutations with only small cycles,* available online on prof. Walsh's webpage:

http://capone.mtsu.edu/dwalsh/3cycless.pdf

[581] D. Walsh, *Counting forests with Stirling and Bell numbers,* available online on prof. Walsh's webpage:

http://capone.mtsu.edu/dwalsh/stirfort.pdf

[582] W. Wang, *Riordan arrays and harmonic number identities,* Comp. Math. Appl. 60 (2010), 1494-1509.

[583] Yi Wang, *Linear transforms preserving log-concavity,* Linear Algebra Appl. 359 (2003), 161-167.

[584] Yi Wang, Y.-N. Yeh, *Log-concavity and LC-positivity,* J. Combin. Th. Ser A 114 (2007), 195-210.

[585] Ch.-Y. Wang, I. Mező, *Some limit theorems with respect to constrained permutations and partitions,* Monatsh. Math. 182 (2017), 155-164.

[586] Ch.-Y. Wang, I. Mező, *Some limit theorems with respect to constrained permutations and partitions – II,* manuscript

[587] C.-Y. Wang, P. Miska, I. Mező, *The r-derangement numbers,* Discrete Math. 340(7) (2017), 1681-1692.

[588] S. de Wannemacker, *On 2-adic orders of Stirling numbers of the second kind,* Integers 5(1) (2005), #A21.

[589] S. de Wannemacker, T. Laffey, R. Osburn, *On a conjecture of Wilf,* J. Combin. Th. Ser. A 114 (2007), 1332-1349.

[590] B. M. M. de Weger, *Equal binomial coefficients: some elementary considerations,* J. Number Theory 63(2) (1997), 373-386.

[591] B. M. M. de Weger, *A binomial Diophantine equation,* Quart. J. Math. Oxford Ser. 186(2) (1996), 221-231.

[592] H. Wegner, *Stirling numbers of the second kind and Bonferroni's inequalities* Elem. Math. 60 (2005), 124-129.

[593] H. Wegner, *Über das Maximum bei Stirlingschen Zahlen, zweiter Art,* J. Reine Angew. Math. 262/263 (1973), 134-143.

[594] E. W. Weisstein, *Integer Sequence Primes*, webpage: `http://mathworld.wolfram.com/IntegerSequencePrimes.html`.

[595] T. P. Whaley, *Postorder trees and Eulerian numbers*, Acta Inform. 28(7) (1991), 703-712.

[596] E. G. Whitehead, *Stirling number identities from chromatic polynomials*, J. Combin. Theory Ser. A 24 (1978), 314-317.

[597] H. S. Wilf, *Generatingfunctionology*, Academic Press, 1994.

[598] H. S. Wilf, *The asymptotic behavior of the Stirling numbers of the first kind*, J. Combin Th. 64(2) (1993), 344-349.

[599] G. T. Williams, *Numbers generated by the function e^{e^x-1}*, Amer. Math. Monthly 52 (1945), 323-327.

[600] K. S. Williams, *The recurrence relation $a_n = (An + B)a_{n-1} + (Cn + D)a_{n-2}$*, Crux Mathematicorum 20 (1994), 91-96.

[601] S. Wolfram, *A New Kind of Science*, Wolfram Media, 2002.

[602] J. Wolstenholme, *On certain properties of prime numbers*, Quart. J. Pure Appl. Math. 5 (1862), 35-39.

[603] B.-L. Wu, Y.-G. Chen, *On certain properties of harmonic numbers*, J. Number Theory 175 (2017), 66-86.

[604] B.-L. Wu, Y.-G. Chen, *On the denominators of harmonic numbers*, C.R. Acad. Sci. Paris, Ser. I 356 (2018), 129-132.

[605] Ai-M. Xu, *Extensions of Spivey's Bell number formula*, Electron. J. Combin. 19(2) (2012), #P6.

[606] Ai-M. Xu, Z. Di Cen, *A unified approach to some recurrence sequences via Faà di Bruno's formula*, Comput. Math. Appl. 62(1) (2011), 253-260.

[607] Ai-M. Xu, T.-H. Zhou, *Some identities related to the r-Whitney numbers*, Integral Trans. Spec. Funct. 27(11) (2016), 920-929.

[608] Y. Yang, *On a multiplicative partition function*, Electron. J. Combin. 8 (2001), #R19.

[609] Sh.-L. Yang, H. You, *On a connection between the Pascal, Stirling and Vandermonde matrices*, (note) Discrete Appl. Math. 155 (2007), 2025-2030.

[610] Q. Yin, Sh. Hong, L. Yang, M. Qiu, *Multiple reciprocal sums and multiple reciprocal star sums of polynomials are almost never integers*, J. Number Theory 195 (2019), 269-292.

[611] P. T. Young, *Congruences for Bernoulli, Euler, and Stirling Numbers*, J. Number Theory 78 (1999), 204-227.

[612] Ya-M. Yu, *Bounds on the location of the maximum Stirling numbers of the second kind*, Discrete Math. 309(13) (2009), 4624-4627.

[613] J. Zhao, S. Hong, W. Zhao, *Divisibility by 2 of Stirling numbers of the second kind and their differences*, J. Number Theory 140 (2014), 324-348.

[614] B.-X. Zhu, *Some positivities in certain triangular arrays*, Proc. Amer. Math. Soc. 142(9) (2014), 2943-2952.

[615] B.-X. Zhu, H. Sun, *Linear transformations preserving the strong q-log-convexity of polynomials*, Electron. J. Combin. 22 (2015), #P3.27.

[616] Q. Zou, *The log-convexity of the Fubini numbers*, Trans. Comb. 7(2) (2018), 17-23.

Index